INDUSTRIAL HEALTH

Prentice-Hall Series in Environmental Sciences
Granville H. Sewell, Editor

INDUSTRIAL
HEALTH

Jack E. Peterson

Marquette University

Prentice-Hall, Inc.   Englewood Cliffs, New Jersey 07632

Library of Congress Cataloging in Publication Data

PETERSON, JACK E (date)
   Industrial health.

   Includes bibliographical references and index.
   1. Industrial hygiene.   2. Occupational diseases.
3. Industrial toxicology.   I. Title.   [DNLM: 1. Industrial medicine.   2. Occupational diseases.
3. Poisoning.   WA400 P485i]
RC967.P47   613.6′2   76-47650
ISBN   0-13-459552-1

10   9   8   7   6   5   4   3   2   1

Printed in the United States of America

PRENTICE-HALL INTERNATIONAL, INC., London
PRENTICE-HALL OF AUSTRALIA PTY. LIMITED, Sydney
PRENTICE-HALL OF CANADA, LTD., Toronto
PRENTICE-HALL OF INDIA PRIVATE LIMITED, New Delhi
PRENTICE-HALL OF JAPAN, INC., Tokyo
PRENTICE-HALL OF SOUTHEAST ASIA PTE. LTD., Singapore
WHITEHALL BOOKS LIMITED, Wellington, New Zealand

CONTENTS

v

2

Laboratory Determination of Toxicity 9

3

Gases 29

4

Metals and Metalloids   *47*

5

Pneumoconioses   *61*

6

Organic Solvents   *75*

7

Monomers and Polymers   105

8

Pesticides   121

9

Sensitization and Dermatitis   141

10

Carcinogenesis   157

11

Abnormal Pressure   175

12

Noise   *191*

13

Biothermal Stress   *221*

14

Nonionizing Radiation   *241*

15

Ionizing Radiation   259

PREFACE

Books, including this one, tend to give the reader an impression that the world is replete with immutable facts and unchanging ideas and that there is only one best way of doing almost anything. The world does change, however, and many immutable facts change right along with it despite their representation by the unchanging written word. This book does not contain a list of threshold limit values (TLV's) for chemicals or for any physical agents. Threshold limits published by the American Conference of Governmental Industrial Hygienists (ACGIH) are updated annually, but this book is not. Few things are so out of date as last year's TLV list. Similarly, there is a positive dearth of analogous data for radioactive materials, air and water pollutants, or legislated standards. Instead, in this text, underlying principles are emphasized, assigning much of their application to current publications such as TLV's, Hygienic Guides, ANSI standards, and scientific journals, all well designed to further the constant battle against information obsolescence.

This book began as the outline of a survey course taught mainly to senior and graduate civil engineering students and ended as an expanded version of that outline fleshed by the ideas of many people. It is, in systematic form, a collection of useful fundamentals and principles behind man's attempt to evaluate and control hazards that abound in his environment. Because they are most immediate, occupational hazards are stressed, ranging from those of chemical toxins to those of various energy forms. Many of the materials and energies are found also in the home, farm, or garden.

The reader is assumed to have some familiarity with science and scientific methodology but no particular background beyond that is necessary. A review of organic chemical nomenclature important in the field is presented in Appendix A, and a glossary of terms mainly from medicine and physiology appears in Appendix B.

Many people contributed to this book but only a few can be thanked. Jim Mellender devised the algorithms and wrote the programs that allowed me to computerize my file of abstracts and then to search it with extreme rapidity. Both computers used for this purpose belong to the Medical College of Wisconsin; much of the abstract file was developed during my employment with that institution.

Marquette University personnel and especially those in the College of Engineering were particularly encouraging throughout all phases of manuscript preparation. A special debt is owed to my students, who helped shape the lecture outlines with questions and comments; to Dean Ray Kipp and Civil Engineering Department chairmen Al Zanoni, and, later, Bill Murphy, who helped ease the burdens of my academic life to allow more time for writing.

At home my wife and two sons not only helped by taking over many of my husbandly and fatherly chores but participated actively in the book production process. Writing could not have been done without their aid.

Much of the industrial hygiene I know can be traced back to The Dow Chemical Company and especially to the man who was responsible for my choice of this above all other fields, Harold Hoyle. Both he and Warren Cook, my friend and thesis advisor at the University of Michigan, taught me well so that the mistakes herein are mine, not theirs.

J. E. P.

INDUSTRIAL
HEALTH

1

ENTRY
AND TOXIC ACTIONS
OF CHEMICAL SUBSTANCES

Toxicity is a property of matter. It is a biological property in much the same way that mass is a physical property. It is an extrinsic, as opposed to intrinsic, property because toxicity is a function of the amount of material. (Mass is an extrinsic physical property.) Toxicity is the ability of a material to injure a living organism by other than mechanical means. The toxicity of a material is not altered by the way it is handled, by its temperature, or by its physical state, but the toxic hazard posed by the material may be greatly influenced by these and by many other external factors. The *toxic hazard* of a material is the likelihood of injury and an evaluation of hazard must consider, in addition to toxicity, physical and chemical properties, physical state, method of handling, and other factors that influence the probability of contact with significant amounts of the material.

Of course not all hazards are toxic hazards. Hazards arise from various forms of energy and from mechanical sources as well as from toxicity. Mainly on the basis of historical precedent, mechanical hazards such as falling, tripping, caught in or between, and fire/explosion are considered to

1

be the province of people who call themselves safety engineers, safety professionals, or safety specialists, and that field is generally called *safety*. All other hazards, and in particular those associated with materials, pressures, and energies encountered by man in his occupations, fall into the field called *industrial hygiene*.

Routes of Entry

While the inherent toxicity of a material is not dependent on how that material gets into the body unless the material is altered chemically by entry, the route of entry is normally specified in expressions of absolute or relative toxicity. This is so because the efficiency of absorption of materials by the body varies with the route of entry, and injection with a hypodermic needle is the only way of being certain that 100% of a dose has been absorbed. Even though drug abuse may be an important industrial problem, injection of industrial chemicals is not. Instead, attention must be focused on entry by the routes of ingestion, skin absorption, and inhalation.

Ingestion

Workmen rarely eat or drink the materials they handle, even in a candy factory, and therefore ingestion is an important route of entry only in cases where the *acute* (short term or single dose) or *chronic* (long term or repeated dose) toxicity by this route is very high. For materials with high oral toxicities, the small amounts that may be transferred to the mouth from the hands on a sandwich or a cigarette may be enough to be cause for concern. This is so with most radioisotopes, some of the metals, and many of the organophosphate insecticides. In general, if the material has an acute oral LD_{50} (lethal dose for 50% of the animals fed) of 1.0 mg/kg or lower, precautions against oral contact should be taken if ingestion is a possible route of entry.

Note that the word *poison* has been avoided. Poison has legal connotations as it is defined in several laws that need not concern us here. Toxicologists and environmental health engineers usually avoid using the word *poison* except in circumstances required by law (labels, for instance) or to mean simply *too much*.

Skin Absorption

Rather than being an absolute barrier, man's skin acts as a selective filter. Some materials such as water-soluble salts are almost absolutely excluded but others such as phenol, hydrogen cyanide, and aniline are readily absorbed through the intact skin. Intact skin is more of a barrier than is abraded or lacerated skin, indicating that the skin does offer at least some resistance to penetration of all materials. Even though some materials penetrate the skin very rapidly and efficiently, most do not; only actual experimentation is likely to determine whether or not skin absorption toxicity of a material is high enough to make this route of entry a potentially important one.

Inhalation

The lungs were designed by nature to provide an extremely efficient method of getting gases into and out of the body. Although the lungs are best equipped to handle oxygen and carbon dioxide, any material dispersed in air is provided an excellent opportunity to diffuse into the bloodstream if that material reaches the *alveoli* (the tiny sacs in the lungs where the actual gaseous exchange takes place). Inhalation, then, is usually an efficient way of introducing a gas or vapor, or even a solid particulate material, into the body; this is just as true for a man filling a drum with carbon tetrachloride as it is for a patient being anesthetized with ethyl ether.

Man breathes from 4 to 10 cubic meters (m^3) of air (140 to 350 ft^3) during an 8-hr shift and therefore can inhale a considerable amount of a material even if that material is present in a rather low concentration in air. Furthermore, many materials are hazardous to inhale in concentrations that neither smell bad nor irritate the respiratory tract. For all of these reasons, and because of the direct route available from lungs to target tissues through the blood, inhalation is the most important route of contact with materials handled in industry.

Mode of Action

The action taken by the material on the body and/or the action taken by the body because of contact with the material can range from nil to a complex network of interrelated activities. Inert gases such as helium and methane pass through the lungs and into the bloodstream but apparently do nothing more than dissolve in body fluids unless their concentration is sufficient to dilute oxygen to hazardous levels. Similarly, many materials appear to have no effect after ingestion or gross skin or eye contact. Such materials are regarded as being physiologically inert (not *nontoxic*) because their toxicity is so low. Nevertheless, in every case, if the dose is large enough, the concentration is high enough, or the contact is prolonged enough, some effect will be experienced.

Action at the Site of Contact

All irritants have an action (irritation) at the site of contact. In many cases that action is severe enough and rapid enough to warn of a contact that may be dangerous if the warning is neglected. On the other hand, fibrogenic dusts that are not irritating in the usual sense of the word and some other materials also act at the site of contact. Action at the site of contact does not necessarily preclude absorption into the body, although hyperemia, edema, erythema, and so forth (see Appendix B for definitions) arise from the body's defenses against this kind of material.

Absorption into Blood

Whether or not there is any action on the skin, lungs, or stomach, the material may be absorbed into the bloodstream from the site of contact. This never happens with perfect efficiency and, therefore, some sort of a distribution coefficient may be determined if enough is known about the absorption process.

The distribution coefficient is usually calculated on the basis that an equilibrium has been attained. When materials are being absorbed into the body, processes at equilibrium are usually much less important than is the rate of absorption, and that rate is usually variable and not known (whether or not the distribution coefficient is known). (See Chapter 3 on carbon monoxide.) Furthermore, the distribution coefficient may be modified by anything that has an effect on the barrier to which the coefficient applies. Nutritional state of the body, presence of other materials, and action or aftermath of disease are only a few variables that may alter distribution coefficients (and rates of absorption). Therefore, the distribution coefficients must be used with caution.

Action on Blood

Under most circumstances, the blood simply transports material that has been absorbed. Some substances, however, have an action on the blood itself as exemplified by carbon monoxide and aromatic amines. These materials act on hemoglobin to reduce the ability of that protein to carry oxygen from the lungs to other tissues. Direct action on the blood other than interference with hemoglobin is very unusual except for some living disease entities.

Being carried throughout the body by the blood, the material has an opportunity to be absorbed into all other tissues. Even though one organ or organ system may be the most affected, the material usually becomes well distributed throughout the body.

Absorption into Tissue

For any other tissue in the body, a second distribution coefficient usually can be calculated to represent absorption from the blood. As with the more external coefficient, care must be taken in the use of an internal coefficient that may be modified by exercise, nutritional state, temperature, and so forth. Furthermore, rate is again usually more important to the toxic action than is the condition at equilibrium.

Some materials are preferentially absorbed into the body's fat depots. These are materials that are good fat solvents or are very soluble in body fat (*lipid*). The main toxic action of a material usually is not exerted on lipid but, instead, the fat is simply a place where the material is stored to be released slowly over a relatively long period of time. The importance of this storage and release to the toxic potential of the material is a function of the amount of fat present as well as of the solubility of the material in fat. These are good reasons why the toxicity of a material may vary widely from animal to animal and from species to species.

If a measure of the total amount of a toxic agent in the body (the *body burden*) at any time is plotted against the duration of a continuous exposure, the resulting curve, which may be nearly linear at first, always approaches equilibrium asymptotically. This generalization applies regardless of the route of entry; the solubility of the material in body fluids or fat; or the rate of uptake, reaction (*detoxification*), or excretion, provided only that the exposure is essentially continuous and not fatal. The generalization is also true if *organ* or *tissue* is substituted for *body*. Straight-line approximations never can be more than just that—approximations.

Once the material is in the body, it tends to disappear in time by reaction with materials already present or by excretion in the breath, urine, feces, sweat, hair, or nails. Some of the material may be stored in tissues such as fat (most of the fat-soluble solvents, for instance) or in bone (the bone-seeking heavy metals), to be eliminated later if at all. Disappearance of the toxic agent may be a very complex phenomenon, with several different "half-lives," each for a different cause. Nevertheless, once exposure has ceased, the body burden tends to diminish at a relatively rapid rate at first, to approach zero or a steady state asymptotically at some later time. Disappearance (or decay) curves tend to be hyperbolic or exponential in nature.

Site of Action

Usually one organ or one organ system will be more sensitive to the material or may absorb more of the material than any other, and action on that organ or system may completely overshadow any other effects. The major site of action may be the kidneys, liver, heart, brain, or even a small portion of one of these organs such as the glomerulus in the kidney. Or, a whole system may be attacked such as the action of benzene on the blood-forming (*hematopoietic*) system or of ethyl ether on the central nervous system (*CNS*). Or, the attack can be localized in a system, such as the action of many of the organophosphates on processes taking place at neural synapses. Perhaps most often, the toxin may interfere with one or more enzymes or enzyme systems, causing effects that may be manifested in many almost inexplicable ways. Even though action at one site may be the major effect of the material, secondary actions may well take place even though they are ignored because they are (or are thought to be) relatively unimportant.

In most cases where the toxicity of a material has been investigated, a general indication of the site of major action of that material is possible. Knowing that site, however, is far from knowing just how the material actually causes the injury. For example, carbon tetrachloride has long been known as a material able to cause severe injury to the liver and kidneys. Despite the hundreds of thousands of man-hours expended, the exact pathway of that toxic process is still disputed. And, even today, the reason why ethyl ether causes unconsciousness is not known, nor the reason why aspirin relieves a headache or reduces a fever.

Toxic Action

Although toxicity is a property of matter, it is a biological property, implying that the variability of response will be much greater than that found in physical systems if only because biological systems are incredibly complex when compared to any physical system. One of the peculiarities of toxicity is that it may differ from individual to individual in addition to differing from species to species. Furthermore, most experimentation with toxicity must be done with animals instead of with man. One of the largest problems the toxicologist must face is how well he can predict the effect of a material on man when he has data only from animals, plants, or protista.

Sometimes after much hard work coupled with a bit of luck a toxicologist is able to put together all the puzzle pieces to form a picture of just how a chemical exerts its toxic action. Sometimes the hard work seems only to add complications to an already complex situation, but each bit of understanding does tend to allow the next one to come a little easier. One of the benefits of the accumulating pool of toxicological knowledge has been a better understanding of the reasons why different species of animals differ in their responses to a toxin. This kind of understanding has enabled better and better predictions of human response from animal data.

One of the earliest observations was that irritants tend to have similar actions on all species of animals tested and hence prediction of their effects on man became almost routine. Surprises still occur, however, especially when the irritant needs water (sweat) in order to do its worst to the skin and the species used for testing does not sweat (as is usually the case) or when the material is a sensitizer and not a primary irritant. Nevertheless, and despite the surprises, animal testing usually enables a very good prediction of the effects of irritant materials on man.

When the absorbed material acts by interfering with one or more enzyme systems, animal testing will enable good predictions if the same pathway is followed in man, but a good prediction of toxicity from one species to another does not imply that the toxicology is the same in both species. Animal species may be different physiologically and, therefore, the mode of action of a toxin may not be the same in all species. One recognition of this fact is the use of animals that are physiologically similar to man by industrial toxicologists. It is unusual for an industrial toxicologist to use an amphibian, a bird, a reptile, an insect, or a tree as a test organism although fish may be used to test industrial waste waters. Instead, he almost invariably chooses a mammal as high on the evolutionary scale as space and budget will permit. He wants organs, organ systems, enzymes, and physiology to resemble those of man as closely as possible so that the mode of entry and especially the mode of action of the toxin will parallel that in man.

2

LABORATORY DETERMINATION OF TOXICITY

The science of toxicity is called *toxicology*. Toxicology is the study of the effects of chemicals on biologic systems, with emphasis on the mechanisms of harmful effects and of the conditions under which those harmful effects occur. Note that neither *man* nor *animal* is specified in this definition or in the definition of toxicity. It is perfectly proper to investigate the toxicity of a material to fruit flies or annual bluegrass, and both are used as test organisms regularly.

Most basic to the study of toxicity is the concept that the effect of a chemical on a biologic system is and will be proportional to the size of the dose of that chemical. This concept was refined by Orfila in 1815 (in the 16th century Paracelsus said "...it is the dose which makes the poison.") and is called the *dose-response* relationship. An extension of the concept is that for all materials there is a dose small enough to have no harmful effect at all and therefore, for all materials, there exists a *threshold* dose that must be exceeded for injury to occur. Both the generalized dose-response relationship and the threshold concept have been demonstrated innumerable times, but some theorists now feel that there may be no real threshold for effects such as carcinogenesis and mutation. Nevertheless, arguments for dose-related effects and for a threshold are strong regardless of the nature of the affecting material or energy [1].* Controversy about happenings and effects at the low end of the dose-response curve will likely continue because absolute proof is quite difficult to obtain. This is so because when the magnitude of the effect becomes smaller, an ever-smaller percentage of subjects will experience that effect, requiring more and more experimental subjects to be exposed before the effect can be demonstrated. In essence, as the magnitude of the effect (or its incidence) tends toward zero, the number of experimental subjects needed to demonstrate the effect tends toward infinity.

Industrial Toxicology

Industrial toxicologists are concerned with the effects of industrially encountered materials on man. The toxicologists commonly experiment on animals but usually are interested in the results only insofar as those findings can be translated into a determination or a prediction of the effect on man. Industrial toxicologists usually confine their efforts to the effects of materials and leave the effects of energies such as sound and radiation to others who specialize in those fields. The industrial toxicologist may not be employed by industry but may work in a government laboratory, at a university, or for a consulting firm. His interests, not his place of employment, determine his profession.

With a material at hand to study, the toxicologist will need or will determine the following information:

A. General chemical and physical properties of the material. Usually molecular configuration, physical state, boiling and melting points, vapor pressure, and chemical reactivity are necessary, not only because they may be components of the handling hazard, but also because they can be used to identify the tested chemical positively. Also, for this reason, a chemical analysis is often obtained. Occasionally other information will be needed, depending on the potential uses of the substance. For instance, if a plastic film is to be used as a food wrap, the toxicologist might have to determine what additives were used in the manufacture of the film as well as the probability of those additives being leached out of the plastic and into various kinds of foods.

B. Procedures for detecting and determining the toxic agent. Even established analytical procedures may not suffice for determining a material

---

*References to the literature are collected at the end of each chapter.

in the micro and submicro concentrations typical of toxins in the air or in body fluids. Then, too, the toxicologist is faced with the problem of finding and quantitating the material in the presence of many other substances that may be present in much higher concentrations. The analytical problem is being solved rapidly with newly developed methods, but occasionally this new ability has posed severe problems for the toxicologist. One such problem is that he is no longer sure that there is such a thing as a concentration of *zero*. Recognition of this uncertainty (which has always existed) has raised philosophical and legal problems that the toxicologist must confront even if he would like to ignore them.

C. The mode of action of the toxic material. At the very least, the toxicologist is expected to determine the more obvious signs of injury. (*Signs* are observable things such as stupor, irritated eyelids, a limp, or a broken fingernail. *Symptoms* are characteristics of the disease experienced by the individual such as an upset stomach, a headache, or a far-out wild "trip." The toxicologist can observe signs of injury in experimental animals or plants, but he can determine symptoms only if his subjects are human.) In most cases, more information, such as the organ or organ system most susceptible to attack, will be found and any lesions ascribed to the toxin will be designated. Very occasionally, a breakthrough is made after the application of much work or much luck or both, and the complete mechanism by which toxic injury occurs can be detailed.

D. The dose-response relationships. Putting dose (the actual amount given) and response on a quantitative basis is what distinguishes the toxicologist from the magician or tribal medicine man. The toxicologist is usually very precise about dosages (doses per unit of body weight) and expresses them as so many grams or milligrams per kilogram of body weight or as so many parts vapor per million parts of an air-vapor mixture (ppm) for so many minutes, hours, or days. He is not usually so precise about response unless the response is death, in which case the exact number of subjects that experienced the response will be noted. Many attempts have been made to quantitate responses such as by indicating the degree of eye or skin irritation on a numerical scale [2, 3], but none is universally accepted. Furthermore, those scales of response that are used are almost always subjective in nature and thus depend on the experience and judgement of the observer. Nevertheless, quantitated or not, the response to a given dose will normally be well described, and thus the dose-response relationship will have been defined.

Dose-Response

No two biologic entities, even within the same species, react exactly alike to an insult from a toxic chemical agent. This variability of response is even greater between species than it is within a given species. It is the basic problem confronting the toxicologist in his choice of experimental subjects, especially when he must extrapolate his data to species entirely different from that used for his experiments.

To enable a comparison of the toxicity of materials over long time spans, the toxicologist may use standardized strains or varieties of his experimental subjects. When he doses a laboratory rat today with a material, the results will be similar to those observed when that same material was given to rats of the same inbred, genetically stable strain 2 decades ago [4]. And, of course, the results he obtains will be similar to those found by his colleague at a different location using the same variety of laboratory rat [5]. The more he uses the same strain of the same species, the more he learns about how that subject reacts to different chemical stresses. The better he knows his subjects, the better any extrapolations of the data are likely to be.

Statistical Expressions of Data

Suppose compound A were administered orally to groups of rats, each group being large enough so that the percentage dying from each dose was known accurately. Suppose further that the first group of rats was given 1.0 mg of compound A per kilogram of body weight (mg/kg) and for each succeeding group the dosage was increased, the final group receiving 100 mg/kg.

Without further thought, we would probably just look at the number (or percentage) that died at each dosage level. We would expect that the lower amounts would kill fewer rats than would the higher ones. That is, 4 mg/kg would result in a smaller percentage dead than would 8 mg/kg. Finally, we would expect that with a high enough dosage all animals in that group would die. The name given to this kind of reporting is *cumulative frequency*.

Suppose, instead, that we examined the increase in the number dead caused by a specified increase in the dosage level. Offhand, we probably would not be able to guess whether or not an increase from, say, 4 to 8 mg/kg would kill more rats than would, say, an increase from 8 to 16 mg/kg. Nevertheless, this sort of *incremental* effect can be examined statistically. Incremental frequency of response is, in fact, the basis for the standard (Gaussian) curve of error. (Incidentally, doubling—or nearly doubling—the amount given to each successive group of animals is standard practice for reasons that should become apparent.) Consider the following (rounded) data for compound A:

Dosage	Percent Dead	
(mg/kg)	Cumulative	Incremental
1	3	3
2	10	7
4	26	16
8	46	20
15	72	26
30	87	15
50	96	9
100	100	4

Incremental data are plotted against dose in Fig. 2-1. The resulting histogram is skewed, indicating that at high dosage levels the same increment of dosage has much less effect than at low levels or, conversely, that a small incremental increase in dosage has a much greater effect at low than at high levels. This even "feels" right because, after all, an increase of 1 mg/kg at a dosage of 1 mg/kg actually doubles the dose, while at a level of 100 mg/kg an increase of 1 mg/kg is only 1%, which should have less effect. Saying that equal *percentage* increments should have roughly the same effect at both high and low dosage levels is equivalent to saying that a logarithmic transformation of the data would be expected to normalize the curve. Such a transformation was used for Fig. 2-2, where the same data

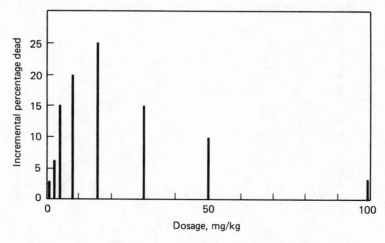

Figure 2-1 Incremental response vs. oral dosage for compound A.

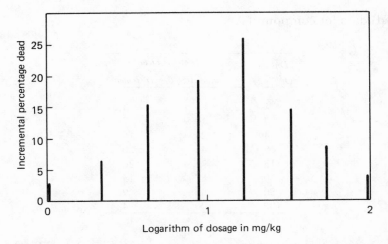

Figure 2-2 Incremental response vs. logarithm of oral dosage for compound A.

are plotted. The resulting histogram looks much more like the normal distribution than does the one in Fig. 2-1.

Interpolating from one dosage to another is very difficult on graphs such as those in Figs. 2-1 and 2-2. Plots of cumulative (instead of incremental) response provide a much better opportunity to fit a smooth curve to the data. Figures 2-3 and 2-4 show the data plotted in this manner. A logarithmic transformation of dosage (Fig. 2-4) causes the curve to approach being a straight line, but an S shape is still apparent.

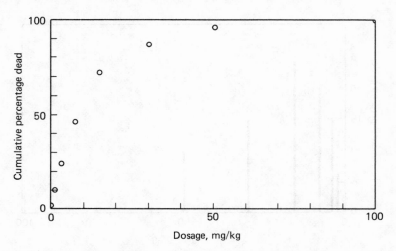

Figure 2-3 Cumulative response vs. oral dosage for compound A.

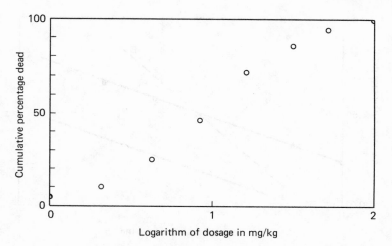

Figure 2-4 Cumulative response vs. logarithm of oral dosage for compound A.

Fortunately, graph paper is available that allows a plot of the probability of cumulative response versus either the dosage or the logarithm of the dosage. Furthermore, if the data are normal, the resulting curve on that graph paper will be a straight line, allowing easy interpolation. Cumulative death rate data for compound A are plotted on arithmatic probability paper in Fig. 2-5 and on log normal paper in Fig. 2-6 (note that neither 0% nor 100% can be plotted in either case). A straight line fits the data points in Fig. 2-6 reasonably well as usually is the case for experiments of this

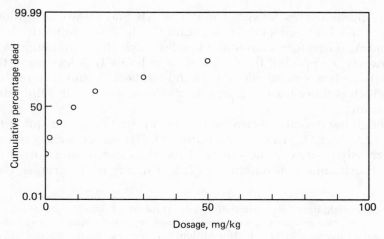

Figure 2-5 Arithmatic-probability plot of dose-response for compound A.

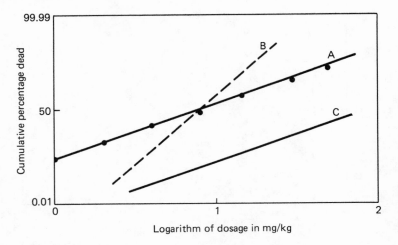

Figure 2-6 Log probability plot of dose-response for compound A.

kind. The line passes through the point (50%, 8.0 mg/kg) and therefore the dosage expected to kill 50% of the animals is 8.0 mg/kg.* Accordingly, that dose (lethal dosage for 50%) is designated the LD_{50} (or, more accurately, the acute oral LD_{50} for that experimental animal). Obviously, the same graph could be used to estimate the LD_{99}, the LD_{10}, or any other expressions of the lethality of the compound.

Interpretation of Data

Dosage-response curves for other materials (B and C) are also plotted in Fig. 2-6. For all dosage levels, compound C kills fewer animals than does compound A. Therefore, compound C is less toxic than compound A. Consider, however, compound B. At low dosage levels B is less *potent* than A, but at high levels it is more potent even though both A and B have the same LD_{50}. Which is more toxic? There is, of course, no simple valid answer to that question [6].

Although the toxicity of two materials can be compared quantitatively by using LD_{50} (or LD_{10}, LD_{99}, or even the LD_0 [7]) values, such a comparison may be grossly in error if the slopes of the dosage-response curves are not similar. Furthermore, although not evident in any of the graphs, the slope

---

*The best straight line (by eye or analytically) through all the data points *must* be used. Connecting each point to the next with straight lines or using a simple or complex curve through the data points is incorrect. If a straight line does not fit the data at all, either a mistake has been made or another technique such as direct determination of the geometric mean must be used.

of the dosage-response curve for a material may not be the same for all kinds of response, further complicating the matter of comparing the toxicity of different materials.

Selection of Subjects

The industrial toxicologist has several criteria that must be met when he chooses animals for his experiments. Most important is that the animals have metabolic processes similar to those in man when results of the tests will be used to predict toxicity to man. The only animal with a metabolism identical to man's is man and therefore man would appear to be the ideal test species. This is not so. Aside from overriding humanitarian considerations, a man is much more expensive to feed and house than almost any other animal and furthermore man lives at a slow rate. In 2 years a mouse or a rat can live its natural lifetime (on experiment or not), while a man would have lived only a few percent of his during the same period of time. Also, if allowed to breed, the mouse or rat family will be represented by several generations in 2 years (and fruit flies will have had several tens of generations in that time). Man is used experimentally when no other animal will yield the correct information and then only when the safety of the experiment has been demonstrated by previous animal or inadvertent human exposures.

Experiments with Animals

There is no one ideal experimental species and even if there were, most toxicologists would still use several animal species for important experiments to determine between-species differences. Criteria for experimental animals vary with the purpose of the experiment, but in general the animals used should be physically small, inexpensive, easily handled, easily fed with inexpensive food, not dangerous to man by direct attack or by carrying man's dangerous diseases, and (if possible) of a standardized strain of the species used. Ideally, the animals should breed easily in captivity and also be clean, neat, and at least not offensive to smell.

Standardized strains of mice and rats are now easily available and these animals tend to dominate in toxicity testing. Others used extensively include rabbits, guinea pigs, hamsters, gerbils, and dogs. For special purposes cats or swine may be ideal; for others, monkeys or apes are chosen. Birds and amphibians are rarely used.

To protect experimental animals from needless cruelty or inhumane treatment, rules have been established by several organizations. Those of the American Physiological Society are representative:

Only animals that are lawfully acquired shall be used in this laboratory, and their retention and use shall be in every case in strict compliance with state and local laws and regulations.

Animals in the laboratory must receive every consideration for their bodily comfort; they must be kindly treated, properly fed, and their surroundings kept in a sanitary condition.

Appropriate anesthetics must be used to eliminate sensibility to pain during operative procedures. Where recovery from anesthesia is necessary during the study, acceptable technic to minimize pain must be followed. Curarizing agents are not anesthetics. Where the study does not require recovery from anesthesia, the animal must be killed in a humane manner at the conclusion of the observations.

The postoperative care of animals shall be such as to minimize discomfort and pain, and in any case shall be equivalent to accepted practices in schools of Veterinary Medicine.

When animals are used by students for their education or the advancement of science, such work shall be under the direct supervision of an experienced teacher or investigator. The rules for the care of such animals must be the same as for animals used for research.

Many journals where the results of animal experiments are published have a policy of not accepting papers unless the author(s) can state that these or similar rules have been followed.

Experiments with Man

Long before a science of toxicology existed, man was the usual experimental subject and, most often, the subject was the experimenter himself or possibly his family and/or close friends and colleagues. Even today this selection of experimental subjects continues to be favored, especially when the risk is not negligible. The risks of human experimentation should be as low as possible but nevertheless some experiments are more hazardous than others; those persons who know most about the experiments tend to assume the greatest risks. The World Medical Association formulated a code of ethics in 1964, governing the use of humans in all kinds of experiments, including those in toxicology. This code is known as the *Declaration of Helsinki* [8], and its provisions relating to this field are:

I. BASIC PRINCIPLES

1. Clinical research must conform to the moral and scientific principles that justify medical research and should be based on laboratory and animal experiments or other scientifically established facts.

2. Clinical research should be conducted only by scientifically qualified persons and under the supervision of a qualified medical man.

3. Clinical research cannot legitimately be carried out unless the importance of the objective is in proportion to the inherent risk to the subject.

4. Every clinical research project should be preceded by careful assessment of inherent risks in comparison to foreseeable benefits to the subject or to others.

5. Special caution should be exercised by the doctor in performing clinical research in which the personality of the subject is liable to be altered by drugs or experimental procedure.

III. NON-THERAPEUTIC CLINICAL RESEARCH

1. In the purely scientific application of clinical research carried out on a human being, it is the duty of the doctor to remain the protector of the life and health of that person on whom clinical research is being carried out.

2. The nature, the purpose and the risk of clinical research must be explained to the subject by the doctor.

3a. Clinical research on a human being cannot be undertaken without his free consent after he has been informed; if he is legally incompetent, the consent of the legal guardian should be procured.

3b. The subject of clinical research should be in such a mental, physical and legal state as to be able to exercise fully his power of choice.

3c. Consent should, as a rule, be obtained in writing. However, the responsibility for clinical research always remains with the research workers; it never falls on the subject even after consent is obtained.

4a. The investigator must respect the right of each individual to safeguard his personal integrity, especially if the subject is in a dependent relationship to the investigator.

4b. At any time during the course of clinical research the subject or his guardian should be free to withdraw permission for the research to be continued.

The investigator or the investigating team should discontinue the research if in his or their judgement, it may, if continued, be harmful to the individual.

Experimental Methods

The work of an industrial toxicologist may be directed toward range-finding studies, chronic inhalation studies, chronic feeding studies, proper labeling of products, satisfying various rules and regulations promulgated by states or nations, or any combination of these major areas along with, perhaps, more basic studies such as those on the mechanism of action of materials. No matter what part of the overall job is his, he works mainly to the end of predicting the effects of industrial chemicals and other materials on people.

Range-Finding

Range-finding is a term applied to a series of tests designed to be relatively inexpensive and yet to provide basic data on the acute toxicity of materials by various routes of exposure. Range-finding tests may consist of some or all of the following:

A. Acute oral (rat or mouse)
B. Acute eye irritation (rabbit)

C. Skin irritation (rabbit or guinea pig)
D. Acute vapor inhalation (rat or rabbit)

In some organizations, all new materials that are to be made, handled, processed, or produced are submitted for range-finding tests. These experiments are designed to alert the industrial physician to the hazards faced by his potential patients and to give him a chance to decide on methods of treating victims well prior to any exposure. In addition, data from these tests allow the industrial hygienist to prepare advice on proper methods of hazard control based on experience with similar materials.

Controls

For any type of toxicity testing at least two groups of animals are used—experimental and control. The control group is as close to identical with the experimental group as is practical, but it does not participate in any experiments. Instead, the subjects in this group are treated in all ways in a manner similar to the way the experimental subjects are treated, but without exposure to any toxins. Control groups are necessary in biological experiments because of the many unanticipated events that can occur. For instance, some sort of infection may become established in a colony of animals. If so, the controls are likely to be affected too, and when the experiment is complete, the experimental animals will be compared with control animals who have had similar adverse effects unrelated to the toxin.

Common adverse effects need not be limited to infection but can take the form of generalized exposure to air or water pollution, to vitamin- or mineral-deficient food, to pests (fleas, etc.) that might become established in a colony, or even to the materials used to control the pests. Control animals must be genetically similar to experimental animals so that when the experiment ends, rational comparisons can be made. Control animals are carefully selected to be similar in age, sex or sex ratio, body weight, and number to the experimental animals.

Another way of solving the control problem is to do before and after studies using each subject as its own control. With this method, especially careful attention must be paid to the requirements for subsequent validity of data statistics. In particular, randomization of experiments must be well done and complete.

Growth Rate

Young adult animals are preferred as subjects for most toxicity testing. Young animals cost less than adults if only because they have consumed less food, housing, and care. More important, these animals are old enough to have passed the hazards of infancy but are still young enough to be growing fairly rapidly. Growth rate can be one of the most sensitive indices of toxic effect. A decrease in growth rate (*Note:* not necessarily a weight loss) may be seen prior to the development of any other sign of intoxication. The average growth rate of a group of animals may well differ from the norm for that species and a determination of whether or not the difference is due to the toxin can only be made by comparison with controls.

Acute Toxicity Tests

Ingestion. To determine acute toxicity via the oral route, a large group of animals is divided into several smaller groups, usually so that the average body weight in each of the smaller groups is about the same. Then each animal in the smaller groups is given the same dosage of material (in grams, milligrams, or milliliters per kilogram of body weight), usually by stomach tube. A progression of dose based on 2, e, $\sqrt{10}$ or 10 is usually used so that each group of animals receives some multiple of the dose given to the preceding group. Actual doses depend on a guessed value of the LD_{50} and are designed to be both smaller than and larger than that guessed value. If the guess was wrong, the experiment usually must be repeated, but this time the toxicologist will have a better idea of the doses to use.

The laboratory rat is the animal normally used, and each group receiving the same dosage will consist of from five to ten animals. Rodents are good animals for this type of test because they cannot vomit. If a dog or cat, for instance, were dosed by stomach tube with a material that caused nausea, it would promptly disgorge the contents of its stomach. A rat cannot do this and once in the stomach the material will stay there to do all the damage it is capable of doing. Mice are rodents and cannot vomit, but mice are not quite so predictable as rats in their behavior toward toxic materials so industrial toxicologists usually prefer rats for most acute oral testing.

Eye irritation. Effects of substances on the eye are ordinarily determined with rabbits [9]. These animals have large eyes that react to most materials in a manner similar to that of a man. In addition, rabbits are not aggressive and even if the material hurts, they are not likely to attack the experimenter.

A drop of the material (or a drop of the material suspended in an inert liquid) is instilled into each of the rabbit's eyes. One eye is then washed thoroughly with water about a minute after contact. Both eyes are examined periodically, first for signs of injury and later for signs of healing if injury has occurred.

This test not only gives information about the irritating potential of the material and how damaging it can be to the eye but also tells the toxicologist whether or not washing with water is likely to be helpful if contact does occur. Washing with water is not harmful if a foreign substance has been introduced into the eye and most often is found to be helpful. Washing is likely to reduce the severity of injury and also to assure prompt healing. The efficiency of a thorough washing with water is not predictable by a knowledge of the solubility of the toxin in water except in a very rough way. Some insoluble materials are washed from the eye very effectively by the mechanical action of the water and a true solution need not be formed.

Many tests have shown that if a wash with water is not effective, then washing with other materials is not likely to be effective either. Especially in the case of acids and bases, the temptation is to provide a weak solution of a neutralizing chemical as the antidote for eye contact. Aside from the fact that the eye is a poor place in which to conduct chemical reaction experiments, diluting and flushing with water is usually more effective than neutralizing. This is so for at least two reasons. First, plenty of water is almost always easily available, and eye washing can be continued easily for the recommended 15 min with flowing fresh water. Such is usually not the case with neutralizing solutions. Second, in almost all areas where eye contact can and does occur, water is much more freely available than is any kind of specific antidote and the time lost in searching for a bottle of antidote while at least partially blind may be the difference between sight and blindness.

Even though the actions to follow after eye contact with an irritating material are now fairly standard, eye irritation testing is still conducted. This is mainly so that the industrial physician can be alerted on the basis of data rather than guess and also to allow a recommendation to be made concerning the type of eye protection that should be used.

Special eye protection may not be necessary if pure water is being handled at atmospheric pressure. On the other hand, if the job involves possible eye contact with concentrated sulfuric acid, the eyes must be completely protected. And, it is almost inevitable that the more complete the protection, the more uncomfortable it is to wear. Therefore, in order to avoid "crying wolf" when such a cry is not necessary, environmental health engineers try to be as reasonable as they can in their specification of the kind of eye protection to be worn.

In most cases, absolutely complete eye protection is recommended only where eye contact is possible with a substance that will cause at least some loss of sight. Complete eye protection is provided only by an absolute barrier such as chemical worker's goggles or a full-face gas mask. On the other hand, the protection offered by safety spectacles is considered adequate if contact is possible and if such contact is not likely to result in an injury severe enough to cause loss of time from work. For cases between the two extremes, side-shield safety glasses, louvered goggles, or face shields are recommended.

Skin irritation and absorption. Skin irritation testing has three phases: repeated, acute, and hypersensitization. Rabbits are normally used for the first phases [9], and the guinea pig or another species may be used for sensitization experiments. Repeated irritation work is done by applying the material to abraded and nonabraded portions of the shaven belly or back. The test area is then covered with a bandage to prevent the animal from licking the area as well as to keep the material in place; the animal is re-caged to be examined a day later. If no effect is found, new applications are made daily until the test is finished, usually in 2 or 3 weeks.

If signs of irritation are present after the first 24-hr application or subsequently, then (depending on the severity) acute testing may be done. For this, the rabbit is held immobile in a stock. Several patches of the material are applied, to be removed at selected time intervals. As before, the skin is examined later for evidence of effect, and the minimum time required to produce the effect is noted. The efficacy of washing the skin with plenty of flowing water after timed contact may also be determined.

During repeated skin applications the animal(s) may become ill or die. Unless there has been an obvious opportunity for ingestion, the material is then assumed to have been absorbed through the skin. Illness or death during repeated skin irritation experiments is the usual signal for skin absorption testing. For this test, the whole trunk of the animal (again usually the rabbit) is clipped and shaven. The material is applied in graded doses to small groups of animals and the shaven area is covered with a bandage and an impermeable plastic or rubber film. The number dying in each group following the application is noted and the acute LD_{50} for skin absorption is calculated.

If there is some basis to suspect that the substance may be a skin sensitizer, a skin sensitization test may be called for [10, 11]. Usually the guinea pig is the animal used, although no animal species responds to sensitizers in a manner identical to that of man [12]. (Consider people who break out with a poison ivy rash after petting an exposed dog while the dog shows no ill effects whatsoever.) Whether guinea pigs or human volunteers are used, the subjects are first given a sensitizing dose by skin contact, followed by a period of a week or two with no contact. At that time, a challenge dose, small in amount or concentration compared to the initial dose, is administered, possibly in graded amounts, and the number or percentage of those responding is noted. (See Chapter 9.)

If an experimental animal shows sensitization, then man is usually assumed also to be susceptible. On the other hand, if animals are not sensitized to the material, man still may be and the only certain test is with a sufficient number of human subjects. What constitutes a sufficient number is quite controversial and may range from a few tens of people to a few thousands, depending on the potential use for the material and the possible consequences of sensitization from overexposure.

Vapor inhalation. Acute vapor inhalation toxicity is determined by having a group of four or five animals (usually rats) inhale air saturated with the vapor of the material for periods of time ranging up to 7 or 8 hr. If the animals survive the longest exposure, no other tests are conducted. If that one test should prove fatal to some of the animals, however, other exposures will be planned, still to the saturated vapor but for shorter durations until a duration is found that allows survival of all the animals. A rough approximation of the concentration of material inhaled by the animals is found by dividing the weight of material evaporated by the volume of air that flowed through the evaporation flask.

Results of the acute vapor inhalation test are used mainly by industrial hygiene engineers to plan emergency procedures necessary in case of a large spill of the substance or other situations where very high vapor concentrations may be encountered. The engineer must be able to estimate the exposure duration to essentially saturated atmospheres that will still allow survival and to plan that no human exposure will approach that duration under any circumstances.

Particulate inhalation. Acute inhalation tests are rarely conducted with aerosols because of the extreme difficulties involved in setting up and maintaining air suspensions of dusts, mists, or fumes where the concentration and particle size range are both known and constant [13, 14]. As an alternative, the results of acute oral tests can be used to predict the effects of acute overexposure to an airborne particulate. This is possible and practical because most of the mass of inhaled particulate material reaches the stomach instead of the depths of the lungs and toxicity is related to the mass (dose) absorbed. The filtration/impaction apparatus in the nose and throat removes large particles from the airstream and those particles are subsequently swallowed. Even material that gets deep into the bronchi is removed by cilliary action that sweeps it up to the throat.

Subacute Testing

Range-finding data may be adequate for many purposes, but if the material is likely to be in potential contact with more than a certain number of people, more extensive testing may be necessary and justified. Although *sub* means *under* or *below*, tests involving more time than that associated with acute but less than that with chronic are called *subacute* (*sub* refers to the dose used and its effects, not to the time involved).

The tests described for skin irritation actually qualify under the term *subacute* because they involve several contacts. For inhalation studies, subacute implies that animals will be exposed to known, precisely controlled concentrations of an air contaminant for short periods of time, perhaps daily, for a few days or weeks [15, 16]. Subacute skin absorption and feeding studies could be conducted, but they rarely are.

The number of people who must be at risk before this type of testing is justified is not constant but varies with circumstances as well as with the potential severity of the hazard. In general, the more subtle the hazard, the more serious the consequences of overexposure, or the greater the number of people at risk, the more likely subacute testing will be done. Every decision of this sort necessarily must include the economic picture and if a material seems to have the potential for contacting a large or an ever-increasing number of persons, more extensive testing may be done even if other considerations do not seem to provide adequate justification.

Chronic Testing

For a material that may contact thousands of workmen, customers, and/or the general public, comprehensive testing is economically justified. In most circumstances, such testing will take the form of chronic feeding studies and/or chronic inhalation studies. In either case, the objective is to determine the subtle rather than the gross manifestations of toxicity and to find a level of no effect.

Chronic feeding. Chronic feeding studies may range in duration from 90 days (and be called subacute) to 2 years, depending on the objective of the study. Where there are legal requirements to be satisfied (for example, chemicals that may contact food), these studies may be done simply to obey the law and to provide data to the Food and Drug Administration or another agency. Other studies may well be instituted with a more nearly scientific objective in mind, but the result of either type of experiment will be a set of data that, when analyzed, may call for supplementary studies to elucidate a facet of oral toxicity.

A chronic feeding study is started by choosing groups of experimental and control animals from a larger colony. From the start of the experiment, each group is fed a diet containing a prespecified amount of the material to be tested, ranging from background (control animals) to a level high enough certainly to cause some effects. At the conclusion of the experiment, all the animals are sacrificed and examined for evidence of gross effects as indicated by the appearance and weight of the major organs; then tissues from those organs are examined microscopically to determine any changes that have occurred at that level of organization.

Most chronic oral studies now include tests to determine whether or not the material has any adverse effects on a pregnant female or on the fetuses she carries. *Teratogenic* effects (those on the fetus) have long been known, but the thalidomide incident that surfaced in 1962 caused much more searching for this type of effect than had been done up to that time. Now chronic oral tests may be done not only on one but on several generations of animals, all of them exposed to the same concentration of the material in their diet.

The number of animals at each dietary level and the number of dosage levels used may be variable or they may be set by legal requirements. In any case, one objective of the study will certainly be to gather as much information per dollar spent as possible. Obviously, the fewer animals used per dosage and the fewer the dosage levels, the less the total cost. On the other hand, animals are variable in their responses to any toxin, and a sufficient number must be used so that the statistics will be believable after the test is over.

Other factors that must be considered include the duration of the experiment (how long is chronic?), the number of animal species to use and which ones, and whether or not teratogenic experiments should be conducted. All the animals must be housed, fed and watered, and cared for during the experiment; at the end, all must be sacrificed and carefully examined. Obviously, costs can mount very rapidly, and the overall price tag for a chronic feeding experiment conducted by or for industry is at least a few tens of thousands of dollars and may run as high as a few hundred thousand if there are many complications.

Chronic inhalation. Chronic inhalation (vapor or aerosol) experiments have all the complications of chronic oral experiments along with several others. Incorporating a chemical into food or water at precise levels is not particularly difficult to do under most circumstances, but setting up and maintaining a known concentration in air for several months can be very difficult—and nearly impossible for a particulate. In order to know the concentration of a material in air over a period of time, the air must be monitored periodically and preferably with an automatic instrument which will not break down when the experiment is half done (or perhaps worse, nearly done) and which will give true, reproducible results day after day. In addition, the animals must be housed in some kind of a sealed chamber while they are being exposed [17, 18] and usually there must be at least one such chamber for each different atmospheric concentration (including zero for the controls). Exposure chambers are, of course, separate from the normal housing of the animals unless continuous exposures are contemplated (rather than the normal 7 or 8 hr/day).

Inhalation exposure chambers are not inexpensive, nor do they operate themselves. For the usual 6- to 12-month exposure duration the chambers must be under constant surveillance by well-trained people because the consequences of an abrupt rise of the air contaminant concentration past the acutely lethal level, for instance, may be disasterous.

The desired result of chronic exposure by any route is the precise determination of the greatest exposure (dose or concentration) without adverse effect upon any of the experimental subjects [19]. For gases, vapors, and aerosols, finding that level may eventually lead to the setting of control concentrations such as threshold limit values (ACGIH) or acceptable concentrations (American National Standards Institute—ANSI) after confirmation by human exposures. In some cases, the control concentrations so found may be adopted by official regulation agencies in the United States such as the Occupational Safety and Health Administration or the National Institute for Occupational Safety and Health; analogous organizations in other countries may do the same.

REFERENCES

[1]   B. D. DINMAN, "'Non-Concept' of 'No-Threshold': Chemicals in the Environment," *Science* **175** (1972), 495–97.

[2]   A. A. FISHER, *Contact Dermatitis* (Philadelphia, Pa: Lea & Febiger, 1967), p. 27.

[3]   J. H. DRAIZE, "Dermal Toxicity," *Appraisal of the Safety of Chemicals in Foods, Drugs, and Cosmetics* (Austin, Tex.: Association of Food and Drug Officials of the United States, 1959), p. 51.

[4]   C. S. WEIL, C. P. CARPENTER, J. S. WEST, and H. F. SMYTHE, JR., "Reproducibility of Single Oral Dose Toxicity Testing," *Am. Ind. Hyg. Assn. J.* **27** (1966), 483–87.

[5]   C. S. WEIL AND G. J. WRIGHT, "Intra- and Interlaboratory Comparative Evaluation of Single Oral Test," *Tox. Appl. Pharm.* **11** (1967), 378–88.

[6]   W. J. HAYES, JR., "The 90-day LD_{50} and a Chronicity Factor as Measures of Toxicity," *Tox. Appl. Pharm.* **11** (1967), 327–35.

[7]   E. R. HERMANN, "Thresholds in Biophysical Systems," *Arch. Env. Health* **22** (1971), 699–706.

[8]   World Medical Association, "Declaration of Helsinki," *Brit. Med. J.* **2** (1964), 177.

[9]   C. S. WEIL AND R. A. SCALA, "Study of Intra- and Interlaboratory Variability in the Results of Rabbit Eye and Skin Irritation Tests," *Tox. Appl. Pharm.* **19** (1971), 276–360.

[10] W. BURCKHARDT, "Comments Regarding Skin-Sensitivity Tests (Patch Test, Contact Test, Impregnation Test, Combined Test, Alkali Resistance)," *Berufsdermatosen* **18** (1970), 179–88.

[11] B. C. KORBITZ, "Possible Improved Technique of Skin Patch Testing," *Arch. Env. Health* **27** (1973), 409–11.

[12] R. L. POOLE, J. F. GRIFFITH, AND F. S. K. MACMILLAN, "Experimental Contact Sensitization with Benzoyl Peroxide," *Arch. Derm.* **102** (1970), 400–404.

[13] O. G. RAABE, ET AL., "An Improved Apparatus for Acute Inhalation Exposure of Rodents to Radioactive Aerosols," *Tox. Appl. Pharm.* **26** (1973), 264–73.

[14] J. D. MACEWEN, ET AL., "A New Method for Massive Dust Exposures by Inhalation," *Am. Ind. Hyg. Assn. J.* **22** (1961), 109–13.

[15] CHENG-CHUN LEE, J. W. NEBGEN, AND F. J. BERGMAN, "Inhalation Chamber and Method for Monitoring Trifluoroamine Oxide," *Tox. Appl. Pharm.* **13** (1968), 62–66.

[16] E. M. GLAUSER AND S. C. GLAUSER, "Environmental Chamber for the Study of Respiratory Stress in Small Animals," *Arch. Env. Health* **13** (1966), 61–65.

[17] L. T. LEACH, ET AL., "A Multiple Chamber Exposure Unit Designed for Chronic Inhalation Studies," *Am. Ind. Hyg. Assn. J.* **20** (1959), 13–22.

[18] R. G. HINNERS, J. K. BURKART, AND C. L. PUNTE, "Animal Inhalation Exposure Chambers," *Arch. Env. Health* **16** (1968), 194–206.

[19] "The Toxicological Basis of Threshold Limit Values" (A symposium of six papers by several authors), *Am. Ind. Hyg. Assn. J.*, **20** (1959), 341–73.

3
GASES

Strictly speaking, the difference between a gas and a vapor is that a vapor is condensable at room temperature and atmospheric pressure to the liquid state and a gas is not. While that definition is not particularly important to the following discussion, nearly all the materials examined are gases, and most of them are inorganic. Furthermore, these gases cause injury either by asphyxiation or by irritation or both.

Asphyxiants

Toxic Effects

Asphyxiation means suffocation and implies a lack of oxygen at the cellular level. Asphyxiation can result from apnea (drowning, choking), from a decrease in the partial pressure of oxygen inhaled (altitude, dilution), from chemical interference with the transport of oxygen by blood to the cells (hemolysis or the formation of carboxyhemoglobin or methemoglobin), or from chemical interference with the enzyme systems in the cells that utilize oxygen supplied (cyanides, nitriles). Gases and vapors encountered in many areas of the environment can cause asphyxiation by all of these methods.

When speaking of processes that lead to suffocation, many people use the terms *hypoxia* and *anoxia* (and sometimes hypoxemia and anoxemia) interchangeably. All do (or could) mean a lack or a deficiency of oxygen but they are not synonyms. *Hypoxia* refers to a deficiency in the partial pressure of oxygen in the inspired air; *anoxia* refers to a deficiency (not necessarily a total lack) of oxygen in body tissues; the suffix *-emia* refers to blood. Asphyxiation is the end result of severe anoxia whether or not hypoxia is the cause. Substances that act by causing hypoxia are called *simple asphyxiants;* those that exert their effects by chemical reactions are called *chemical asphyxiants*.

The onset of acute anoxia is rarely apparent to the person experiencing this condition unless he has had some training in the recognition of the characteristic subtle symptoms. Although the beginning of anoxia may be signaled by an increase in heart and/or respiratory rate, the trained person usually is aware first of a feeling of mild euphoria. This is likely to be followed by a headache or perhaps by a feeling of great fatigue after mild exertion.

If the anoxia progresses, respiration may become grossly altered and nausea and perhaps vomiting are likely. The feeling of fatigue may increase to the point that the victim collapses, perhaps unconscious, perhaps not. There may be convulsions and gasping respiration until breathing stops. The heart stops a few minutes later.

Nerves are not particularly sensitive to anoxia. Most nerves in the human body can survive complete oxygen deprivation for several hours. On the other hand, the brain is very sensitive; the cerebrum can recover only if the oxygen deprivation has been for less than about 8 min; even 5 min without oxygen can cause some irreversible damage to the cortical centers.

The most sensitive objective sign of anoxia appears to be a diminution of a person's ability to detect slight differences in the brightness of objects [1]. Other components of vision may also be affected, including one called the *visual evoked response* (VER). When a light is flashed into one's eyes, the brain reacts and the response has been evoked by stimulating the sense of vision. When that occurs, a signal typical of the reaction may be detected at the scalp. The signal is very small (microvolts) and detection requires a rather sophisticated signal averaging technique. Changes in the VER have been demonstrated during anoxia caused by carbon monoxide inhalation [2, 3].

Hypoxia

Hypoxia may be caused either by adding a physiologically inert gas or vapor to the inhaled air or by decreasing the oxygen concentration (partial pressure) directly by reaction as in a fire. Materials which have a very low inhalation toxicity and which exert their principal effect by dilution of oxygen (simple asphyxiants) can be classified into four groups:

A. Noble gases: He, Ar, Ne, Xe, Kr
B. Nonmetabolized elements: N_2, H_2
C. Nonreactive compounds: CH_4, C_2H_6, C_3H_8, C_4H_{10}, C_2H_2, CF_4, SF_6, N_2O, etc.
D. The special case of carbon dioxide

Unlike the other materials in this listing, carbon dioxide is not physiologically inert. In fact, it is very active in the body, being the substance that controls the frequency of respiration under most circumstances. In high concentrations, carbon dioxide exerts two principal effects other than mere dilution of oxygen. It causes an increase in the rate and depth of respiration, thus hastening any adverse effects associated with inhalation, and it causes the blood to become more acid than normal, a condition termed *acidosis*. Carbon dioxide is an active rather than a passive cause of anoxia.

Low concentrations of oxygen are encountered in tanks, vats, shipholds, silos, mines, and other poorly ventilated areas. Dilution and chemical or biological oxidation are both rather frequent causes of small disasters in such areas. Almost every season the newspapers carry a story about someone entering a poorly ventilated sewer, for instance, and then collapsing. Often, in rescue attempts, several people will successively enter the same area only to collapse in turn. Finally the fire department will be summoned and a fireman equipped with a supplied-air respirator will enter and remove the victims.

Why does a man who can hold his breath for a minute collapse unconscious after a few breaths of an oxygen-deficient ($< 6\%$) atmosphere?

As it enters the lungs, air at normal atmospheric pressure has a partial pressure of oxygen of about 149 mm Hg. (This is found by multiplying the partial pressure of air by the concentration of oxygen. Air entering the lungs is saturated with water vapor, which has a partial pressure of 47 mm Hg at body temperature. The partial pressure of air is then the barometric pressure minus 47 mm Hg or, at sea level, 713 mm Hg. This times the fraction of oxygen—0.2096—equals 149 mm Hg. See Chapter 11). The partial pressure of CO_2 in ambient air is about 0.2 mm Hg, close to zero. At the break point in breath holding, the partial pressure of oxygen in the lungs is about 70 mm Hg and that of CO_2 is 45 to 50 mm Hg. The lungs, even at the break point, hold a reserve, still supplying O_2 as it is used.

Suppose one inhales air containing 5% O_2. The partial pressure of O_2 in that saturated air is about 35.6 mm Hg. This partial pressure is about the same as that in venous blood and, in fact, to the blood the lungs appear to be oxygen-starved tissue. Instead of adding oxygen to the blood as it passes through, the lungs may even absorb O_2 in a very efficient manner. That blood returns to the heart where it is pumped through the circulatory system, thereby supplying a bolus of oxygen-deficient blood to the brain, which immediately "turns off." The elapsed time from first breath to collapse may be as little as a few seconds if the oxygen concentration is low enough and the victim is breathing hard from exertion, for instance, or from gulping air after breath holding. If the main cause of a low concentration of oxygen is a high concentration of carbon dioxide, the sequence of events will almost always be rapid because of the stimulating effects of excess carbon dioxide.

Transport Interference

Although myoglobin does carry some oxygen in the blood, most is carried by hemoglobin. Heme is a very complex molecule built around iron in the ferrous (reduced or Fe^{2+}) state. When heme is joined to an even more complex molecule, the protein globin, a molecule of hemoglobin (Hb) is formed. One molecule of hemoglobin contains four iron atoms and can carry four oxygen (O_2) molecules. The oxygen is bound loosely (oxyhemoglobin, O_2Hb) and it is easily released when the O_2Hb encounters tissue that has a low partial pressure of oxygen.

Blood has 14 to 16 g of Hb per 100 ml in the average human male. When fully saturated, that much Hb allows each 100 ml of blood to carry about 20.8 ml of O_2. Put another way, each gram of Hb is fully saturated with 1.38 ml of O_2 (the volume of O_2 being measured at standard temperature and pressure, dry—STPD). Some oxygen is carried by physical solution in blood, but at standard barometric pressure the amount is quite small (about 0.3 ml/100 ml of blood) in comparison with that carried by hemoglobin.

Any time the *erythrocytes* (red blood cells) have their hemoglobin altered from the normal forms (Hb and O_2Hb), the ability of the blood to supply oxygen to tissue from the lungs is diminished. The alterations of concern here are the formation of carboxyhemoglobin (COHb) and methemoglobin (MetHb) and the massive destruction of erythrocytes and hemoglobin.

Carboxyhemoglobin. *Carboxyhemoglobin* is like oxyhemoglobin in that carbon monoxide (instead of oxygen) is reversibly bound to the hemoglobin molecule. Each molecule of Hb can combine with four molecules of CO, but the affinity of man's Hb for CO is about 220 times that for O_2. When a COHb molecule encounters a low partial pressure of CO or a high partial pressure of O_2, it will release the bound CO, leaving the hemoglobin none the worse for wear.

Hemoglobin in the blood can carry either O_2 or CO or both on the same molecule, but for every carrying site occupied by CO there is one less site for an O_2 molecule and also Hb carrying one or more CO molecules holds oxygen more tenaciously. Because both CO and O_2 can occupy the same sites on Hb, the amount of COHb in the blood is usually expressed as the percentage saturation, referring to the total capacity of Hb in the blood for either CO or O_2. That is, if the blood has 10% COHb saturation, the maximum amount of oxygen it could carry at that time would be 90% of that carried by Hb saturated with O_2.

Carboxyhemoglobin has an even redder color than oxyhemoglobin does; reduced hemoglobin is a bluish red. Persons suffering from *carboxyhemoglobinemia* (the presence of COHb in the blood) may have a healthy looking skin color that can persist after death.

All of us produce some CO within our bodies at all times, the rate of production averaging about 0.007 ml/min. Most of the CO is released when senile erythrocytes are destroyed; in a hemolytic disease the amount of CO produced can be elevated greatly. As a consequence of *endogenous* (within the body) production of CO, the blood of a healthy nonsmoker who has no exposure to CO contains from 0.4 to about 0.8% COHb saturation [4]. The usual cause for an elevated COHb level is the smoking of tobacco and, not surprisingly, those who inhale the most tobacco smoke are likely to have the highest COHb levels [5, 6].

Carbon monoxide is not metabolized by the body, nor is it excreted by any route other than the lungs; therefore, the total body burden of CO at any time is expressed by a knowledge of the COHb saturation. Coburn, Forster, and Kane [7] formulated a theoretical equation that expresses the relationship of COHb concentrations in blood to several other variables:

$$\frac{A[COHb]_t - B\dot{V}_{CO} - P_{I_{CO}}}{A[COHb]_0 - B\dot{V}_{CO} - P_{I_{CO}}} = \exp\frac{-tA}{BV_b} \tag{3-1}$$

where $A = \bar{P}_{C_{O_2}}/M[O_2Hb]$

$B = 1/D_L + P_L/\dot{V}_A$

$\bar{P}_{C_{O_2}}$ = average partial pressure of O_2 in lung capillaries (mm Hg)

M = ratio of the affinity of Hb for CO to that for O_2

$[O_2Hb]$ = concentration of O_2Hb in blood (ml/ml)
  $= [O_2Hb]_{max} - [COHb]_t$

$[O_2Hb]_{max}$ = $(1.38) \times$ (g Hb/ml blood) (ml/ml)

$[COHb]_t$ = concentration of COHb in blood at time t (ml/ml)

$[COHb]_0$ = initial concentration of COHb in blood (ml/ml)

$\dot{V}_{CO}$ = endogenous production rate of CO (ml/min)

D_L = diffusion capacity of lung for CO (ml/min-mm Hg)

P_L = barometric pressure minus the partial pressure of water vapor at body temperature (mm Hg)

$$\dot{V}_A = \text{alveolar ventilation rate (ml/min)}$$

$$P_{I_{CO}} = \text{partial pressure of inhaled CO (mm Hg)}$$

$$t = \text{duration (min)}$$

$$V_b = \text{blood volume (ml)}$$

exp = 2.7182 ... (the base of natural logarithms raised to the power of the following expression)

Equation (3-1) was verified by a series of human exposures to constant concentrations of CO for varying periods of time [8]. As shown in Fig. 3-1, the fit of the equation to the data points (each the average of two to eight values) is very good.

One of the problems that has beset environmental health engineers for

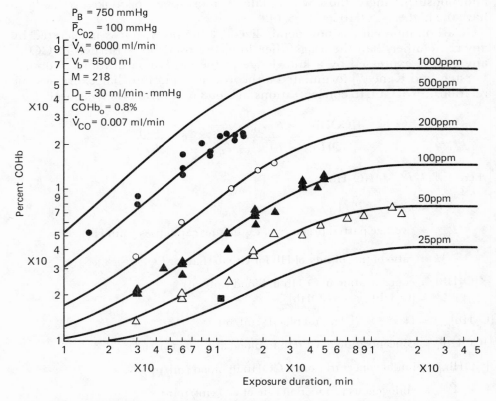

Figure 3-1 Carboxyhemoglobin levels for man as a function of exposure duration and of CO concentration determined by solving the CFK equation.

as long as the profession has existed is that of integrating inhalation exposures. For lack of any better way, the time-weighted average (TWA) has been used. The TWA is found by summing the products of individual exposure durations and concentrations and then dividing by the sum of the durations. That is, if C = concentration and t = time,

$$\text{TWA} = \Sigma \, Ct/\Sigma \, t \qquad\qquad (3\text{-}2)$$

This technique has the distinct advantage of simplicity and it therefore enjoys wide use even though experiment has shown that the toxicity of most materials cannot be so easily expressed.

Figure 3-1 can be used to integrate exposures to carbon monoxide. The graph is entered at the initial COHb percentage saturation. Then for each succeeding interval of constant concentration the appropriate concentration line is followed for the duration of that interval, following lines of constant COHb level between intervals. Unfortunately, this graph cannot be used if the exposure is to a CO concentration of zero or any other value not plotted, nor can it be used if one of the other parameters such as the ventilation rate or barometric pressure is varied. In such cases, Eq. (3-1) must be used and if even a very small computer is available, the equation is the method of choice. By allowing each final COHb level to be the initial level for the next increment of exposure, an exposure of any duration can be summed over any range of inhaled concentrations to result at the end of each step in the actual level of COHb. That concentration can then be converted to percentage saturation by dividing by $[\text{COHb}]_{max}$ and multiplying by 100. Similarly, knowing the total exposure duration and the initial and final COHb levels, the average concentration of CO that was inhaled can be calculated. This is possible even well after the exposure has ceased provided only that levels of all variables except one (the inhaled concentration in this case) are known or can be estimated accurately.

An example can illustrate the latter situation. Assume that a normal man (V_b = 5,500 ml, M = 218) was exposed for 30 min in normal air to CO during which he was moderately active ($\dot{V}_A$ = 20,000 ml/min). As a nonsmoker, his initial COHb level can be assumed to be 0.8% saturation or 0.0016 ml CO/ml blood (a normal Hb of 15 g/100 ml). The barometric pressure was about 750 mm Hg (therefore, P_L = 703 mm Hg and $\bar{P}_{C_{O_2}}$ = 100 mm Hg) and, with no history of hemolytic disease or lung defect, $\dot{V}_{CO}$ and D_L can be approximated as 0.007 ml/min and 30 ml/min-mm Hg. The man was found unconscious and taken ($\dot{V}_A$ = 4,000 ml/min) to a hospital where a blood sample was obtained 75 min after his exposure ended. (During that 75 min his exposure is assumed to be to a CO concentration of zero.) The blood sample at the hospital was found to contain 42% COHb saturation or

0.084 ml/ml. What was the average concentration to which he was exposed?

Two exposure durations must be integrated. First, the 75 min at zero concentration to find the COHb level at the end of the exposure and then the 30 min of exposure to find the average CO concentration. For the first, values for all variables in Eq. (3-1) are known except the initial COHb concentration, which is calculated by substituting in

$$[COHb]_0 = \frac{[COHb]_t}{\exp(-tA/BV_b)} \qquad (3\text{-}3)$$

where, with a little manipulation, $[COHb]_0$ is found to be 0.1071 ml/ml or 51.8%.

Knowing the final COHb level after the exposure (51.8%) and knowing that the exposure was for a duration of 30 min, the average CO concentration inhaled can be calculated. That the concentration must have fluctuated during the exposure is not important, as a rearrangement of Eq. 3-1 enables a calculation of a true (integrated) average exposure concentration.

$$P_{I_{CO}} = \frac{(A[COHb]_0 - B\dot{V}_{CO})\exp(-tA/BV_b) - A[COHb]_t + B\dot{V}_{CO}}{\exp(-tA/BV_b) - 1} \qquad (3\text{-}4)$$

Substituting known levels, the partial pressure of inhaled CO is calculated to be 1.535 mm Hg. At a total pressure of 750 mm Hg, the concentration in parts per million (ppm) = (1.535) (1,000,000)/750 = 2,046 ppm.

The same technique could have been used if the victim had been given oxygen by mask (assume 95% O_2 inhaled) during his trip to the hospital and until the blood sample was obtained by knowing the effect of inhaled oxygen on the average partial pressure of O_2 in lung capillaries [9]. (In this case, $0.95 \times 703 - 49 = 618.85$ mm Hg.) For the same conditions except a finding of 17% COHb saturation at the hospital, his COHb level at the end of the exposure can be found to have been 67.7% saturation, and the average concentration to which he was exposed for the 30 min was 3,024 ppm CO.

Using Eq. (3-1) and altering the level of several of the variables one at a time from nominal values, Table 3-1 was constructed [9].

Methemoglobin. Oxygen is bound loosely to normal hemoglobin so that under conditions of low oxygen partial pressure that exist, for instance, in oxygen-poor tissue, O_2 is given up rather easily from the blood. Methemoglobin, on the other hand, is a form of hemoglobin in which the iron has been oxidized to the ferric (Fe^{3+}) state and in which O_2 is bound very tightly. Even a complete vacuum where the oxygen partial pressure is obviously zero will not cause the release of O_2 from methemoglobin.

Table 3-1. Effect of Various Parameters on COHb Saturation
(Values Are % COHb Levels Calculated from the CFK Equation
by Changing Only the Indicated Parameter.)

	60-Min Exposure					480-Min Exposure				
	ppm CO					ppm CO				
Parameter	8.7	35	50	100	1,000	8.7	35	50	100	1,000
Nominal (Fig. 3-1)	1.08	1.68	2.02	3.16	23.2	1.40	4.25	5.83	10.9	59.8
P_B = 400 mm Hg	1.11	1.62	1.90	2.86	20.0	1.64	4.53	6.15	11.4	65.2
P_B = 600 mm Hg	1.09	1.66	1.99	3.07	22.2	1.49	4.37	5.98	11.2	61.7
P_B = 1,500 mm Hg	1.05	1.70	2.07	3.31	24.8	1.19	3.82	5.27	9.9	55.0
$\dot{V}_A$ = 15 L/min	1.12	2.17	2.77	4.75	37.6	1.41	4.93	6.85	12.8	60.7
$\dot{V}_A$ = 50 L/min	1.16	2.77	3.68	6.69	49.9	1.39	5.15	7.18	13.4	60.8
$\dot{V}_A$ = 100 L/min	1.18	2.99	4.02	7.39	52.8	1.39	5.17	7.21	13.4	60.8
V_b = 1,000 ml	1.33	3.66	4.96	9.22	57.6	1.55	5.35	7.38	13.6	60.8
V_b = 2,000 ml	1.20	2.63	3.45	6.13	46.8	1.54	5.25	7.25	13.4	60.8
V_b = 7,000 ml	1.07	1.55	1.82	2.73	18.9	1.36	3.87	5.28	9.8	58.6
$[COHb]_0$ = 2%	1.93	2.53	2.87	4.00	23.9	1.67	4.50	6.08	11.1	59.8
$[COHb]_0$ = 7%	6.14	6.73	7.06	8.19	27.8	3.00	5.76	7.29	12.2	59.9
D_L = 10 ml/min-mm Hg	1.07	1.50	1.74	2.55	17.0	1.39	3.73	5.05	9.3	57.7
D_L = 50 ml/min-mm Hg	1.09	1.74	2.11	3.34	25.0	1.41	4.37	6.01	11.3	60.1
% O_2 = 10	1.19	1.82	2.18	3.38	24.8	2.25	6.35	8.66	16.1	79.8
% O_2 = 100	0.50	0.90	1.13	1.88	14.5	0.24	0.85	1.19	2.3	19.0
Hb = 10 g/100 ml	1.12	1.98	2.47	4.11	32.0	1.48	4.79	6.62	12.3	60.6
Hb = 20 g/100 ml	1.06	1.52	1.78	2.65	18.1	1.35	3.80	5.17	9.6	58.3

Many materials can cause the formation of methemoglobin in the blood of man. A partial list follows:

Inorganic	*Organic*
Nitrites	Aromatic nitro- compounds
Dichromates	Aromatic amines
Nitrogen trifluoride	Nitroso compounds
Tetrafluorohydrazine	Phenylhydroxylamines
Perchlorates	Aliphatic nitrites

While the mechanism of formation of methemoglobin (MetHb) is not completely known, the evidence suggests that several steps are involved, especially with the aromatic materials, and that the active material(s) may be species as well as compound dependent. In the presence of oxygen and/or oxyhemoglobin, MetHb is slowly reduced to hemoglobin, probably under the influence of an enzyme system. Detoxification can be hastened by the administration of several materials. Methylene blue, a dye and a poor MetHb-former, has been used as has a reducing sugar, glucose.

Persons with methemoglobinemia are likely to have bluish-colored lips and fingernails, a condition called *cyanosis,* after *cyan,* the name of a hue between blue and green. Methemoglobin is an even bluer red than is reduced hemoglobin and the altered blood color is especially evident where the skin is thin over a profusion of capillaries as in the lips, ear lobes, fingernails, etc.

The presence of methemoglobin alters the normal relationship between oxyhemoglobin and tissue in a manner similar to that observed with carboxyhemoglobin. That is, when MetHb is present, O_2Hb holds its oxygen more tenaciously than otherwise. In this respect, MetHb seems to be roughly half as efficacious as COHb and, therefore, anoxia is somewhat less severe with MetHb than with the same percentage saturation of COHb.

Apnea

Of all the commonly encountered gases, only hydrogen sulfide (H_2S) has the ability to halt the process of respiration. Upon exposure to a high concentration (several hundred to a few thousand ppm) of H_2S, the first effect is an increase in the frequency of breathing with, perhaps, a decrease in the depth of each inhalation. After a few breaths, depending on the concentration, breathing stops (*apnea*), either preceded or immediately followed by collapse. If the victim is rescued and treated by artificial respiration within a very few minutes of collapse, complete recovery is likely.

In low concentration, H_2S has the odor of rotten eggs familiar to every freshman chemistry student. Olfactory fatigue is rapid even at low concentrations and an exposure of a few minutes to 10 or 15 ppm is sufficient to inhibit odor sensation completely. In higher concentrations, the first part of the first breath may be sufficient to paralyze the sense of smell so that acutely dangerous concentrations may have no odor.

A concentration of 10 ppm will probably cause some minimal eye irritation after an exposure of several hours. As concentration increases, corneal irritation appears more rapidly with symptoms of "sand in the eyes," a halo around bright lights, and perhaps the impression of a blue fog in dimly lit areas. After a few hours of nonexposure, recovery is complete.

Hydrogen sulfide is rapidly oxidized in the blood so that an exposure to a relatively high concentration is necessary before life is threatened. This gas seems to have the ability to act directly on the respiratory control centers if its concentration in blood is high enough. Concentrations in air greater than 400 ppm are dangerous and concentrations in excess of 600 ppm cause apnea rapidly on inhalation.

As with other asphyxiants, if respiration is reestablished soon enough after collapse to prevent injury to the brain from anoxia, recovery from an acute episode will be complete. There are no effects of chronic exposure that are significantly different from those associated with acute exposure.

Enzyme Inhibition

Cyanide ion (CN^-) has the ability to inhibit the action of several enzyme systems in the animal body. Among the most sensitive of these are cytochrome oxidase and other enzymes that control oxidation at the cellular level. When these systems are inhibited by CN^-, they can no longer use oxygen supplied by the blood and those cells most sensitive to anoxia die quite rapidly. Because oxygen carried by the blood is not used, venous blood takes on the appearance of arterial blood, giving a glow of ruddy good health to the victim. Action by CN^- is rapid; symptoms include those common to anoxia: weakness, headache, confusion, and occasionally nausea and vomiting prior to collapse.

Hydrogen cyanide (HCN) is the most common occupational source of CN^- even if the salts of this acid are being used, because the CO_2 in air is sufficiently acid to release HCN from its salts. Another cause of cyanide intoxication is acrylonitrile, $CH_2{=}CHCN$, which releases CN^- in the body.

HCN is very rapidly absorbed from air by the lungs and is more slowly absorbed from air by the skin. The liquid is rapidly absorbed through the intact skin but it does not irritate so that its warning properties are poor. Cyanides and nitriles have a characteristic odor said by some detective story authors to be that of "bitter almonds." The vapor of acrylonitrile is somewhat irritating to breathe in acutely dangerous concentrations but that of HCN is not. HCN poses an extreme fire hazard with a flash point of $-18°C$ (0°F) and explosive limits of 5.6 to 40% in air.

A concentration of 10 ppm HCN (or 20 ppm acrylonitrile) will probably not cause any symptoms even from an exposure of many hours. Two to four times that level will cause minimal symptoms in several hours. A concentration of 100 to 135 ppm HCN can be fatal after 30 min; 250 to 300 ppm causes immediate collapse and death if the victim is not rescued and treated. Recovery from an acute episode is complete if brain damage is avoided and there are no chronic effects different from the acute ones.

The class of cyanides is one of the very few for which specific antidotal procedures are available. Methemoglobin is much less toxic than is cyanide and it reacts rapidly with CN^-, removing it from the bloodstream. The antidote is then to form methemoglobin as rapidly as possible. This is done by inhaling amyl nitrite [$H_3C{-}(CH_2)_3{-}CONO$] and/or by injecting sodium nitrite ($NaNO_2$). Sodium thiosulfate ($Na_2S_2O_3$) reacts in the blood with CN^- to produce sulfite (SO_3^{2-}) and thiocyanate (SCN^-) ions, neither of which shares the high toxicity of CN^- so that sodium thiosulfate may also be injected as an antidote to cyanide poisoning.

Other than training personnel in the use of antidote kits, the only effective means of controlling the toxic hazard of HCN is to design equipment and procedures so that contact is minimized. Gas mask manufacturers produce a canister that is specific for HCN and, of course, supplied-air and self-contained masks are useful. Respiratory protection is not sufficient, however; if a dangerous concentration of HCN must be entered, a plastic airtight suit must be worn to avoid skin absorption.

Irritants

Gases that irritate on contact with the respiratory tract may be grouped according to the location of the irritation produced when the material is encountered in the usual industrial situation. Some materials produce irritation mainly in the throat and bronchi (the upper respiratory tract); others cause irritation in the depths of the lung (the lower respiratory tract). Between these extremes is a third group: "whole lung" irritants.

Upper Respiratory Tract

Gases that irritate the upper respiratory tract (URT) are characterized by their relatively great solubility in water and by the fact that hydrolysis is not a necessary prerequisite to their action. Materials in this category most commonly encountered include ammonia and ammonium hydroxide (NH_3 and NH_4OH), the halogen acids (HF, HCl, HBr, and HI), and the sulfur oxides (SO_2 and SO_3). All of these materials are very irritating to moist tissue and all are quite soluble in water, hence their action on the surface of mucous membrane.

These gases are highly toxic, but none is very hazardous to handle. They are sufficiently irritating to warn of their presence well before acutely toxic concentrations are inhaled. In fact, under most circumstances these materials will drive people out of an area well before dangerous amounts are encountered. In high concentrations the URT irritation is so pronounced that the bronchi may go into spasm and close so completely that breathing is impossible. Lung injury from inhaling one of these gases is extremely rare and is likely to be encountered only where someone is trapped so that escape is impossible.

Ammonia and sulfur dioxide are peculiar in that some people become conditioned (hardened or acclimatized) and that other people appear to develop a true hypersensitization to them. In either case the hazard is thereby increased. When a person becomes tolerant of high concentrations, the possibility of deep lung irritation is definitely increased. Hypersensitivity, on the other hand, can lead to an asthmatic-type response and inhaling a high concentration could well be fatal (see Chapter 9).

Hydrogen chloride is the only one of the URT irritants known to have a chronic effect, namely, erosion of tooth enamel. Doubtless, other acidic materials in this group could cause much the same kind of effect if exposures were prolonged to sufficiently high concentrations.

Lower Respiratory Tract

Those inorganic gases that act primarily on the lower respiratory tract (LRT) or the deep lung are those with low but appreciable water solubility or they hydrolyze in the water of mucous membrane to produce a material that is a strong irritant. In commonly encountered concentrations, these materials may not be particularly irritating to breathe but hours later may cause intense irritation deep in the lungs. Because of the typical delayed action, LRT irritants are quite hazardous to handle. Carbonyl chloride, $COCl_2$, (phosgene) was used as a war gas in World War I, causing thousands of casualties.

Nitrogen dioxide is the anhydride of nitrous and nitric acids and is one of the very few colored gases, having a reddish-brown hue that varies in intensity with concentration and with the length of the light path through which it is viewed. When the thickness of the gas-air mixture is a few centimeters or meters, the concentration must be on the order of hundreds of ppm for the color to be seen. When the thickness of the layer is on the order of kilometers (over a city after the formation of photochemical smog), however, concentrations of hundredths of a ppm can color the air noticeably.

Nitrogen dioxide (NO_2) is not very soluble in water, and the formation of nitrous (HNO_2) and nitric (HNO_3) acids is slow, accounting for the typical delayed action seen after inhaling moderately high concentrations of this material. If concentrations are high (several tens of ppm to a few hundred ppm), there will be some acute irritation of eyes and nose but that irritation is far from being sufficiently strong to drive one out of the area.

Carbonyl chloride ($COCl_2$, phosgene) and carbonyl bromide ($COBr_2$) have low water solubility and hydrolyze to the corresponding halogen acid (HCl or HBr) rather slowly. Their delayed action after inhaling relatively low concentrations may result from the parent compound and not the corresponding acid [10]. When the concentration is higher (a few tens of ppm), the whole respiratory tract as well as the eyes may be irritated, the intensity of the irritation being at least somewhat proportional to the concentration. Even though these materials are considerably more irritating acutely than is nitrogen dioxide, the irritation is still not sufficient to warn of concentrations capable of causing a severe delayed response.

In many references to the toxicity of the carbonyl halides, especially phosgene, the odor is said to be that of new-mown hay. This description may have been useful when, in World War I, phosgene was used as a war gas and a large percentage of the soldiers came from farms, but today relatively few people have smelled new-mown hay. Instead, a better description of the odor might be that of fresh grass clippings, an odor most people are familiar with whether or not they have lived on a farm.

The carbonyl halides are encountered industrially when they are used as solvents or when they are formed by thermal decomposition of a halogenated hydrocarbon. *Pyrolysis* (thermal decomposition) of chlorinated or brominated hydrocarbons typically takes place in a fire, when the vapors of these materials encounter red-hot metallic surfaces or when welding is being done near a source of halogenated hydrocarbon vapor such as a degreaser [11]. In almost all cases of pyrolytic formation of a carbonyl halide the corresponding halogen acid is formed in much greater concentrations and therefore acts as a warning against overexposure. This is not always true, however, and whenever temperatures high enough to produce a red heat and halogenated hydrocarbon vapors in concentrations of a few ppm or more may coexist, the presence of a carbonyl halide must be suspected.

Phosgene and other lung edema producers such as ozone and nitrogen dioxide may produce this effect by action on specific receptor sites in the lung rather than by generalized irritation [12]. There may also be significance in the fact that hexamethylenetetramine injections have been found to prevent lung edema formation in men overexposed to phosgene [13].

Typical LRT delayed irritation and pneumonitis have been caused by inhalation of mercury vapor in high concentrations [14]. The picture seen is similar to that caused by the volatile metal carbonyls and halides (see Chapter 4).

Whole Respiratory Tract

Inorganic irritants which are intermediate in water solubility and which may become more irritating after a latent period tend to affect neither the URT nor the LRT preferentially and therefore are said to be *whole lung* or *whole respiratory tract* irritants. The most commonly encountered members of this class are the halogens (fluorine, F_2; chlorine, Cl_2; bromine, Br_2; and iodine, I_2) and ozone (O_3).

In low concentrations these materials may be irritating to the eyes and URT and therefore act somewhat like typical URT irritants. In higher concentrations, however, the intensity of acute irritation is not so severe as with the URT irritants and hence voluntarily tolerated exposures may cause serious irritation deep in the lung. There may even be a somewhat delayed response typical of the LRT irritants.

Chlorine, bromine, and iodine are all colored gases: chlorine being yellow-green; bromine, reddish-brown; and iodine, violet. Pure fluorine is colorless. Ozone forms a deep blue liquid when condensed but is colorless as a gas, at least in commonly encountered concentrations.

The ability of the halogens to hydrolyze in water is inversely proportional to molecular weight. That is, iodine (with the highest molecular weight) does not hydrolyze appreciably but fluorine actually decomposes water violently to form several products, among which are HF and OF_2. Both chlorine and bromine hydrolyze to give the corresponding hydro- -ic and hypo- -ous acids, probably accounting for the increased severity of response occasionally seen some time after the cessation of an irritating overexposure. The toxicity of fluorine is not materially different from that of HF although fluorine is somewhat more reactive and hence more irritating.

Ozone is a peculiar gas. Within the body (and elsewhere, also) it acts as the bi- free radical it is. The consequences of this action are widespread, giving rise to what has been termed the *radiomimetic* properties of ozone [15, 16].

Ionizing radiation appears to exert its harmful effects upon the body by ionizing substances (chiefly water) within the body and by causing the formation of free radicals (see Chapter 15). Free radicals have the ability to react with any other substance regardless of whether such a reaction would occur under normal circumstances. In the process of reacting, free radicals not only alter the molecules with which they react but usually cause the chain formation of other free radicals.

Genetic materials [DNA (deoxyribonucleic acid) and RNA (ribonucleic acid)] are not particularly reactive compounds and under most circumstances are not altered by the chemical processes taking place in cells. Free radicals can and do react with DNA and RNA, however, thereby causing alterations in the genetic makeup of the cell. The signs and symptoms of overexposure to O_3 are similar to those associated with ionizing radiation (see Chapter 15) except, of course, that O_3 is also a local irritant of the whole respiratory tract.

Animal experiments have shown that even a brief exposure to O_3 can protect against the irritant effects of a subsequent exposure [15]. Although this effect is limited to lung irritation (edema formation), a sublethal exposure has been found able to protect completely against a normally lethal exposure some weeks later. Furthermore, later experiments have shown that this immunity is not limited to ozone—that an exposure to ozone can protect against exposure to many other irritants and vice versa. Materials sharing this cross-tolerance are, in general, whole lung irritants that cause edema.

Several materials used as antioxidants in foods and elsewhere have also been found to protect against one or more of the manifestations of ozone toxicity [17, 18, 19] as have some general enzyme-inducing agents [20]. Information of this kind is developed much more to aid in eliciting mechanisms of injury than to find antidotes. Knowledge of toxic action always aids further experimentation and usually also forms the best base from which to develop protective measures of many kinds.

Whenever any of the irritating gases is condensed to the liquid state, even momentary skin or eye contact with that liquid is likely to result in severe irritation. With the exceptions of iodine, bromine, and phosgene, all the materials discussed in this section boil at temperatures well below the freezing point of water. Therefore, in addition to the irritation hazard, all of them are capable of causing frostbite upon contact of the liquid with the skin surface. A light plastic film which may be impermeable to the liquid and which therefore will protect against irritation will not protect against the frostbite hazard.

Fluorine, hydrofluoric acid, and fluorides pose a skin and eye contact hazard that is qualitatively different from that of most of the irritants. Irritant fluorides react with protoplasm in such a manner that the resulting products of reaction are quite insoluble in water and body fluids. Those reaction products thus remain firmly fixed to tissue and are difficult or impossible to remove except by *debridement* (cutting away tissue). Irritation by one of these materials that at first seems to be nearly superficial may thereby result in a chronic, ulcerating abscess. In the eye, these materials can result in blindness that even corneal transplants are unable to cure because of the slow translocation of fluoride ion from surrounding tissue to the cornea, resulting in renewed opacification.

Asphyxiating Irritants

Arsine (AsH_3) [21] and stibine (SbH_3) and perhaps germane (GeH_4) are in a class by themselves. These gases are formed whenever the parent metals contact nascent hydrogen. They are whole lung irritants capable of causing pulmonary edema that, in fact, may well be the cause of death in some cases. In addition, however, these materials can produce a massive destruction of hemoglobin in the blood, a process called *hemolysis*. The body tries to rid itself of the hemolyzed blood, but the mass of material to process may be so great as to overload the liver and kidneys, resulting in pain, *jaundice* (a yellowish coloration of the skin), and *hematuria* (blood in the urine). Because the debris resulting from hemolysis is filtered from the blood, the hemoglobin content of the blood is reduced, thus causing a reduction in the amount of oxygen that can be carried to tissue. If the victim dies, the cause of death is very likely to be anoxia.

REFERENCES

[1]   R. A. McFarland, F. J. Roughton, M. H. Halperin, and J. J. Niven, "The Effects of Carbon Monoxide and Altitude on Visual Thresholds," *J. Aviation Med.* **15** (1944), 381–94.

[2]   C. Xintaras, et al., "Brain Potentials Studied by Computer Analysis," *Arch. Env. Health* **13** (1966), 223–32.

[3]   M. J. Hosko, "The Effect of Carbon Monoxide on the Visual Evoked Response in Man," *Arch. Env. Health* **21** (1970), 174–80.

[4]   A. Kahn, et al., "Carboxyhemoglobin Sources in the Metropolitan St. Louis Population," *Arch. Env. Health* **29** (1974), 127–35.

[5]   J. R. Goldsmith and S. A. Landaw, "Carbon Monoxide and Human Health," *Science* **162** (1968), 1352–59.

[6]   R. D. Stewart, et al., "Carboxyhemoglobin Levels in American Blood Donors," *J. Am. Med. Assn.* **229** (1974), 1187–95.

[7]   R. F. Coburn, R. E. Forster, and P. B. Kane, "Considerations on the Physiology and Variables that Determine the Blood Carboxyhemoglobin Concentration in Man," *J. Clin. Inv.* **41** (1965), 1899–1910.

[8]   J. E. Peterson and R. D. Stewart, "Absorption and Elimination of Carbon Monoxide by Inactive Young Men," *Arch. Env. Health* **21** (1970), 165–71.

[9]   J. E. Peterson and R. D. Stewart, "Predicting the Carboxyhemoglobin Levels Resulting from Carbon Monoxide Exposures," *J. Appl. Physiol.* **39** (1975), 633–38.

[10] T. Nash and R. E. Pattle, "The Absorption of Phosgene by Aqueous Solutions and its Relation to Toxicity," *Ann. Occup. Hyg.* **14** (1971), 227–33.

[11] M. H. Noweir, E. A. Pfitzer, and T. F. Hatch, "Decomposition of Chlorinated Hydrocarbons: A Review," *Am. Ind. Hyg. Assn. J.* **33** (1972), 454–60.

[12] A. R. Gregory, "Inhalation Toxicology and Lung Edema Receptor Sites," *Am. Ind. Hyg. Assn. J.* **31** (1970), 454–59.

[13] P. Stavrakis, "The Use of Hexamathylenetetramine (HMT) in Treatment of Acute Phosgene Poisoning," *Ind. Med. Surg.* **40** (1971), 30–31.

[14] J. Milne, A. Christophers, and P. DeSilva, "Acute Mercurial Pheumonitis," *Brit. J. Ind. Med.* **27** (1970), 334–38.

[15] H. E. Stokinger, "New Concepts and Future Trends in Toxicology," *Am. Ind. Hyg. Assn. J.* **23** (1962), 8–19.

[16] A. N. M. Nasr, "Biochemical Aspects of Ozone Intoxication: A Review," *J. Occup. Med.* **9** (1967), 589–97.

[17] J. N. Roehm, J. G. Hadley, and D. B. Menzel, "Antioxidants vs. Lung Disease," *Arch. Intern. Med.* **128** (1971), 88–93.

[18] B. D. Goldstein, et al., "p-Aminobenzoic Acid as a Protective Agent in Ozone Toxicity," *Arch. Env. Health* **24** (1972), 243–47.

[19] B. L. Fletcher and A. L. Tappel, "Protective Effects of Dietary Alpha-Tocopherol in Rats Exposed to Toxic Levels of Ozone and Nitrogen Dioxide," *Env. Res.* **6** (1973), 165–75.

[20] B. D. Goldstein and O. J. Balchum, "Modification of the Response of Rats to Lethal Levels of Ozone by Enzyme-Inducing Agents," *Tox. Appl. Pharm.* **27** (1974), 330–35.

[21] B. A. Fowler and J. B. Weissberg, "Arsine Poisoning," *N. E. J. Med.* **29** (1974), 1171–74.

4

METALS
AND
METALLOIDS

A metal is an element, crystalline when solid, characterized by ductility, opacity, electrical conductivity, and a metallic luster, especially when freshly fractured. A metalloid, on the other hand, is a material that is intermediate in character between metals and nonmetals. Metalloids have some but not all the properties of metals.

Other than sharing the general properties of metals, members of the metalloid group have little in common. Melting point, density, toxicity, vapor pressure, and other properties can be found to vary across most of any possible range. With a few exceptions, then, these materials are usually studied individually with emphasis on those most likely to be encountered or on those with the highest hazard potential. Study in that manner may be pursued in recent journal articles, Hygienic Guides, ANSI standards, and in books such as those written by Patty [1] and Browning [2].

Some similarities do exist between members of subsets of the set metals/ metalloids. While most similarities that could be found are of little use to the student of industrial hygiene (consider valence states, heat capacity, or atomic weight, for instance), a few have some value. In particular, group-ings on a basis such as similar toxic effects or similar tendency to form highly volatile compounds can be useful.

In forming the groups of similar materials that follow, no attempt was made to include those materials usually studied most intensively. This col-lection of similarities is intended to supplement, not replace, the more con-ventional approach.

Dietary Essentials

Thirteen metallic elements presently are known to be essential in the diet of man. They are calcium, cobalt, copper, chromium, iron, magnesium, man-ganese, molybdenum, phosphorus, potassium, selenium, sodium, and zinc. Several other metals such as boron, vanadium, and zirconium are almost always found in some tissues when sought but have no known uses within the body. As further research may well find more essential metals, the list of 13 cannot be regarded as complete.

Toxicity of the essential elements varies considerably. Cobalt, copper, chromium, phosphorus, and selenium all are high in toxicity; chromium is a known carcinogen and cobalt is strongly suspected of having a similar ability. Zinc and manganese are fairly toxic metals, but the remainder of the essential elements are low in toxicity, especially by the oral route in ionized forms. The alkali metals, sodium, potassium, and calcium, can dis-place oxygen from water, leaving hydrogen, which may be ignited by the heat of the reaction. These metals and their hydroxides are extremely cor-rosive to all tissue. Nutritional necessity and degree of toxicity have no cor-relation [3].

Organometallic Compounds

The term *organometallic* refers to those materials formed from metals and organic molecules but which are neither chelates nor salts of organic acids. Because of the wide variety of organic molecules coupled to metals and be-cause such compounds can be made with nearly all metals, only a few gen-eralizations are useful. For example, organometallic compounds usually have a toxicity that is different in kind in some major aspects from that of the metal and may also be different in degree. Volatility is a component of hazard and organometallic compounds that are more volatile than their parent metals may well be more hazardous to handle.

Tetraethyl lead and tetramethyl lead have been produced in larger amounts than any other organometallic compounds. Both materials were used mainly as antiknock compounds in gasoline and both are considerably more toxic and more hazardous to handle than is lead itself. Both, for instance, are absorbed through the intact skin in acutely toxic amounts, while no inorganic lead compound is so absorbed. Even in very small amounts, the lead alkyls can have devastating effects on the *central nervous system* (CNS) [4]; inorganic lead compounds mainly affect the gastrointestinal (GI) tract, blood-forming system (and blood), and muscles, especially in the wrists and ankles, manifestations of effects on the peripheral nervous system.

The aluminum alkyls and the aluminum halo-alkyls (for instance, tri-ethyl aluminum and diisobutyl aluminum chloride) are used in large amounts as catalysts, especially in the production of polyethylene and other polyolefins. Many of these compounds are *pyrophoric* (burst into flame spontaneously) in moist air or in contact with the skin or eyes.

Several organomercurials have commercial importance as preservatives of seeds and woods. In general, the toxicity of these materials is considerably greater than that of metallic mercury. In some cases, human fatalities have resulted from the accidental ingestion of grain treated with a mercurial pesticide [5] and poisonings have occurred from eating the meat of animals that have eaten treated grain [6]. Furthermore, aquatic microorganisms have been found that are able to alkylate metallic mercury (usually discharged from mercury cells used for production of sodium hydroxide and chlorine by electrolysis) to methyl mercury, which is then concentrated by fish and other seafoods [7]. Eating fish so contaminated has caused outbreaks of poisoning in several countries; the first to be recognized was in Japan [8].

Organomercurials used as antiseptics, germicides, and diuretics are generally less toxic than mercury as well as being less volatile and considerably less hazardous to handle. Solubility in body fluids plays a large role in determining the toxicity of these materials, by whatever route of administration.

Volatile Hydrides

Many of the metals and metalloids can combine with hydrogen to form one or more hydrides. Most of the compounds so formed are solids at room temperature and, with the exception of lithium hydride, have little commercial importance. A few of the metals and metalloids, however, form hydrides that are liquids or gases at room temperature and therefore may pose vapor inhalation hazards.

The hydrides of arsenic (arsine, AsH_3), antimony (stibine, SbH_3), and germanium (germane, GeH_4) result whenever nascent hydrogen contacts inorganic arsenic, antimony, or germanium. Of all the volatile hydrides, only overexposure to these three results in massive hemolysis, which may be fatal because the blood is no longer capable of supplying adequate oxygen to tissues (see Chapter 3).

Hydrogen selenide (H_2Se), hydrogen telluride (H_2Te), and phosphine (PH_3) are all considerably more volatile than their parent metals (metalloids) but are not significantly more toxic. Hydrogen selenide and hydrogen telluride are irritants formed by the action of nascent hydrogen on the metals; phosphine is formed by the contact of phosphorus with hot alkali or by the contact of phosphides with water. Phosphine causes lung irritation although it is not particularly irritating to the eyes and upper respiratory tract.

The boron hydrides have been used as high-energy fuels. All of them are much more toxic than boron. Pentaborane (B_5H_9) is probably the most toxic of the group; its effects are mainly on the CNS [9]. Overexposure to diborane (B_2H_6) causes irritation of the whole lung in addition to a metal fume fever-like syndrome. Chronic exposure, on the other hand, is more likely to involve the CNS. Both diborane and pentaborane are pyrophoric in moist air. Decaborane ($B_{10}H_{14}$) is a solid that hydrolyzes slowly in water. Both acute and chronic overexposure produce neurotoxic symptomatology.

Tin forms a gaseous hydride (SnH_4) called stannane, which is more toxic than arsine and affects the CNS. Gallium and silicon form volatile hydrides called, respectively, gallanes and silanes. The toxicology of these materials has not been investigated very thoroughly.

Sensitizers

Many of the metals (metalloids) are capable of sensitizing either the skin or the respiratory tract (or both) of man [10]. The best known of the metals for its ability to sensitize is nickel. "Nickel itch" is relatively common among those who plate other metals with nickel and the wearing of nickel-plated objects may cause a response in those persons susceptible to nickel sensitization [11]. Many stainless steels contain nickel, and even nickel coins have caused problems.

Other metals known to sensitize the skin include beryllium [12], gold [13], cobalt, hexavalent chromium [14], mercury [15], and thallium. Platinum in complex salts is a skin sensitizer although the metal itself may not have that ability. Copper [16] and tellurium may be sensitizers, but if so they are not particularly active.

Zirconium has long been known to be almost completely inert physiologically. Its ability to sensitize the skin of man was only found after some zirconium salts (mainly the lactate) had been used widely in underarm deodorants.

Selenium dioxide dust can cause a sensitization of the eyelids known as *rose eye*. Several compounds of arsenic are primary skin irritants and some inorganic arsenicals have sensitized the skin of workers.

For a discussion of the phenomenon of hypersensitivity, see Chapter 9.

Volatile Carbonyls

The combination of a metal with carbon monoxide to form a compound called a *metal carbonyl* is relatively common. Most of the carbonyls, however, are solids at room temperature and few of them have any importance, either economically or toxicologically. Four metals (iron, nickel, osmium, and ruthenium) form carbonyls, however, that are liquid at room temperature. These materials are all considerably more toxic than are the parent metals and because of their volatility they are considerably more hazardous to handle. All are capable of causing an immediate or delayed pneumonitis following inhalation.

Volatile carbonyl halides, hydrocarbonyls, and nitrosoyl carbonyls also may be formed regardless of whether the metal carbonyl is volatile. The toxic properties of such materials are not well known.

Iron pentacarbonyl ($Fe[CO]_5$) forms at room temperature when carbon monoxide is held at high pressure in contact with iron or steel [17]. This condition is attained when CO is confined in a steel compressed gas cylinder. Investigation has shown that under reasonable conditions concentrations as high as a few thousand ppm of the carbonyl may be formed. Fortunately, $Fe(CO)_5$ is probably the least toxic of all the carbonyls and, even more fortunately, this material is adsorbed by activated charcoal very readily while CO is not, providing a good means of purifying the CO.

Nickel carbonyl ($Ni[CO]_4$) is formed and then decomposed during the Mond process for refining nickel. Persons working in this process have been exposed to nickel carbonyl vapor. Furthermore, the incidence of lung cancer among Mond process workers has been found to be considerably higher than that in an unexposed population. Because of these facts and some animal experiments, nickel carbonyl has been held to be a carcinogen, although whether the carbonyl itself or the finely divided nickel formed upon its decomposition is the true agent is not yet known [18].

Metal Fume Fever

Fume is an aerosol of solid particles formed by condensation, often accompanied by some oxidation. The syndrome known as *metal fume fever* [19, 20] is a collection of symptoms that have been likened to those of malaria or influenza. Typically, the victim will have had at least 2 days (a weekend) of nonexposure. He will then come to work, be overexposed, and a few hours later develop chills and fever, nausea, cough, bone aches, and other symptoms of the disease. A few hours after that the fever will break with profuse sweating; approximately 24 hr after the onset, no trace of the disease will remain.

Metal fume fever has been caused by inhaling the fumes of antimony, arsenic, beryllium, cadmium, cobalt, copper, iron, lead, manganese, mercury, nickel, silicon, tin, and zinc. In addition, the typical syndrome has also been caused by inhaling the fume from polytetrafluoroethylene and perhaps similar materials [21].

Metal (or polymer) fume fever is most likely to result when the fume is freshly formed, and the available evidence strongly indicates that the syndrome may be the body's response to the inhalation of very small particles. The cause then may be a sufficient concentration of sufficiently small particles within the lung. Although the exact mechanism is still unclear, an antibody-antigen type of reaction (see Chapter 9) appears to be responsible.

The Burton Line

Several metals are capable of causing a punctate dark-colored line to form around the edges of the gingival mucosa. This line or series of dots is most likely to appear after excessive exposure to one of five or six metals that have dark-colored insoluble sulfides and, furthermore, is more likely to be found in a poorly cared-for mouth. The line is likely to be close to or allied with diseased gum tissue and/or diseased teeth and is found at the edge where the gum meets the tooth and between the teeth; it is rarely found elsewhere in the mouth.

Burton line deposits have been formed after exposure to cadmium (yellow dots), bismuth, copper, and lead (purple dots), as well as mercury and zinc (blue dots). The dots may appear black in any case except that of cadmium. Presence of the line is diagnostic of an exposure to any of these materials but does not of itself indicate that an overexposure has occurred. The line appears to be formed by the precipitation of metal sulfide along the junction of teeth and gum, where bacteria can thrive under anaerobic conditions.

Teratogenic Effects

Routine toxicological evaluation of materials did not include an investigation of *teratogenic* effects (fetal, or birth, defects) until after the mid 1960s. For that reason, a discussion of this effect of metals and metalloids or of any other materials cannot be complete or exhaustive. At this writing, the following materials have been shown by one or more investigators to cause birth defects in animals or humans: arsenic [22], cadmium [23], indium [24], lead [25], alkyl mercury [26], strontium, and thallium [27].

Methyl mercury was shown in the late 1960s and early 1970s to be the agent responsible for malformed fetuses, abortions, and malformed children in Japan and elsewhere. While these have been the most dramatic incidents associated with a metal, others have probably occurred.

The ubiquitous chelating agent, ethylenediaminetetra-acetic acid, EDTA, causes congenital malformations in rats when injected or ingested during days 6 to 21 of gestation. Administration of a diet rich in zinc to the pregnant females prevents this teratogenic effect [28].

Mental Effects

Many metals (metalloids) exert one or more effects on the nervous system but in only a few cases have those effects been of a nature to affect thought processes.

Mercury has long been known to be capable of causing severe mental upsets following prolonged overexposure. The mad hatter of the 19th century, characterized well in *Alice in Wonderland,* was a relatively common victim. He inhaled gross amounts of mercuric nitrate during a process called *carroting* in which felt was made from animal hair. The beaver hat fad was responsible for the prevalence of this disease. Characteristic of the syndrome is a pathologic shyness in which the victim can be literally paralyzed by the unexpected appearance of a stranger.

Chronic manganese intoxication, manganism, is characterized by many grotesque signs including a mask-like (or doll's) face and the mental syndrome of wild hysterical laughter, euphoria, and speech defects including muteness. In general, persons subject to this disease are those who have been grossly overexposed to manganese during its mining and purification. Although it is not highly toxic, manganese appears to be quite hazardous to mine and refine.

Thallium is an extremely toxic metal and even a single overexposure, not immediately fatal, can result in mental disorders to the point of psychosis. Chronic overexposure can result in similar signs and symptoms but usually not so severe.

Rather than causing psychoses, lithium and its salts have been used in the treatment of such diseases. Lithium carbonate and citrate are sedatives, but their toxicity is so high that they are not used widely [29].

A teratogenic effect of lead upon mental processes has been demonstrated in lambs. Pregnant ewes fed sufficient lead during the gestation period delivered lambs that about 1 year after birth were slower to learn visual discrimination than controls or lambs from mothers fed less lead [30].

Fuming Halides

Several metals and metalloids combine with one or more halogens to produce a liquid. Many of these liquids react with moist air in such a way that they visibly *fume*. Compounds having this propensity include boron tribromide, trichloride, and trifluoride (BBr_3, BCl_3, and BF_3); arsenic trichloride $(AsCl_3)$; germanium tetrachloride $(GeCl_4)$; phosphorus tribromide, trichloride, oxychloride, and thiochloride $(PBr_3, PCl_3, POCl_3,$ and $PSCl_3)$; silicon tetrabromide and tetrachloride $(SiBr_4$ and $SiCl_4)$; vanadium tetrachloride (VCl_4); tin tetrachloride $(SnCl_4)$; and titanium tetrachloride $(TiCl_4)$. The tetrachlorides of vanadium, tin, and titanium have been used as mordants in the dyeing industry and to increase the weight of the dyed fabrics.

The fume that forms above these liquids was long assumed to consist mainly of the hydrohalic acid and the metal oxide (for instance, hydrochloric acid and titanium dioxide from $TiCl_4$). More recent investigations, however, have shown that the fume may be the hydrate of the parent molecule, for instance, $TiCl_4 \cdot 5 H_2O$. Subsequently, hydrolysis may ensue in the air, but that reaction usually takes place over a fairly long period, typically hours or days. The tetrachlorides of both tin and titanium have been used in the production of smokes for warfare as well as for tracing air currents in ventilation work.

All the fuming metal halides are irritating to the eyes and to the respiratory tract. The severity of irritation depends on the metal as well as on the stability of the metal halide hydrate.

Antagonists

Potentiation and antagonism of the toxic effects of one metal by another appear to be rare occurrences. On the other hand, most studies of toxicity are not designed to search for the effects of two or more substances and therefore both potentiation and antagonism may be much more common than would be expected from the number of examples that follow.

Sodium is antagonistic to the action of lithium on the kidney (said to be its primary target organ). That is, the nephrotoxicity of lithium becomes more and more evident as dietary sodium is reduced. Furthermore, with an increased intake of sodium, the experimental animal is able to tolerate an increased intake of lithium. This is true antagonism that apparently works, in this case, because of a competition between the two ions for reabsorption into the bloodstream from nephrons.

Copper, a metal highly toxic to animal life [31], is antagonistic toward the toxic action of molybdenum in cattle and fowl and to the action of zinc in rats. In both cases the antagonism appears to be developed at the enzyme level.

Both zinc [32] and selenium [33] appear to inhibit teratogenic and other toxic effects of cadmium, and cadmium may exert some of its toxic effects by competing with the essential elements copper, cobalt, and zinc. Zinc [34] and selenium [35] both reduce teratogenic effects of mercury.

Calcium is a muscle depressant and barium is a muscle stimulator. These ions are mutually antagonistic. Sodium arsenate and arsenite prevent hemolysis and liver injury from selenium by maintaining succinic hydrogenase levels of the liver. Selenium, on the other hand, inhibits the teratogenic effect of arsenic [33].

Because of the phenomenon of antagonism, one investigator decided to determine if pretreatment by a metal would alter the toxicity of that metal. In other words, he wanted to see if a metal could be antagonistic to itself [36]. He found that this was indeed the case with silver, arsenic, cadmium, mercury, indium, lead, manganese, and tin in mice. Furthermore, pretreatment with any metal in that series had the effect of increasing the amount required for lethality for any other metal in the series. Pretreatment had no effect with copper, nickel, selenium, or thallium and it increased toxicity of barium, chromium, iron, and zinc.

Magnesium ion protects rats against the seizures caused by acute exposure to high pressure oxygen; manganese and zinc protect against lung damage caused by those exposures. Calcium inhibits the protective action of magnesium [37].

Carcinogens

Chromium in its hexavalent form has long been known to cause lung cancer in some overexposed persons [38]. Even in this case, however, the cancer has usually appeared only after many years of chronic overexposure sufficient to have caused perforation of the nasal septum long before the establishment of cancer [39].

Nickel carbonyl is formed in the Mond process for the purification of nickel and is associated with an increased incidence of lung cancer among Mond process workers. Recent evidence appears to indicate that nickel may be a carcinogen in the ionic form and as the metal; therefore, combination with carbon monoxide is not necessary for carcinogenesis.

Arsenic is known to cause perforation of the nasal septum and epidemiologic evidence [40] indicates that arsenic, like chromium, is capable of causing lung cancer.

Other metals suspected of being carcinogens but without overwhelming evidence as yet include beryllium [41], cobalt, cadmium, titanium, and lead. Uranium can cause cancer at many sites; whether the cancer is caused by the chemical toxicity of the metal, by its emitted ionizing radiation (see Chapter 15), or by a combination of these effects along with those of cigarette smoking is not known. Lung cancer among uranium miners, however, appears predominantly among those who smoke cigarettes [42].

Recent developments appear to indicate that cigarette smoking may be a contributing factor in the development of bronchial cancer from chromates [43] as well as from uranium and asbestos. On the other hand, selenium has been shown to inhibit the development of cancer in some cases [44].

Bad Odors

A few metals and metalloids are associated with stenches. The odor may come from the metal, its oxide, or its hydride (or all three) or from the breath and/or sweat of a person exposed to the material and hence from an organometallic compound.

Osmium was named because of the foul odor associated with its oxide (OsO_4); *osme* is the Greek word for odor.

Selenium and tellurium are found together and either tellurium alone or selenium as well is able to give the breath, sweat, and urine of exposed persons a garlic-like odor that is attributed to formation of the methyl selenide or telluride. Because tellurium causes this effect in extremely low doses and because selenium is contaminated with small amounts of tellurium in most places where it is encountered, there is still some doubt about whether pure selenium is capable of causing the effect. Selenium dioxide, however, is said to have the odor of rotten horseradish.

A garlic odor is associated not only with selenium and tellurium but also with antimony (the hydride and oxide) and with arsenic, especially the hydride. Phosphorus and phosphine both have foul odors as do the hydrides of boron and silicon.

Miscellaneous Oddities

Silver is a metal with a rather low toxicity but one where exposures must be kept as close to zero as possible. The reason is that any silver compound taken into the body is subsequently reduced to metallic silver and deposited. If that metallic silver is near the skin surface, the skin above takes on a bluish-gray cast given the name *argyria*. Argyria can be a disfiguring industrial disease as there is no way to reverse its course. The disease causes no discomfort and no illness; it simply makes its victim an object of curiosity, scorn, and ridicule.

Excessive intake of phosphorus can cause a particularly disfiguring disease called *phossy jaw*. This disease is a necrosis of bone, particularly of the lower jaw or mandible. The necrotic lesions cause the bone to become malformed, resulting in a grotesque appearance. Progress of the disease is facilitated by poor dental care and by diseased teeth.

Summary

This collection of similarities between metals (metalloids) has skipped lightly over some materials and dwelt more heavily on others with little or no regard to the numbers of people exposed or to the real hazards involved. Beryllium [45] and cadmium [46] have hardly been mentioned, for instance.

Furthermore, one facet of similarity does not imply others; study of the separate materials is a necessary adjunct to knowledge.

REFERENCES

[1] F. A. PATTY, *Industrial Hygiene and Toxicology,* 2nd rev. ed., D. W. Fassett and D. D. Irish, ed. (New York: Interscience Publishers, Inc., 1963), vol. II.

[2] ETHYL BROWNING, *Toxicity of Industrial Metals* (London: Butterworths, 1961).

[3] D. B. LOURIA, M. M. JOSELOW, AND ANN A. BROWDER, "The Human Toxicity of Certain Trace Elements," *Ann. Intern. Med.* **76** (1972), 307–19.

[4] A. D. BEATTIE, M. R. MOORE, AND A. GOLDBERG, "Tetraethyl-Lead Poisoning," *Lancet* **2** (July 1, 1972), 12–15.

[5] L. K. A. DERBAN, "Outbreak of Food Poisoning Due to Alkyl-Mercury Fungicide on Southern Ghana State Farm," *Arch. Env. Health* **28** (1974), 49–52.

[6] P. E. PIERCE, ET AL., "Alkyl Mercury Poisoning in Humans: Report of an Outbreak," *J. Am. Med. Assn.* **220** (1972), 1439–42.

[7] J. J. VOSTAL AND T. W. CLARKSON, "Mercury as an Environmental Hazard," *J. Occup. Med.* **15** (1973), 649–56.

[8] K. KOHIMA AND M. FUGITA, "Summary of Recent Studies in Japan on Methyl Mercury Poisoning," *Toxicology* **1** (1973), 43–62.

[9] G. J. LEVINSKAS, ET AL., "Chronic Toxicity of Pentaborane Vapor," *Am. Ind. Hyg. Assn. J.* **19** (1958), 46–53.

[10] N. American Contact Dermatitis Group, "Epidemiology of Contact Dermatitis in North America," *Arch. Derm.* **108** (1973), 537–40.

[11] T. L. WATT AND R. R. VAUMANN, "Nickel Earlobe Dermatitis," *Arch. Derm.* **98** (1968), 155–58.

[12] N. KRIVANEK AND A. L. REEVES, "The Effect of Chemical Forms of Beryllium on the Production of the Immunologic Response," *Am. Ind. Hyg. Assn. J.* **33** (1972), 45–52.

[13] M. L. ELGART AND R. S. HIGDON, "Allergic Contact Dermatitis to Gold," *Arch. Derm.* **103** (1971), 649–53.

[14] L. H. JANSEN AND L. BERRENA, "Hypersensitivity to Chromium Compounds," *Dermatologica* **137** (1968), 1–16.

[15] L. J. MIEDLER AND J. D. FORBES, "Allergic Contact Dermatitis Due to Metallic Mercury," *Arch. Env. Health* **17** (1968), 960–64.

[16] V. P. BARRANCO, "Eczematous Dermatitis Caused by Internal Exposure to Copper," *Arch. Derm.* **106** (1972), 386–87.

[17] R. S. BRIEF, R. S. AJEMIAN, AND R. G. CONFER, "Iron Pentacarbonyl: Its Toxicity, Detection, and Potential for Formation," *Am. Ind. Hyg. Assn. J.* **28** (1967), 21–30.

[18] A. P. WEHNER, ET AL., "Chronic Inhalation of Nickel Oxide and Cigarette Smoke by Hamsters," *Am. Ind. Hyg. Assn. J.* **36** (1975), 801–10.

[19] M. OHMOTO, ET AL., "Toxicity of Electric Arc Welding Fumes—10 Experimental Studies on Welder's Fever," *Jap. J. Ind. Health* **16** (1974), 103–11. Abstracted in *Ind. Hyg. Digest* **38**; Abst. No. 1411/74 (Nov., 1974).

[20] D. S. ROSS, "Welders' Metal Fume Fever," *J. Soc. Occup. Med.* **24** (1974), 125–29.

[21] W. D. Kuntz and C. P. McCord, "Polymer-Fume Fever," *J. Occup. Med.* **16** (1974), 480–82.

[22] R. D. Hood and Sally L. Bishop, "Teratogenic Effects of Sodium Arsenate in Mice," *Arch. Env. Health* **24** (1972), 62–65.

[23] S. Ishizu, et al., "An Experimental Study on Teratogenic Effect of Cadmium," *Ind. Health* **11** (1973), 127–29.

[24] V. H. Ferm and S. J. Carpenter, "Teratogenic and Embryopathic Effects of Indium, Gallium, and Germanium," *Tox. Appl. Pharm.* **16** (1970), 166–70.

[25] R. M. McClain and B. A. Becker, "Teratogenicity, Fetal Toxicity, and Placental Transfer of Lead Nitrate in Rats," *Tox. Appl. Pharm.* **31** (1975), 72–82.

[26] J. W. Spann, et al., "Ethyl Mercury *p*-Toluene Sulfoanilide: Lethal and Reproductive Effects on Pheasants," *Science* **175** (1972), 328–31.

[27] J. E. Gibson and B. A. Becker, "Placental Transfer, Embryotoxicity, and Teratogenicity of Thallium Sulfate in Normal and Potassium-Deficient Rats," *Tox. Appl. Pharm.* **16** (1970), 120–32.

[28] Helen Swenerton and Lucille S. Hurley, "Teratogenic Effects of a Chelating Agent and their Prevention by Zinc," *Science* **173** (1971), 62–64.

[29] F. Y. Aoki and J. Ruedy, "Severe Lithium Intoxication," *Can. Med. Assn. J.* **105** (1971), 847–48.

[30] T. L. Carson, et al., "Slowed Learning in Lambs Prenatally Exposed to Lead," *Arch. Env. Health* **29** (1974), 154–56.

[31] S. R. Cohen, "A Review of the Health Hazards from Copper Exposure," *J. Occup. Med.* **16** (1974), 621–24.

[32] H. A. Schroeder, et al., "Vascular Reactivity of Rats Altered by Cadmium and a Zinc Chelate," *Arch. Env. Health* **21** (1970), 609–14.

[33] R. E. Holmberg, Jr., and V. H. Ferm, "Interrelationships of Selenium, Cadmium, and Mammalian Teratogenesis," *Arch. Env. Health* **18** (1969), 873–77.

[34] T. F. Gale, "The Interaction of Mercury with Cadmium and Zinc in Mammalian Embryonic Development," *Env. Res.* **6** (1973), 95–105.

[35] H. E. Ganther, et al., "Selenium: Relation to Decreased Toxicity of Methylmercury Added to Diets Containing Tuna," *Science* **175** (1972), 1122–24.

[36] H. Yoshikawa, "Preventive Effect of Pretreatment with Low Dose of Metals on the Acute Toxicity of Metals in Mice," *Ind. Health* **8** (1970), 184–91.

[37] M. W. Radomski and J. D. Wood, "Effect of Metal Ions on Oxygen Toxicity," *Aerosp. Med.* **41** (1970), 184–91.

[38] P. E. Enterline, "Respiratory Cancer among Chromate Workers," *J. Occup. Med.* **16** (1974), 523–26.

[39] L. Hanslian, et al., "Upper Respiratory Tract Lesions from Chromic Acid Aerosol," *Pracovni Lekarstvi* **19** (1967), 294–98. Abstracted in *Ind. Hyg. Digest* **32** (Oct., 1968), 19.

[40] M. Gerald, et al., "Respiratory Cancer and Occupational Exposure to Arsenicals," *Arch. Env. Health* **29** (1974), 250–55.

[41] T. F. Mancuso, "Relation of Duration of Employment and Prior Respiratory Illness to Respiratory Cancer among Beryllium Workers," *Env. Res.* **3** (1970), 251–75.

[42] V. E. Archer, et al., "Uranium Mining and Cigarette Smoking Effects on Man," *J. Occup. Med.* **15** (1973), 204–11.

[43]   S. LANGARD AND T. NORSETH, "A Cohort Study of Bronchial Carcinomas in Workers Producing Chromate Pigments," *Brit. J. Ind. Med.* **32** (1975), 62–65.

[44]   D. V. FROST, "The Two Faces of Selenium—Can Selenophobia Be Cured?," *Crit. Rev. Tox.* **1** (4) (1972), 467–514.

[45]   L. B. TEPPER, "Beryllium," *Crit. Rev. Tox.* **1** (1972), 235–59.

[46]   D. F. FLICK, H. F. KRAYBILL, AND J. M. DIMITROFF, "Toxic Effects of Cadmium: A Review," *Env. Res.* **4** (1971), 71–85.

5

PNEUMOCONIOSES

Many years ago *pneumoconiosis* was coined to mean the presence in the lungs of dust whether or not the dust caused any ill effects. In 1971 the Fourth International Conference of Experts on Pneumoconiosis [1] met and re-defined the word to mean *the accumulation of dust in the lungs and the tissue re-actions to its presence.* They then defined *dust* to be *an aerosol composed of solid inanimate particles.* (An *aerosol* is a dispersion of solid or liquid particles in a gaseous medium.) While this definition of dust excludes living entities such as pollen grains or bacteria, it does include most particulates of vegetable or animal origin as well as inorganic and man-made organic materials.

Fibrogenic Materials

A pneumoconiosis with permanent scarring of the lungs may be caused by fibrogenic dust such as silica or asbestos or by a severe reaction, an altered tissue response, to a dust that is not normally fibrogenic. *Fibrogenic* simply means that fibrous bodies of collagen form in place of normal tissue. The result is similar to scar tissue that forms in the place of a deep laceration. Fibrous tissue in the lung is nonfunctional and takes the place of functional tissue; it tends to inhibit gaseous exchange in proportion to its surface area. This kind of response is the reaction of the lung to chronic irritation; the peculiarity of crystalline silica and asbestos is that these materials are not irritants in the usual sense but still are fibrogenic.

Many theories have been advanced concerning the means by which silica and asbestos cause the proliferation of fibrous tissue in the lungs. None of the theories gives a completely satisfactory explanation of why some of these materials are fibrogenic and others similar in size, shape, and chemical makeup are not. This problem is being investigated actively by several groups around the world.

Disability as the result of a fibrogenic disease usually takes a period of many years to develop. Furthermore, neither silicosis nor asbestosis causes fatalities directly. Instead, they cause shortness of breath (*dyspnea*) and decreased chest expansion with a reduced physical work capacity in proportion to the disease severity.

Free Crystalline Silica

All *free crystalline silica* (or crystalline free silica) is SiO_2 but not all SiO_2 is free crystalline silica. An *amorphous* (noncrystalline) form of SiO_2 is produced commercially by the combustion of ethyl silicate and diatomaceous earth, opal, and tripoli are natural forms of amorphous silica. The amorphous varieties do not cause fibrosis except perhaps from massive inhalation [2, 3]; instead, if the particle size is small enough, they may cause an acute disease similar to metal fume fever. The tendency of free crystalline silica to cause fibrosis is related to its crystallinity [4].

Crystalline silica exists in three forms: quartz, tridymite, and cristobalite. These varieties of SiO_2 differ in several physical properties such as refractive index, density, thermal behavior, and X-ray diffraction pattern; these may be used to distinguish them analytically. Toxicologically, cristobalite and tridymite are about equal in their fibrogenic activity, quartz is less active. Nevertheless, because quartz is by far the most prevalent form, most cases of silicosis have been caused by its chronic inhalation.

Diatomaceous earth, or fuller's earth (a *fuller* is a person who treats cloth to make it seem more "full" to the hand), is a deposit of the shells of microscopic diatoms that inhabited the now-vanished oceans where the deposits are found. The shell material is chiefly amorphous silica and, as mined, diatomaceous earth usually contains less than 2% crystalline silica. A refining process (calcining) that subjects the earth to high temperature, however, causes the transformation of as much as 60% of the amorphous variety to cristobalite [5]. Calcined diatomaceous earth is the main source of possible human exposures to cristobalite.

Quartz is almost ubiquitous on the earth's surface. In addition to being the main component of beach sand, it exists as a contaminant or main component of granite, sandstone, many clays, some limestones, and even some slate, talc, and soapstone. It is used in abrasives, refractories, ceramics, paints, and even in some fertilizers as well as in many manufacturing processes. Amethyst, cat's eye, bloodstone, onyx, agate, and rock crystal are among several semiprecious gems that are almost pure quartz.

Silicosis has the distinction of being probably the oldest occupational disease [6]. One of man's first full-time occupations must have been that of miner or tool and weapon maker and the first tools and weapons were made of stone. Quartz is present in many of the rocks that are suitable for hand-forming into tools such as knives, scrapers, and heads for spears and arrows. Cavemen who mined the rock or made tools must have been afflicted with silicosis if they lived long enough. The occupational origin of silicosis was recognized at least as early as 3,000 b.c., although then the true cause was not known. Nevertheless, even at that time most hard-rock miners had a short life expectancy. Until the industrial revolution, silicosis was a disease mainly of those who worked in hard-rock or hard-coal mines; furthermore, most miners were afflicted. Agricola in 1556 wrote that some women in the Carpathian mountains married as many as seven miners in succession, each miner in turn dying because of lung disease.

Prior to about the middle of the 20th century, tuberculosis (TB) was one of the most potent killers of men. This disease was endemic in most populations; hence avoiding contact with infected persons was nearly impossible. Silicosis increases susceptibility to TB and tends to make any case of TB more severe. The combined disease, silico-tuberculosis, was present in many of those who had advanced silicosis and was usually the cause of death. Uncomplicated silicosis kills rarely if ever, but silico-tuberculosis can be a fulminating, rapidly fatal disease.

The industrial revolution brought increased opportunities for the inhalation of all kinds of dusts and, in addition, more dust in the air. Dust is an aerosol produced mainly by mechanical action. More power available for working rock invariably meant that more dust would be produced. Explosives and mechanized mining equipment tremendously increased dust concentrations in mines because improvements in ventilation and other means of dust control did not keep pace with the development of mining tools. Much the same can be said for other areas of industry where dust was produced, from ore reduction to tool sharpening.

The prevalence of silicosis in Europe and in the United States probably peaked in the early decades of the 20th century although the problems of industrial disease reporting prevent an accurate assessment. On a worldwide scale the peak prevalence may well occur sometime in the 21st century unless the mistakes made by the now developed countries are avoided by the developing nations.

Abrasive (sand) blasting probably accounts for most potentially hazardous exposures to crystalline silica today in the United States [7]. Even though respiratory protective hoods are universally worn by men doing the blasting, the protection offered by the hoods is often far from complete. One means of reducing the hazards of this practice is to use materials other than sand as the abrasive.

Because of the prevalence of silicosis in many industries, people have sought ways and means either of strengthening man's resistance to crystalline silica or of curing silicosis once it has become established. The most popular prophylactic (preventive) treatment has been the inhalation of aluminum or aluminum oxide powder [8], but other materials such as potassium carbonate [9] have been used. Steroids have been used therapeutically, as has been polyvinylpyridine-*N*-oxide [10, 11], but no treatment has proved to arrest or reverse the course of silicosis completely.

In the 1960s and 1970s in the United States [12] and Great Britain [13], one of the most pressing occupational disease problems was *coal-worker's pneumoconiosis,* or CWP. Also called *anthracosis* and *black-lung disease,* its cause was obscure although obviously related to the inhalation of coal dust. Although the symptoms of CWP are similar to those of silicosis, one of the signs of the disease, namely the altered appearance of lungs on X ray, is quite different. Not all coals are able to cause this potentially disabling disease, and those able to are not alike in that ability [14]. Crystalline silica is present in most of the coal usually associated with CWP or in the rock that must be worked along with the coal seams, but silica is also present in some coals not associated with the disease. Nevertheless, for the most part the incidence and severity of CWP appear to be proportional to the extent of chronic crystalline silica inhalation. Complicated CWP or *progressive massive fibrosis* (PMF) results from an altered tissue response to coal dust [15] that is not normally fibrogenic. The cause of the altered response is obscure, but metal content of the coal may play a role [16].

Early environmental health surveys of workers exposed to the dust of crystalline silica showed that neither the incidence of silicosis nor its severity were directly related to the mass of dust per unit volume of air inhaled. Such information must have been shocking at first to those who were aware of dose-response phenomena in toxicology. Experiment showed that the disease was related to the number of crystalline silica particles per unit volume of air inhaled rather than to their mass. Hindsight makes the reason for this obvious. In any aerosol, most of the mass of material (and hence the weight of a contaminant in an air sample) is concentrated in a few large particles. When the aerosol is a dust, most of the large particles will either drop out of the suspension through gravity or, if inhaled, will be removed from the air by impaction on moist mucous membranes in the nose and throat. The large particles never reach the lungs; only the small ones do.

Particles with unit density and a diameter of 10 μm have a negligibly small but predictable probability of reaching the lungs. With decreasing diameter the probability of reaching the lungs increases, as does the deposition in the lungs until a diameter of about 1 μm is reached. Then while the particles reach respiratory bronchioles and alveoli in ever-increasing numbers, the probability of deposition and of toxic effect decreases until the particles begin to approach the size of molecules. Where the main site of toxic action of an inhaled material is the deep lung, the hazard is proportional to the amount of material deposited in the deep lung. Most suspended aerosol particles are smaller than 10 μm in diameter and hence the inhalation hazard is related to the number of particles per unit volume of air sampled, not to the total mass found in an air sample.

For many years air samples were obtained with midget impingers and were analyzed by using a light-field microscope to count the number of particles visible in each of several size ranges. Fortunately, responding to impetus provided by the U.S. Atomic Energy Commission, samplers were developed in the 1960s that use a mechanical system to separate the sample into nonrespirable and respirable fractions. The nonrespirable fraction is discarded and the respirable fraction is simply weighed [17, 18]. Epidemiologic studies have shown that the silicosis hazard is proportional to the mass per unit volume of respirable (*respirable* being very carefully defined) dust and thus the simple, accurate procedure of weighing could substitute for the many problems of dust counting if the proportion of crystalline silica in the sample were known.

Asbestos

The word *asbestos* has an even broader generic meaning than does *silica* because it is applied to materials that have little in common except fibrous nature and mineral derivation [19]. Chemically, all varieties of asbestos are silicates, but other elements may be present in the molecule in addition to silicon and oxygen. Most of the mineral forms called *asbestos* are derived from two varieties of rock, serpentine and amphibole, which also exist in nonasbestos forms. Serpentine, for instance, has the same chemical composition ($H_4Mg_3Si_2O_9$) as chrysotile, the most widely used variety of asbestos, but serpentine has no fibrous structure. Or rather, serpentine that does have a fibrous structure is called *chrysotile*.

Amphibole types of asbestos vary widely in chemical composition as well as in other properties. Those with some commercial importance are crocidolite, anthophyllite, amosite (a variety of anthophyllite), tremolite, and actinolite. Elements that must be present are silicon and oxygen; others such as Al, Ca, F, Fe, H, K, Mg, Na, and Ti may also be present in the mineral. Amphibolic minerals that do not have the fibrous structure typical of asbestos include arfvedsonite, cummingtonite, glaucophane, hornblende, and riebeckite.

Asbestos has long been known capable of inducing a fibrotic reaction in lung tissue, a pneumoconiosis called *asbestosis*. For many years, some authorities argued that only chrysotile had this particular toxic property, but further experience with the less well-known forms of asbestos showed that all varieties cause fibrosis despite their widely varying composition and physical properties [20, 21, 22]. Furthermore, rocks with identical composition and physical properties (except being nonfibrous) are not capable of causing fibrosis. Fibrous glass, with dimensions similar to at least some asbestos varieties, is also incapable of causing a fibrotic reaction as is asbestos with fiber lengths less than about 5 μm, leaving investigators with little upon which to base a theory of the toxicology of asbestos. Of course, not having much upon which to base a theory has not noticeably curtailed hypothesis and theory formation. At any one time at least two or three theories have been in vogue, each with its own supporters [19, 23, 24].

Some people exposed to asbestos have, years later, developed a cancer called *mesothelioma*, indicating that the cancer arises in mesothelial tissue, which is the type used by the body for linings or sacs. Both the pulmonary and abdominal cavities are enclosed in sacs of mesothelial tissue and mesotheliomas have arisen in both areas subsequent to minor asbestos inhalation. Furthermore, the incidence of mesothelioma in the unexposed population is extremely low.

In the lung, any fibrous material may be coated with a collagen-like substance rich in iron, often giving the fiber a "dumbbell" appearance upon microscopic examination [25]. These curiously shaped structures were first called "asbestos" or "asbestosis" bodies because of the obviously asbestos-like fiber at the core [26]. Then, when chrysotile asbestos could not be found in the core when it should have been, the name was changed to "ferruginous bodies," indicating that they contain iron [27]. Nevertheless, the scientific community was shocked to hear that pathologists were finding ferruginous bodies in ever-greater percentages of lungs examined at autopsy. In fact, reports have appeared that ferruginous bodies are found in essentially 100% of the lungs examined [28, 29, 30], even of people with no known exposure to asbestos, and that most may indeed be formed around a core of asbestos [31].

Fibrous materials or asbestos-like fibers are found in nearly all air samples obtained in urban areas to evaluate particulate air pollution [32]. The reason for this appears to be that since the early 1900s the amount of asbestos in our environment has been increasing rapidly. Asbestos is used in so many products, ranging from brake shoes for automobiles to insulation for ventilation ducts, that probably everyone in the civilized world has inhaled at least some of this material. That, coupled with the fact that decades after a minimal exposure a person (especially one who smokes) has a reasonably good chance of developing a mesothelioma or other cancer [33], has caused asbestos to be seen in a new light. Laws have been passed and rules have been made to limit occupational and environmental exposure but the nagging suspicion remains that perhaps most people who live long enough might be afflicted with a cancer caused by asbestos. That problem has not been completely resolved.

Persons who are routinely exposed to asbestos because of their occupation ought to have a much greater cancer risk than the general population. Therefore, intensive epidemiological and industrial hygiene surveys of such groups were begun in the 1960s [34, 35]. The persons most closely watched were insulation workers such as pipe coverers because their trade was known to be associated with gross inhalation exposure to asbestos, especially chrysotile. Several surveys showed that these workmen indeed had a high incidence and prevalence of asbestosis and also a much higher than normal death rate from mesothelioma of the pleura or peritoneum and lung cancer [36, 37]. Some data, however, appeared to indicate that an asbestos exposure alone might not cause cancer even though that exposure increased the cancer incidence among cigarette smokers [38, 39, 40, 41]. Although complicated by a high prevalence of cigarette smoking among insulation workers, prospective studies should eventually provide at least some of the more important answers to questions raised about the real hazard of asbestos inhalation.

Nonfibrogenic Materials

Most aerosols man inhales do not irritate lung tissue in a manner that results in fibrosis. Instead, their presence in the lung results in a noncollagenous pneumoconiosis where the alveolar structure remains intact. Because the alveoli are not changed by the presence of these dusts, a nonfibrogenic pneumoconiosis is at least potentially reversible. The classification *nonfibrogenic* includes materials as hard and sharp as diamond and as soft as gossamer; in fact, essentially all solids except asbestos and crystalline silica. Nonfibrogenic materials may cause many effects on the respiratory tract (or the body), including the violent and possibly fatal reaction known as anaphylactic shock, but they do not usually cause fibrosis. All, of course, cause pneumoconioses, some of which have special names.

Inorganic Materials

Silicates, other than asbestos, are present in almost all dust but rarely have any noticeable effect unless the exposure is overwhelming. On the other hand, most silicates can be contaminated with varying amounts of crystalline silica, especially quartz. Talc, for instance, is a silicate that can be inhaled over a prolonged period of time with little effect other than a characteristic lung X-ray pattern diagnostic of *talcosis* [42]. On the other hand, some powders used for dulling the shine on milady's nose or helping to keep baby bottoms smooth and dry have been found to be contaminated with surprisingly large amounts of quartz even though supposedly being essentially pure talc [43]. Moreover, the mineral talc is occasionally associated with tremolite or anthophyllite, which may contaminate the commercial product [44, 45]. Other silicates such as kaolin, soapstone, mica, and feldspar are often encountered and often contain quartz as an impurity.

Some of the metals and metal oxides give rise to named pneumoconioses. A partial list is found in Table 5-1.

Table 5-1 Some Named Pneumoconioses and Causes

Pneumoconiosis	*Agent*
Siderosis	Iron/iron oxides
Baritosis	Barium oxide/sulfate
Shaver's disease	Aluminum oxide (?)
Berylliosis	Beryllium and compounds
Stannosis	Tin/tin oxide

The pneumoconioses associated with iron, barium, and tin are called *benign* because they have been found to have no effect other than a characteristic appearance on an X-ray film. Beryllium, on the other hand, is well known to be capable of causing an acute pneumonitis and, more seriously, a chronic *granulomatous* (granules rather than fibers form in place of normal tissue) lung disease that can be fatal years after termination of an exposure.

Aluminum and its oxide are usually thought to be innocuous in the lung but Shaver's disease, associated with production of abrasives from fused aluminum oxide, has been fatal. The causative agent is not known. Purposeful inhalation of aluminum (or aluminum oxide) powder has been practiced by several groups of workers in an attempt to protect against silicosis where that disease was an especially prevalent occupational hazard. Although some experimental data appear to indicate that prophylaxis is possible in this manner, exposure control (see Chapter 16) is almost always the better route to follow. Nevertheless, such chronic purposeful inhalation of aluminum has not caused any reported ill effects. Whether aluminum powder inhalation protects against silicosis is still disputed.

Many metals and/or their oxides and/or other inorganic compounds are capable of causing metal fume fever (see Chapter 4), which is not primarily a lung effect, and/or pneumonitis, which is. The metals best known for causing the acute lung irritation known as *pneumonitis* are cadmium, beryllium, and manganese; each of these has caused many deaths by this mechanism.

Of all the metals, only cobalt [46] and possibly manganese [47] and cerium [48] are associated with occasional fibrotic reactions in the lung following prolonged exposure. In fact, at one time some of the effects of asbestos were suspected of being caused by cobalt, present in most asbestos as a contaminant added from the metal equipment used in asbestos production [49]. Even though altered tissue reactions to cobalt may result in fibrosis, asbestos alone is considerably more effective in this regard. Cobalt contamination has been shown to be relatively unimportant.

Portland cement is produced as a powder with a very small particle size. When it is dispersed into the air, many of the particles are small enough to reach the lungs where they can be quite irritating. Some portland cement contains crystalline silica as an impurity, but concentrations of that material are usually low and silicosis is not at all likely even from massive chronic overexposure to cement dust. On the other hand, some cements contain appreciable amounts of chromium, which is an irritant and potential carcinogen. All portland cements are alkaline and most of the mild skin and respiratory tract irritations from these materials are probably caused by the alkalinity. Unless one becomes sensitized to one of the components of cement (chromium, for instance), the chronic inhalation hazard is probably not very much greater than the acute inhalation hazard.

Fibrous glass has properties in common with both crystalline silica and asbestos. That is, all fibrous glass has a high percentage of SiO_2 and, of course, is fibrous in nature as is asbestos. Fibrous glass does not cause a fibrotic reaction in the lung, however, even after chronic inhalation of concentrations very high compared to those considered safe for either asbestos or crystalline silica [50]. Glass is an amorphous material. In some respects, glass can be considered to be a supercooled liquid, and thus it does not contain any crystalline silica at all. Furthermore, the diameter of most glass fibers is large in comparison with the fibers of asbestos.

Each fiber of fibrous glass is an individual entity. That is, the material is essentially monofilamentary, while asbestos is composed of bundles of fibers as well as monofilaments. Fibrous glass fibers are solid; some asbestos fibers are hollow. These physical facts may partially explain why fibrous glass does not cause fibrosis but asbestos does [23]. Finally, fibrous glass is not a silicate, but all asbestos varieties are silicates. Chemical nature of the fibers probably also plays an important role in the development of disease.

Organic Materials

Almost all wood dusts are capable of producing an allergic reaction in a small percentage of the human population. Such reactions may take the form of either lung or skin sensitization, or both, although an asthmatic reaction is the more serious disease because it can be life-threatening. The common woods such as fir, pine, spruce, oak, and maple cause much less trouble than do the fine exotic woods such as western red cedar [51] even though many more people are exposed to the common woods.

Natural organic materials other than wood can sensitize some people, although the sensitization may be the result of ingestion or skin contact rather than dust inhalation. Offenders include all grains; cotton; wool; flax; soy, castor, coffee, and cocoa beans; tobacco; tea; and many others. Proteolytic enzymes and the dusts associated with handling cotton and bagasse (sugar cane residue) have been the most frequent industrial offenders.

Proteolytic enzymes and in particular those derived from *Bacillus subtilis* called *subtilisins* began to appear in presoaks and family wash products in the late 1960s [52]. Although these products have caused few problems among users, they have been responsible for respiratory tract-related illness among workers [53]. Whether the problems are caused by a true sensitization or by some other action of the enzymes has been the subject of much research, most of which supports the sensitization theory [54].

Dust arising from equipment used to process raw cotton, flax, or soft hemp into thread has caused a disease called *byssinosis* that is characterized by dyspnea during or following the first day at work after two or more days with no exposure [55]. Furthermore, this acute phase can merge into a more chronic phase where the dyspnea is so extreme as to cause physical disability if exposures are prolonged over many years. Final stages of the disease may include chronic bronchitis and emphysema; the resultant strain on the right side of the heart has been fatal [56].

The specific cause of byssinosis (if there is one cause) is not known although the fiber (cotton, flax, or hemp) itself does not appear to be the culprit. Instead, inhalation of dust arising from waste associated with the whole plant (rather than with the fiber) appears to be responsible. One interesting suggestion is that the predominant cause may be a proteolytic enzyme in concentrations too low to be found by ordinary analytical methods. Symptoms and etiology of byssinosis are similar to those associated with subtilisins, tending to enhance that theory. On the other hand, byssinosis lacks some of the features of a true sensitization such as skin response to potential offenders, thereby confusing an already perplexing situation [57].

After sucrose has been extracted from sugar cane, the residue, which consists of plant fiber and is called *bagasse,* must be disposed of. Bagasse has been used for everything from thatch on huts to fuel for boilers to base for a crude paper and, if handled while still fresh, causes few problems. If this material is stored for any length of time before it is used, however, workers handling it may contract a disease called *bagassosis,* with signs and symptoms similar to those of byssinosis or "detergent worker's lung." In this case, the disease is probably caused by a mold on the fiber and not by the fiber itself and, in that respect, is similar to the more common (in the United States) "farmer's lung," caused by handling moldy hay [58, 59].

Most man-made polymeric materials are physiologically inert and are not any more troublesome than are the common woods. However, many plastics contain plasticizers added to modify certain properties of the material, and the plasticizers may not be inert. Furthermore, incompletely cured plastics may contain monomers (see Chapter 7) that can be sensitizers or the cause of diseases related to the direct toxic action of the materials.

If the substances inhaled do not cause pulmonary fibrosis and do not sensitize the lung, they are usually thought of as physiologically inert (not *nontoxic*) if they are without direct toxic action. In such cases, the main action of the dust is to load the clearance mechanisms (mainly ciliary action and phagocytosis) of the respiratory tract. An overexposure may overload these mechanisms even to the point where the cilia cease to function and then physical blockage of airflow may be a factor. Also, if the clearance mechanisms become inoperative from exposure to an inert material, an exposure to a normally innocuous amount of a *noninert* material could have serious consequences. For these reasons, limits must be placed on exposure to any dust, no matter how inert experiment and experience prove it to be. As a practical matter, dust concentrations in excess of about 10 mg/m³ or 30 mppcf (millions of particles per cubic foot of air) usually indicate poor practice and excessive material loss. They should be avoided for many reasons, including health.

A partial list of materials generally recognized as being *nuisance* dusts is presented in Table 5-2.

Table 5-2 Nuisance Dusts

Aluminum oxide	Mangesium oxide
Calcium carbonate	Marble
Calcium sulfate (gypsum)	Portland cement
Emery	Silicon carbide
Glass	Starch
Graphite	Sugar (sucrose)
Iron oxide	Tin oxide
Kaolin	Titanium dioxide

REFERENCES

[1] "Pneumoconiosis Redefined," *Brit. Med. J.* **2** (1972), 552.

[2] W. KOLSTERKOTTER, "Studies of Animals Experimentally Inflicted with Silicosis through Inhalation of Amorphous Arc Silic Acid," *Arch. Hyg. Bakteriol.* **150** (1966), 542–57. Abstracted in *Ind. Hyg. Digest* **32** (Jan., 1968), 13.

[3] R. PROCHAZKA, "Dust Measurements in the Immediate Vicinity of Electric-Arc Furnaces for Ferro-Alloys: Measurement at Ferrosilicon Furnaces," *Staub Reinhalt Luft* **31** (1971), 8–16. Abstracted in *Ind. Hyg. Digest* **36** (Sept., 1972), 9.

[4] H. Brieger and P. Gross, "On the Theory of Silicosis. III. Stishovite," *Arch. Env. Health* **15** (1967), 751–57.

[5] A. Franzinelli, et al., "A Contribution to the Study of Fossil Flour Pneumoconiosis," *Med. Lavoro.* **62** (1971), 258–71. Abstracted in *Ind. Hyg. Digest* **36** (May, 1972), 5.

[6] H. V. Brown, "The History of Industrial Hygiene: A Review with Special Reference to Silicosis," *Am. Ind. Hyg. Assn. J.* **26** (1965), 212–26.

[7] A. W. Blair, "Abrasive Blasting Protective Practices Study," *Am. Ind. Hyg. Assn. J.* **34** (1973), 61–65.

[8] W. B. Dix, "Aluminum Powder and Silicosis Prevention," *Can. Mining J.* **92** (1971), 35–42.

[9] R. K. Melnichenko, et al., "Potassium Carbonate Inhalation for the Prevention of Pneumoconiosis," *Gigiena i Sanit.* **32** (1967), 119–23. Abstracted in *Ind. Hyg. Digest.* **32** (Sept., 1968), 23.

[10] R. Pott, et al., "The Resorption of Collagen in Experimental Silicosis after Treatment with Polyvinylpyridin-*N*-oxide," *Beitr. Klin. Tuberk.* **141** (1970), 259–67. Abstracted in *Ind. Hyg. Digest* **35** (June, 1971), 16.

[11] J. L. Kaw, et al., "Studies of Quartz Cytotoxicity on Peritoneal Macrophages of Guinea Pigs Pretreated with Polyvinylpyridine-*N*-oxide," *Env. Res.* **9** (1975), 313–20.

[12] W. K. C. Morgan, et al., "The Prevalence of Coal Workers' Pneumoconiosis in U.S. Coal Miners," *Arch. Env. Health* **27** (1973), 221–26.

[13] S. Rae, "Pneumoconiosis and Coal Dust Exposure," *Brit. Med. Bull.* **27** (1971), 53–58.

[14] W. S. Lainhart, "Roentgenographic Evidence of Coal Workers' Pneumoconiosis in Three Geographic Areas in the United States," *J. Occup. Med.* **11** (1969), 399–408.

[15] C. A. Soutar, et al., "Circulating Antinuclear Antibody and Rheumatoid Factor in Coal Pneumoconiosis," *Brit. Med. J.* **3** (1974), 145–47.

[16] D. V. Sweet, et al., "The Relationship of Total Dust, Free Silica, and Trace Metal Concentrations to the Occupational Respiratory Disease of Bituminous Coal Miners," *Am. Ind. Hyg. Assn. J.* **35** (1974), 479–88.

[17] G. Knight, et al., "Development of a Dust Sampling System for Hardrock Mines Based on Gravimetric and Quartz Assessment," *Am. Ind. Hyg. Assn. J.* **35** (1974), 671–80.

[18] T. F. Tomb and H. N. Treaftis, "A New Two-Stage Respirable Dust Sampler," *Am. Ind. Hyg. Assn. J.* **36** (1975), 1–9.

[19] L. J. Cralley, et al., "Research on Health Effects of Asbestos," *J. Occup. Med.* **10** (1968), 38–41.

[20] H. Weill, et al., "Lung Function Consequences of Dust Exposure to Asbestos Cement Manufacturing Plants," *Arch. Env. Health* **30** (1975), 88–97.

[21] L. O. Meurman, et al., "Mortality and Morbidity among the Working Population of Anthophyllite Asbestos Miners in Finland," *Brit. J. Ind. Med.* **31** (1974), 105–12.

[22] G. K. Sluis-Cremer, "Asbestos in South African Asbestos Miners," *Env. Res.* **3** (1970), 310–19.

[23] P. Gross and R. A. Harley, "The Locus of Pathogenicity of Asbestos Dust," *Arch. Env. Health* **27** (1973), 240–42.

[24] H. C. Lewinsohn, "Health Hazards of Asbestos: A Review of Recent Trends," *J. Soc. Occup. Med.* **24** (1974), 2–10.

[25] E. S. Flowers, "Relationship between Exposure to Asbestos, Collagen Formation, Ferruginous Bodies, and Carcinoma," *Am. Ind. Hyg. Assn. J.* **35** (1974), 724–29.

[26] F. D. Pooley, "Asbestos Bodies, Their Formation, Composition and Character," *Env. Res.* **5** (1972), 363–79.

[27] P. Gross, R. T. P. deTreville, and M. N. Haller, "Pulmonary Ferruginous Bodies in City Dwellers. A Study of Their Central Fiber," *Arch. Env. Health* **19** (1969), 186–88.

[28] M. D. Utidjian, Paul Gross, and R. T. P. deTreville, "Ferruginous Bodies in Human Lungs," *Arch. Env. Health* **17** (1968), 327–33.

[29] J. Bignon, et al., "Incidence of Pulmonary Ferruginous Bodies in France," *Env. Res.* **3** (1970), 430–42.

[30] Paul Gross, et al., "Mineral Fiber Content of Human Lungs," *Am. Ind. Hyg. Assn. J.* **35** (1974), 148–51.

[31] A. M. Langer, I. J. Selikoff, and A. Sastre, "Chrysotile Asbestos in the Lungs of Persons in New York City," *Arch. Env. Health* **22** (1971), 348–61.

[32] P. P. Holt and D. K. Young, "Asbestos Fibres in the Air of Towns," *Atm. Env.* **7** (1973), 481–83.

[33] A. L. Reeves, et al., "Inhalation Carcinogenesis from Various Forms of Asbestos," *Env. Res.* **8** (1974), 178–202.

[34] J. L. Balzer and W. C. Cooper, "The Work Environment of Insulating Workers," *Am. Ind. Hyg. Assn. J.* **29** (1968), 222–27.

[35] J. E. Dunn, Jr. and J. M. Weir, "A Prospective Study of Mortality of Several Occupational Groups," *Arch. Env. Health* **17** (1968), 71–76.

[36] J. Lieben, "Malignancies in Asbestos Workers," *Arch. Env. Health* **13** (1966), 619–21.

[37] M. Borow, et al., "Mesothelioma Following Exposure to Asbestos: A Review of 72 Cases," *Chest* **64** (1973), 641–48.

[38] J. Lieben and H. Pistawka, "Mesothelioma and Asbestos Exposure," *Arch. Env. Health* **14** (1967), 559–68.

[39] I. J. Selikoff, E. C. Hammond, and J. Churg, "Asbestos Exposure, Smoking, and Neoplasia," *J. Am. Med. Assn.* **204** (1968), 106–12.

[40] I. J. Selikoff, E. C. Hammond, and J. Churg, "Carcinogenicity of Amosite Asbestos," *Arch. Env. Health* **25** (1972), 183–86.

[41] P. E. Enterline and V. Henderson, "Type of Asbestos and Respiratory Cancer in the Asbestos Industry," *Arch. Env. Health* **27** (1973), 312–17.

[42] Benjamin Weiss and E. A. Boettner, "Commercial Talc and Talcosis," *Arch. Env. Health* **14** (1967), 304–08.

[43] L. J. Cralley, et al., "Fibrous and Mineral Content of Cosmetic Talcum Products," *Am. Ind. Hyg. Assn. J.* **29** (1968), 350–54.

[44] M. Kleinfeld, J. Messite, and A. M. Langer, "A Study of Workers Exposed to Asbestiform Minerals in Commercial Talc Manufacture," *Env. Res.* **6** (1973), 132–43.

[45] M. Kleinfeld, et al., "Mortality Experiences among Talc Workers: A Follow-Up Study," *J. Occup. Med.* **16** (1974), 345–49.

[46] A. O. Bech, "Hard Metal Disease and Tool Room Grinding," *J. Soc. Occup. Med.* **24** (1974), 11–16.

[47] S. H. Zaidi, et al., "Experimental Infective Manganese Pneumoconiosis in Guinea Pigs," *Env. Res.* **6** (1973), 287–97.

[48] F. Heuck and R. Hoschek, "Pulmonary Changes in Cerium Pneumoconiosis, a Previously Unknown Occupational Disease," *Fortschr. Gebiete Roentgenstrahlen u. Nuklearmed.* **106** (1967), 489–502. Abstracted in *Ind. Hyg. Digest* **32** (June, 1968), 16.

[49] L. J. Cralley and W. S. Lainhart, "Are Trace Metals Associated with Asbestos

Fibers Responsible for the Biologic Effects Attributed to Asbestos?", *J. Occup. Med.* **15** (1973), 262–66.

[50] P. E. ENTERLINE AND V. HENDERSON, "The Health of Retired Fibrous Glass Workers," *Arch. Env. Health* **30** (1975), 113–16.

[51] T. ISHIZAKE, ET AL., "Occupational Asthma from Western Red Cedar Dust (*Thuja Plicata*) in Furniture Factory Workers," *J. Occup. Med.* **15** (1973), 580–85.

[52] M. L. H. FLINDT, "Pulmonary Disease Due to Inhalation of Derivatives of *Bacillus Subtilis* Containing Proteolytic Enzyme," *Lancet* **1** (1969), 1177–81.

[53] N. S. SHORE, R. GREENE, AND H. KAZEMI, "Lung Dysfunction in Workers Exposed to *Bacillus Subtilis* Enzyme," *Env. Res.* **4** (1971), 512–19.

[54] D. C. LITTLE AND J. DOLOVICH, "Respiratory Disease in Industry Due to *B. Subtilis* Enzyme Preparations," *Can. Med Assn. J.* **108** (1973), 1120–25.

[55] J. A. MERCHANT, ET AL., "Responses to Cotton Dust," *Arch. Env. Health* **30** (1975), 222–29.

[56] A. BOUHUYS, J. C. GILSON, AND R. S. SCHILLING, "Byssinosis in the Textile Industry," *Arch. Env. Health* **21** (1970), 475–78.

[57] V. POPA, ET AL., "An Investigation of Allergy in Byssinosis: Sensitization to Cotton, Hemp, Flax and Jute Antigens," *Brit. J. Ind. Med.* **26** (1969), 101–08.

[58] C. E. D. HEARN AND V. HOLFORD-STEVENS, "Immunological Aspects of Bagassosis," *Brit. J. Ind. Med.* **25** (1968), 283–92.

[59] HELEN A. DICKIE AND J. RANKIN, "The Lung's Response to Inhaled Organic Dusts," *Arch. Env. Health* **15** (1967), 139–40.

6

ORGANIC SOLVENTS

Almost all the materials considered in this chapter are organic chemicals, but a few are not (having no hydrogen, neither CCl_4 nor CS_2 can be classed as organic). Almost all are good solvents for organic compounds such as oils and greases, although solvency may not be the main use for the material. Most of the compounds considered are liquids at normal room temperature, and all are colorless when pure. Their toxicity [1] varies from that of the simple asphyxiants to moderately high, and their hazards also vary over a wide range.

Solvents are used for degreasing, for dry cleaning, and in the manufacture of many materials ranging from paints, varnishes, shellacs, and lacquers to rubber and synthetic resins. When not being used as solvents, these materials may function as fuels for automobiles, trucks, airliners, and electric power plants as well as for ships and railroad trains. They may also act as chemical intermediates with or without regard to their ability to put other materials into solution. Because of their utility, some of these materials are encountered by everyone in the civilized world. A knowledge of their toxicity and toxic hazards has great utility.

A simple listing of all materials that can be classed as organic solvents would include several hundred, perhaps several thousand, chemicals. To add to that list synonyms and even brief information about the toxicity and hazards of each would occupy untold pages of text to repeat information with a high rate of obsolescence that is available elsewhere.

Some simplification is necessary. In this chapter, organic solvents are considered in five broad categories plus a sixth for miscellaneous. The catagories are based on chemical structure with little attention paid to fine detail so that a few of the materials discussed are gases or solids and are never used as solvents. Use of the material in this chapter requires a working knowledge of organic chemical nomenclature, a review of which may be found in Appendix A.

Simplification extends to the complete omission of air or body fluid sampling and analytical methods mainly because progress in this area is so rapid that no general text can hope to keep pace. Furthermore, although the time-honored wet chemical methods are giving way to the newer physical/instrumental techniques, the transition is far from complete so that a discussion of each field for each class of compounds would be necessary. For general information on this topic, see Chapter 16.

Aliphatic and Alicyclic Hydrocarbons

The principal effect of the lower aliphatic and alicyclic hydrocarbons on man is narcosis. These materials have anesthetic-like action that, however, is absent in methane, ethane, and propane and only weakly present in butane; it becomes stronger with increasing molecular weight [2]. Although unsaturation and chain branching have very little effect on the toxicity of these materials, ethylene and acetylene are weak anesthetics. Cyclopropane has been used for many years in surgical anesthesia.

Paralleling narcosis, irritation of mucous membranes becomes apparent in the alkanes only at high molecular weights, increasing slightly in proportion to molecular weight [2]. Alkenes and alkynes are similar to the alkanes [3].

The principal hazard of the aliphatic and alicyclic hydrocarbons is that of fire and explosion. Lower members of each homologous series have high vapor pressures and fairly wide explosive ranges. The influence of molecular weight on explosive range in air is illustrated by the following equations for the normal alkanes, derived from data for compounds from methane to dodecane [4] submitted to a computerized regression program:

Lower explosive limit:

$$\% = \frac{1}{0.105 + 0.119 N_c}$$

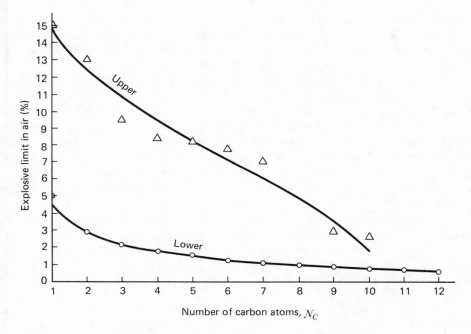

Figure 6-1 Explosive range for normal alkanes in air.

Upper explosive limit:

$$\% = [219 - 94 \ln (\mathcal{N}_c)]^{1/2}$$

where $\mathcal{N}_c$ is the number of carbon atoms in the molecule. These equations and the data points used in their derivation are plotted in Fig. 6-1.

As the molecular weight increases, so in general does the toxicity, but with a consequent decrease in vapor pressure and lipid solubility the toxic hazard (the likelihood of toxic injury) from inhalation eventually decreases to near zero.

Aromatic Hydrocarbons

The first member of the homologous series of aromatic hydrocarbons is benzene, which is also incorrectly called benzol and which may be confused with benzine, a name sometimes given to petroleum naphtha. Benzene is an illustration of the *first-member* rule, which states *the toxicity of the first member of a homologous series is likely to differ qualitatively from the toxicity of other members of the series.*

The toxicity of benzene is indeed different from the toxicity of further members of the simple (or even complex) aromatic hydrocarbons. (That methane, the first member of the aliphatic hydrocarbon homologues, does not differ in kind from further members of the series is one of the several exceptions that "prove the rule.") Benzene is almost unique in its ability to damage the *hematopoietic* or blood cell-forming system of the body as a result of chronic overexposure [5]. A few other materials, none of them simple aromatics, share that ability. Chronic overexposure may result in many kinds of blood disorders or dyscrasias, including *aplastic* (resulting from the nonformation of tissue) *anemia,* low white cell count, low platelet count, and so forth. Symptoms may include dyspnea, anemic appearance, rapid heart rate, low blood pressure, nose bleeds, and generalized weakness leading ultimately to death. Symptoms and the disease itself may be irreversible and may appear weeks or months after cessation of a prolonged exposure.

Neither toluene nor any of the xylenes, ethyl benzene, styrene, or higher members of the series share this form of toxicity [6, 7] with benzene although toluene and the xylenes have been suspected in the past. Apparently the reason for such suspicion was either that all the aromatics were used and handled together so that the others were indicted upon suspicion alone or that the toluene (or xylene) was contaminated with small amounts of benzene, which then caused the problem. Petroleum naphtha, which should contain no aromatics at all, and gasoline, which may, have occasionally been found to contain benzene [8, 9], especially if the boiling range includes 80°C [10].

In general, the aromatic hydrocarbons have pleasant odors in low concentration. The odor threshold for benzene is 2 to 5 ppm; for toluene, 1.5 to 2 ppm; the xylenes, even lower, about 0.3 ppm [11]. They all act on the body in a manner similar to the aliphatic hydrocarbons, only more intensely. All are narcotic, and all are irritating to mucous membranes, more or less in proportion to molecular weight. Benzene is appreciably less narcotic and less irritating than toluene or the xylenes. Styrene (vinyl benzene or phenylethylene) is more irritating than is ethyl benzene, illustrating that unsaturation increases this property in the aromatic hydrocarbons. With those aromatics having higher molecular weights than xylene, irritation becomes self-limiting so that the hazard of inhalation overexposure is small.

Aromatic hydrocarbons pose substantial fire and explosion hazards, but none is a direct substitute for gasoline in automobiles. Instead, these materials have excellent "mixing octane" ratings. When added to gasoline, which is primarily a mixture of aliphatic hydrocarbons, they increase the octane number of the gasoline considerably. In that respect, the aromatic hydrocarbons act as highly branched aliphatic hydrocarbons. By adding larger amounts of aromatics, the octane rating of gasoline can be kept high even though tetraethyl and tetramethyl lead are removed, but with two penalties in addition to the greater hazards associated with the aromatics.

The aromatics usually are more expensive than the aliphatics so the first penalty is cost and, therefore, price. In addition, adding aromatics to gasoline has the effect of increasing the concentrations of photochemically reactive (smog-forming) hydrocarbons and of polynuclear aromatic hydrocarbons (PAH) emitted in the exhaust [12]. As a class, the PAH contain many carcinogens such as benzo[a]-pyrene. The second penalty is then a possible increase in cancer related to exhaust products. As long as catalytic converters and other systems reduce the amounts of PAH and other hydrocarbons emitted by automobile engines, however, the chances appear good that concentrations of carcinogenic PAH in the ambient air will also be reduced even when the aromatic hydrocarbon content of gasoline is increased.

Toluene has moderate narcotic potential and is only moderately irritating. Those circumstances coupled with the fact that toluene is a good solvent led to a problem of almost epidemic proportions in the late 1960s and early 1970s—that of glue sniffing [13]. Children would purchase model cement containing a high percentage of toluene (along with other solvents, mainly ketones), squeeze the tube dry into a paper bag, and then inhale vapors from the bag to get "high." Toluene thus became a drug of abuse. Several fatalities and near-fatalities eventually caused several states to force a change in the solvent composition for glues to include at least one material such as oil of mustard, allylisothiocyanate, unpleasant to inhale in high contrations. Whether this composition change, of itself, really helped is debatable but the problem of glue sniffing declined in severity after a few years.

$$H_2C{=}CH{-}CH \begin{matrix} SCN \\ \\ SCN \end{matrix}$$

Allylisothiocyanate

Halogenated Aliphatic and Aromatic Hydrocarbons

Halogenated hydrocarbons do not occur naturally at least in detectable concentrations; all are man-made. Because chemical processing costs money, halogenated hydrocarbons are more expensive than their unhalogenated counterparts except when a complex hydrocarbon is formed by a synthesis that includes a halogenated hydrocarbon as an intermediate. Chlorine is the least expensive of the halogens and, therefore, chlorinated hydrocarbons are those most frequently encountered. Other halogens are found associated with organic compounds only when the chlorinated material will not serve the purpose. After chlorine, bromine is most common, followed by iodine as a weak third.

Fluorine is qualitatively different from the other halogens and fluorinated hydrocarbons are qualitatively different from their chlorinated, brominated, or iodinated analogs in many respects, including toxicity and hazard. In some circles, therefore, fluorine is not regarded as being a halogen at all. Nevertheless, in this chapter fluorine is a halogen. Fluorine usually is found in hydrocarbons in combination with chlorine or, less often, with bromine.

The halogen with the largest atomic number, astatine, has no commercial importance and neither do any of its organic derivatives.

General Rules

Halogenated hydrocarbons are good fat solvents but usually are not so good in this respect as unhalogenated analogs with similar evaporation rates.

Adding a single halogen to a molecule does not usually alter the flammable properties of that molecule significantly but the ability of an organic species to burn or explode in air usually vanishes well before all its hydrogens have been replaced by halogens. In this respect, fluorine is most potent and chlorine least, with bromine and iodine intermediate. The ability to extinguish burning fires, however, appears to be proportional to the atomic weight or number of the halogen.

For analogous materials, vapor inhalation toxicity increases in the order fluorinated, unhalogenated, chlorinated, brominated, iodinated. Even though these quantitative differences are found, for the most part the toxicity of the halogenated hydrocarbons does not change in kind with a change in halogen except when fluorinated materials are compared with the others. Fluorine tends to bind the whole molecule together more tightly than do other substituents, making that molecule less susceptible to all kinds of reactions, including those in the body. For that reason, fluorinated materials are usually much less toxic than even unhalogenated counterparts.

When two (or more) different halogens are present in the same molecule, the toxicity of that compound is likely to lie between the toxicities of the analogous compounds containing only one variety of halogen. For instance, the toxicity of a bromochloro- compound will usually be less than that of the dibromo- analog but greater than that of the dichloro-.

All halogenated hydrocarbons with the exception of a few *perfluorinated* (all hydrogens replaced with fluorine) materials cause narcosis. Some are much more effective in this respect than others, but there is no regular simple relationship to chemical structure.

If the effect is narcosis, unsaturation of the molecule (either double or triple bonds) has little or no effect. Ethyl and vinyl chlorides are both weak narcotics, for instance, and both 1,1,2-trichloroethane and trichloroethylene are moderate to strong narcotics. If the toxic effect is other than narcosis, the effect of unsaturation is unpredictable. Narcosis is the only effect of ethyl chloride and vinyl chloride is a moderately strong liver and kidney toxin; furthermore, it is carcinogenic. The difference between ethyl and

vinyl chlorides may be an illustration of the first-member rule, but trichloroethylene has little ability to affect the liver and kidney while 1,1,2-trichloro-
ethane is almost as effective in this regard as carbon tetrachloride. Furthermore, 1,1,1-trichloroethane is a good narcotic but has essentially no effect
on the liver and kidney, even after chronic exposure to preanesthetic levels
[14]. The toxicity of the halogenated hydrocarbons to the liver is augmented
by ethyl and isopropyl alcohols and acetone. This effect occurs only if the
liver is a target organ of the halogenated hydrocarbon and thus occurs with
carbon tetrachloride, chloroform, 1,1,2-trichloroethane, and trichloroethylene but not with 1,1,1-trichloroethane [15, 16].

Many of the halogenated hydrocarbons that have narcosis as a main
effect also sensitize the heart to epinephrine (adrenaline). After cardiac
sensitization, a release of epinephrine into the blood from excitement,
alarm, or other reason may cause *ventricular fibrillation,* a twitching of the
ventricular muscle so rapid that no coordinated contractions can occur.
This effect was especially noted with fluorocarbons having such a low toxicity that high blood levels could be attained without injury to the subject [17].
Further experimentation showed that cardiac sensitization appears to be a
rather general effect of halogenated hydrocarbons but, as with narcosis, the
ability to cause the effect varies in a yet unpredictable manner among these
materials.

Thermal decomposition (pyrolysis) of halogenated hydrocarbon vapors
in air produces the corresponding hydrohalic acid(s) (for example, hydrochloric acid) along with varying amounts of the halogen(s) and of the carbonyl halide(s). Usually the acid is present in high enough concentrations
to provide reliable warning by irritation [18], but under some conditions,
pyrolysis of carbon tetrachloride will produce hazardous concentrations of
carbonyl chloride (phosgene) without adequate warning [19]. A decomposition product even more toxic than phosgene, dichloroacetylene, has been
formed from trichloroethylene in contact with potassium superoxide [20].
Dichloroacetylene may be formed in pyrolysis of several ethane derivatives
and may constitute the major hazard; thus, analysis for phosgene may not
tell the whole story.

Welding, and particularly shielded-arc welding with a combination of
high temperature arc, exposed surfaces of hot metal, and intense ultraviolet
radiation, is particularly able to decompose halogenated hydrocarbon vapors. Even under circumstances where the health hazard is small, the acids
and halogens formed may cause corrosion of nearby metal parts and structures. This is a particular problem in the vicinity of degreasers where ferrous parts may have had a protective oil coating stripped from them. Although welding is an example of good pyrolytic conditions, those conditions
are also present in many direct-fired gas space heaters.

The first-member rule is followed in the halogenated methanes and
ethanes. Methyl chloride and methyl bromide are narcotics and this is the

principal hazard from acute exposure. Chronic overexposure to these materials leads to a progression of preanesthesia signs and symptoms even after the exposure has ceased. Complete recovery may take many months. None of the more fully halogenated lower alkanes shares this kind of CNS effect. The monohalogenated ethanes have narcosis as the only effect of either acute or chronic exposure. More fully halogenated ethanes have other effects in addition to narcosis with the possible exception of 1, 1, 1-trihaloethanes.

All halogenated hydrocarbons that are liquids at room temperature (including aromatics) can defat the skin, leading to the typical solvent dermatitis. Some of the more volatile materials can cause frostbite on contact. Many of the halogenated hydrocarbons are quite irritating to the skin, especially if not allowed to evaporate freely, and a few can be absorbed through the intact skin in acutely toxic amounts.

Specific Compounds

Carbon tetrachloride was once regarded as being almost the perfect oil and grease solvent. Its solvent power was high, it was available in high purity at low cost, its evaporation rate was high enough so that degreased parts dried rapidly (but not too rapidly), and it would not burn. Only gradually did carbon tetrachloride users become aware that this "perfect" solvent was capable of causing severe liver and kidney injury, especially from chronic inhalation overexposure. One of the tetrachloroethanes (acetylene tetrachloride, 1,1,2,2-tetrachloroethane) was found to be an even worse liver and kidney toxin, and other substitutes such as ethylene dichloride, trichloroetylene, and perchloroethylene, while much less toxic, appeared to have the same kind of toxic effect. Therefore, in addition to being known as narcotics (chloroform was one of the first anesthetics), the chlorinated hydrocarbons began to have the reputation of being liver and kidney toxins. Even today, there are knowledgeable people who are quite surprised to find that some halogenated hydrocarbons have little if any ability to injure these two organs.

Methylene chloride (dichloromethane, CH_2Cl_2) does not attack the liver or kidneys and in the 1950s methyl chloroform (1,1,1-trichloroethane) was shown by extensive animal experimentation to share this lack of effect [21]. Furthermore, even though trichloroethylene was found to be mildly toxic in this respect, its hazards were easily controlled and, therefore, these three materials largely replaced carbon tetrachloride in solvent applications. In addition, the chlorofluorocarbons began to be used more extensively, especially where their higher cost could be justified by their very low inhalation toxicities and absence of flammability in air.

In 1970 a group at the Medical College of Wisconsin was conducting human inhalation experiments with carbon monoxide. As part of an attempt to learn more about the variation of carboxyhemoglobin levels in nonexposed people, most of the staff volunteered to donate a blood sample every Monday morning during the summer and fall. Variability of COHb levels was about as expected until one Monday morning when the COHb level of Terrance N. Fisher, M.D., was found to have risen from his normal value of about 1% to about 7%. Despite extensive investigation, no source of CO could be found in his home or automobile. The next Monday his blood was normal, but a week later, his COHb level was again over 7%. After a good deal of thought, Dr. Fisher decided that the only activity the weekends had in common was his use of a paint and varnish remover containing methylene chloride. Despite assurances from others in the group that methylene chloride was not a possible source of the high COHb levels, he became convinced that no other source existed. His conviction was firm enough that he insisted on exposing himself to a known low level of methylene chloride in air to see what effect that exposure would have on his COHb level. On a day when the exposure chamber was not being used for carbon monoxide experiments, a 1-hr exposure to about 200 ppm (the TLV was then 500 ppm) of methylene chloride elevated his COHb level from less than 1% to nearly 2.5% [22]. This was evidence enough to convince others in the group that the lead should be followed.

Subsequent experimentation first with laboratory staff and later with paid volunteers showed that the endogenous conversion of methylene chloride to carbon monoxide was a universal phenomenon and not a peculiarity of Dr. Fisher [23]. Later, other experimenters found that only the dihalo methanes participate in this reaction and that rats and dogs share with man the ability to convert these materials to carbon monoxide in the blood [24, 25]. On the basis of these experiments, the TLV for methylene chloride was reduced even though this material had caused remarkably few problems at a TLV of 500 ppm.

Oxygenated Hydrocarbons

Most of the compounds discussed in this section are aliphatic because only a few oxygenated aromatics are used as solvents. Several oxygenated aromatics that are not solvents are discussed in "Miscellaneous Compounds" on p. 99 and in Chapters 7 and 8.

Aliphatic Alcohols

All aliphatic alcohols cause narcosis. In general, the ability to cause this effect increases with increasing molecular weight.

Methyl, ethyl, and propyl alcohols cause little or no irritation to the skin or to the respiratory tract in commonly encountered concentrations. Beginning with the butanols, eye irritation and then respiratory tract irritation become more and more severe with increasing molecular weight.

Methyl alcohol is the only alcohol with a direct effect upon the optic nerve, another example of the first-member rule. Overexposure, particularly chronic and especially oral, may cause blindness.

As the alcohols increase in molecular weight, their volatility is reduced. Therefore even though the toxicity in general increases with molecular weight, the toxic inhalation hazard tends to decrease.

Flash points of the alcohols from methanol to the butanols range from 11 to 29°C with no obvious simple relationship to molecular weight or structure.

As the carbon chain length of the aliphatic alcohols increases to eight and higher, the toxicity of these compounds tends to decrease but the main effect may no longer be narcosis. Toxicity of the aliphatic alcohols is usually not altered materially by chain branching [26] but is increased by unsaturation. In particular, unsaturation tends to increase the ability of the alcohol to irritate all tissue.

Ethanol is in a class by itself because of its generally low toxicity [27] and because of its use in beverages. Beverage ethanol is taxed in most countries of the world, somewhat restricting its use as an industrial solvent. To avoid beverage tax, ethanol may be denatured by the addition of materials that reduce its desirability as a drink with odor, taste, color, aftereffects, or a combination of these deterents. In addition, denaturants must be difficult to remove by means readily available to the average person.

Unfortunately, denaturants may materially increase the toxicity and/or hazards of the resulting mixture and, therefore, denatured alcohol must be chosen and used with due regard to such potential problems. In particular, those formulas that use benzene as the denaturant (formulas 2B and 12A) should be avoided where there is opportunity for vapor inhalation or skin contact. A tabulation of denaturants and of the physical properties of the 50 or so resulting denatured alcohols can be found in *Lange's Handbook of Chemistry* [28].

Cyclic and Aromatic Alcohols

Cyclohexanol and the methyl cyclohexanols are similar in toxic properties to the aliphatic alcohols with similar molecular weight; benzyl alcohol is considerably more toxic with effects similar to those of phenol, namely respiratory stimulation followed by respiratory and muscular paralysis.

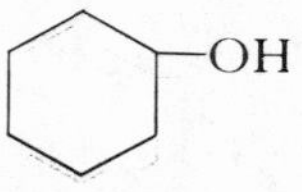

Cyclohexanol 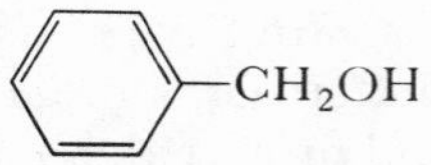   Benzyl Alcohol

   Phenol acts on the CNS (spinal cord), and an overexposure by any route leads to rapid collapse and death, probably by respiratory paralysis. The main industrial hazard from phenol is overexposure resulting from rapid absorption through the intact (or abraded) skin. Phenol on the skin causes irritation and pain, which then subsides because of phenol's analgesic action. If the area contacted is small, the only result may be a depigmented patch of skin. If the area contacted is more than a few square centimeters and the contact is prolonged for more than a few minutes, sudden collapse may occur. Although several specific remedies have been proposed, the best method of removing phenol from the skin is to flood the affected area with large amounts of water along with mechanical scrubbing for several (10 to 15) minutes. Such action can be lifesaving even after gross contact.

   The dihydroxybenzenes, pyrocatechol, resorcinol, and hydroquinone, are absorbed through the skin less readily than is phenol, but skin absorption is still a major factor in hazard evaluation. These materials have toxic actions similar to those of phenol.

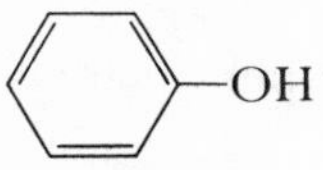

Phenol

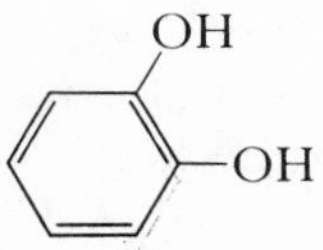

Pyrocatechol

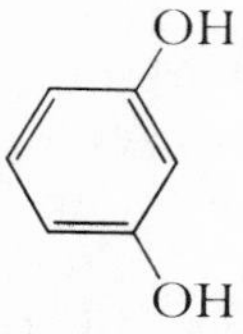

Resorcinol

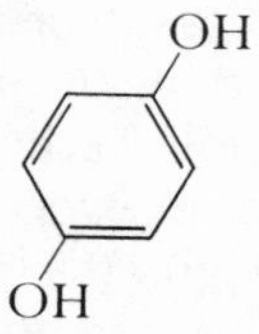

Hydroquinone

Glycols

The word *glycol* is used for any aliphatic hydrocarbon containing two hydroxyl groups not on the same carbon atom. Polyglycols (diethylene glycol, triethylene glycol, dipropylene glycol, polyethylene glycol, etc.), on the other hand, are really glycol ethers but are not named as such:

Ethylene Glycol	$HOCH_2—CH_2OH$
Diethylene Glycol	$HOCH_2—CH_2—O—CH_2—CH_2OH$
Triethylene Glycol	$HOCH_2—CH_2—O—CH_2—CH_2—O—CH_2—CH_2OH$
Polyethylene Glycol	$H(OCH_2CH_2)_nOH$
Polypropylene Glycol	$H(OCH_2CH_2CH_2)_nOH$

All of these materials are low in toxicity by any route. Ethylene glycol can cause injury from chronic (or massive acute) ingestion [29] or inhalation [27]. The lethal dose of ethylene glycol for a man is about 100 ml. As the molecular weight of the polyglycols increases, both the toxicity and the hazards decrease.

In the series of glycols and polyglycols based on ethane, propane, and butane, the propane series has been found to be considerably less toxic than the others. For this reason propylene glycol has been used extensively in pharmaceutical and cosmetic preparations [30] as well as in foods.

Flash points of the glycols are all well over 100°C, indicating only a moderate fire hazard.

Ethers

All the saturated aliphatic ethers produce narcosis if the dose is large enough. In general, irritation and narcotic potential increase with molecular weight, but none is a serious industrial health hazard. Ethyl ether (diethyl ether) is well known as a very serious fire/explosion hazard, especially because of peroxide formation in material that has been stored for extended periods of time [31]. Other ethers, especially when anhydrous, share this tendency.

Cyclic ethers (epoxides) are more irritating and much less narcotic than other aliphatic ethers and in addition may cause skin and/or respiratory tract sensitization as well as liver and kidney effects. Aromatic ethers are low in toxicity and hazard by all routes of administration [32].

Both chloromethyl methyl ether and bis (chloromethyl) ether have been found to be carcinogenic in several species of animals [33, 34]. Both materials cause cancer that originates in the respiratory tract regardless of where or how the dose is administered. Of the two, bis (chloromethyl) ether is the more potent carcinogen [35, 36].

$$ClCH_2—O—CH_3 \qquad ClCH_2—O—CH_2Cl$$

Chloromethyl Methyl Ether   bis(Chloromethyl) Ether

Glycol Ethers

The first member of this homologous series is 2-methoxyethanol (ethylene glycol methyl ether), which has definite toxic effects on the kidney, brain, and blood resulting from inhalation or skin absorption. Effects on the CNS, including mental retardation and neurological symptoms, appear to be peculiar to this first member. Other members of the series (2-propoxy, 2-butoxy, etc., ethanol) injure the kidney and cause hemolysis also. They may have effects on the peripheral nervous system but apparently not the CNS. Of all the lower homologues, 2-ethoxyethanol is the least toxic.

$$H_3C—CH_2—O—CH_2—CH_2OH$$
2-Ethoxyethanol

Ethers of propylene glycol are appreciably less toxic than their ethylene glycol counterparts and therefore find use even in cosmetics where contact may be very prolonged and often repeated.

Flash points range from 35°C (2-ethoxyethanol) up. Only the lower members of the glycol ethers pose much of a fire hazard.

Aldehydes

The most general toxic effect of aliphatic aldehydes is irritation of the skin, eyes, and respiratory tract. This effect decreases with increasing molecular weight but increases with unsaturation of the molecule. General toxicity of aldehydes is related primarily to their irritating properties.

The first member of the homologous series of aliphatic aldehydes, formaldehyde, is capable of causing hypersensitization of the skin following contact with water solutions. This may be an example of the first-member rule, but there is some evidence that acetaldehyde may be a very weak sensitizing agent for man. Higher aldehydes do not appear to share this toxic effect.

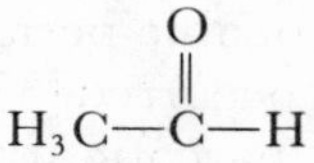

$$\overset{\displaystyle O}{\underset{\displaystyle}{H_3C—\overset{\|}{C}—H}}$$
Acetaldehyde

Both chloral hydrate (a hydrated aldehyde) and paraldehyde (a polymer of acetaldehyde) are capable of causing narcosis and, indeed, this is their principal effect. Other aldehydes may also have narcotic properties, but in general these materials are too irritating to allow sufficient voluntary exposure for the effect.

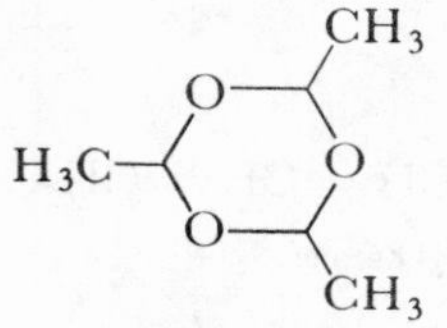

$$Cl_3C—\overset{H}{C}(OH)_2$$

Paraldehyde   Chloral Hydrate

None of the aliphatic aldehydes has been shown to have an effect upon chronic exposure that is different in kind or degree from the effects produced by acute exposure. Any organic injury produced can be attributed to the irritating properties of these materials.

Benzaldehyde is the only aromatic aldehyde with much commercial importance. In the body it is rapidly oxidized to benzoic acid, which in turn is rapidly excreted in the urine. Serious systemic effects from either material have not been found in industry.

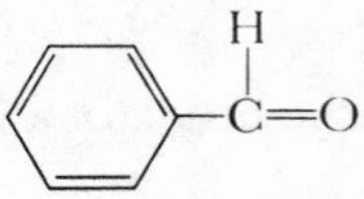

$$\text{C}_6\text{H}_5 - \underset{\underset{}{|}}{\overset{\overset{H}{|}}{\text{C}}} = \text{O}$$

Benzaldehyde

Formaldehyde and acetaldehyde have low flash points (below $-35°C$) and extremely wide explosive ranges. In air, their lower explosive limits are 7.0 and 4.1% and their upper explosive limits are 73 and 55%, respectively. Most of the commercially important aldehydes have flash points below 70°C and several pose a distinct fire hazard.

Ketones

Ketones do not obey the first-member rule except in that the first member, acetone, is far less toxic than any of its homologues. The toxicity of acetone is so low, in fact, that attaining a concentration in air that will be fatal to experimental animals in a reasonable period of time is difficult. Nevertheless, industrial intoxication has occurred [37].

As a class, the ketones are narcotic irritants and the irritation of eyes and nasal passages usually suffices to limit acute exposures to those causing insignificant narcosis. With one known exception, the aliphatic ketones pose no chronic hazard that is significantly different from the acute hazard. The exception is methyl *n*-butyl ketone, which upon prolonged overexposure can cause a peripheral neuritis that may be incapacitating [38]. Although the mechanism of action of the material is not known, experiments with animals indicate that this effect is enhanced by simultaneous exposure to methyl ethyl ketone (MEK, butanone) [39].

$$\text{H}_3\text{C} - \overset{\overset{O}{\|}}{\text{C}} - \text{CH}_2 - \text{CH}_2 - \text{CH}_2 - \text{CH}_3$$

Methyl *n*-Butyl Ketone

Increasing molecular weight of the aliphatic ketones brings an increase in narcotic potential, irritation, and general toxicity but a decrease in vapor pressure. Unsaturation increases the ability to irritate and hence the overall toxicity [40] (but not hazard). Members of this series have pleasant odors as long as the molecule is saturated. Unsaturation causes the odor to become pungent unless the unsaturation is associated with an aromatic ring, in which case the odor again is usually pleasant [11]. Acetophenone, for instance, is used in the perfume industry.

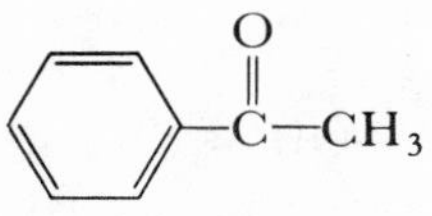

Acetophenone

All the commercially important ketones are potential fire/explosion hazards with flash points that range from $-18°C$ (acetone) to about $85°C$ (isophorone). Lower explosive limits vary from about 1 to 2% in air, while upper explosive limits decrease with increasing molecular weight from 13% for acetone to about 4% for isophorone (3,5,5-trimethyl-2-cyclohexene-1-one).

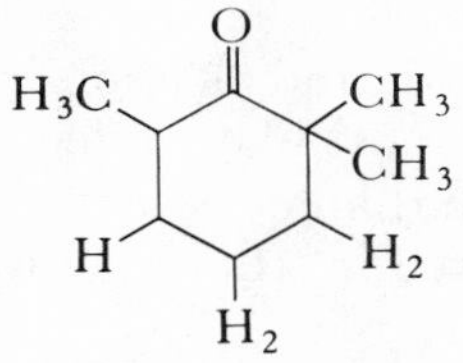

Isophorone

Acids and Acid Anhydrides

The ability of the organic acids and anhydrides to irritate tissue is a function both of molecular weight and of water solubility. As long as the compound is soluble in water, its ability to irritate increases with increasing molecular weight except for the first member of the aliphatic series, formic acid. Formic acid is probably the most severely irritating of all the organic acids, with the possible exception of some halogenated acids, and is the principal component of ant venom. At high molecular weights the acids and anhydrides are no longer water-soluble and their irritation potential drops accordingly.

Unsubstituted aliphatic acids do not sensitize man, nor do they inhibit any enzyme systems as a primary toxic effect. Iodoacetic and fluoroacetic acids, however, are very active enzyme inhibitors and this effect has led to some remarkable advances in enzyme research. Iodoacetic acid can cause hypersensitization.

The first member of the aromatic acid homologues, benzoic acid, is a mild irritant and an occasional skin sensitizer. Being a solid, it is not used as a solvent. Other simple aromatic acids are not sensitizers but phthalic anhydride is. With the exception of hypersensitization, acids and anhydrides rarely have a chronic toxicity that is different from their acute action.

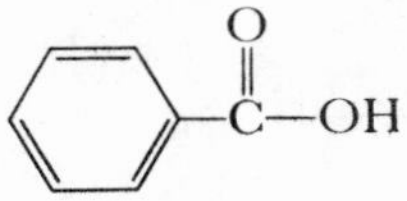

Benzoic Acid 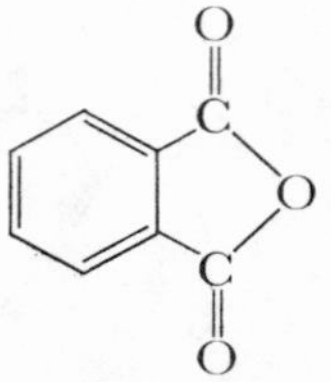   Phthalic Anhydride

Acetic acid has the lowest flash point (40°C) of the aliphatic acids; that of propionic acid is 20°C higher and flash points of the series increase rapidly with an increase in molecular weight.

Esters

Being reaction products of acids and alcohols, esters are the organic equivalents of inorganic salts. The most likely effect of the aliphatic esters is narcosis. Esters are less narcotic than ethyl ether and most chlorinated hydrocarbons but are more narcotic than ethanol, acetone, and aliphatic hydrocarbons such as pentane.

Formates are irritating to the eyes, skin, and respiratory tract; acetates are less so. Most of the esters are more irritating than chlorinated hydrocarbons of analogous structure, probably because of hydrolysis to the alcohol and acid.

None of the aliphatic esters creates much of a problem in normal handling; none has a chronic toxic effect that differs greatly from its acute effects but experimentation has shown some liver injury from acetates [41]. Most esters have pleasant fruity or flowery odors.

Unsaturation in the aliphatics (ethyl acrylate is an example) can increase the irritation potential greatly, perhaps to the point where the ester is recognized as a lachrimator. Furthermore, ethyl acrylate has an extremely low threshold for its characteristic rather unpleasant odor (0.0003 ppm) [11].

Because of their volatility, the esters have relatively low flash points and correspondingly high fire/explosion hazards. The flash point for methyl formate, for instance, is −19°C; that for banana oil (isopentyl acetate) is 32°C.

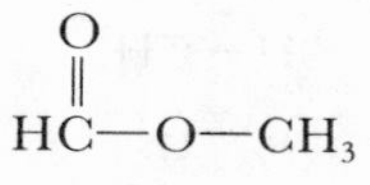

$$\underset{\text{Methyl Formate}}{\overset{O}{\overset{\|}{HC}}-O-CH_3} \qquad \underset{\text{Isoamyl (or Isopentyl) Acetate}}{H_3C-\overset{O}{\overset{\|}{C}}-O-CH_2-CH_2-\overset{\overset{\displaystyle CH_3}{|}}{\underset{|}{CH}}_{\displaystyle CH_3}}$$

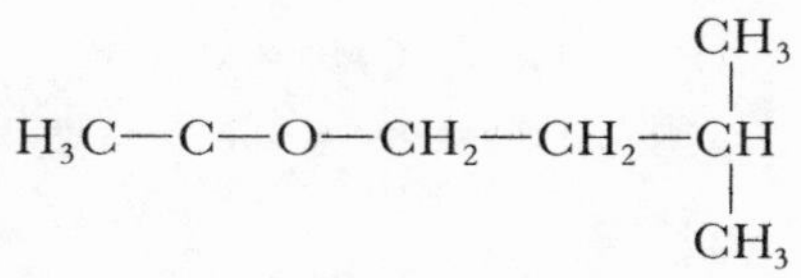

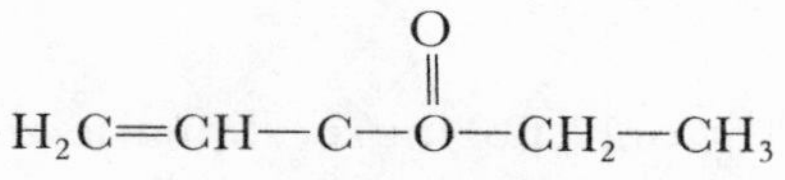

$$\underset{\text{Ethyl Acrylate}}{H_2C=CH-\overset{O}{\overset{\|}{C}}-O-CH_2-CH_3}$$

Nitrogen-Containing Compounds

Compounds containing nitrogen are always toxicologically active and interesting. Interesting because nitrogen does so many peculiar things, even when it is locked into a ring as with pyridine. One of the few really general comments that can be made about these materials is that most have odors and that few smell good. Toxic properties vary from simple irritation to methemoglobin formation to blood vessel dilation.

Mononitro Aliphatic Hydrocarbons

Nitromethane is a mild respiratory tract irritant and a mild narcotic; it is also capable of producing liver and some kidney injury upon prolonged overexposure, usually by inhalation. It does not cause the formation of methemoglobin; it is essentially nonirritating to the skin; it is not absorbed through the skin in acutely toxic amounts.

$$\underset{\text{Nitromethane}}{H_3C-NO_2}$$

Nitroethane is similar in toxicity to nitromethane but, having a higher molecular weight, is more irritating.

$$\underset{\text{Nitroethane}}{H_3C-CH_2-NO_2}$$

Both 2- and 1-nitropropane are still more irritating, but have a similar toxicology with one exception: Both cause methemoglobin formation as do higher members of this series.

$$H_3C-CH_2-CH_2-NO_2 \qquad H_3C-\underset{\underset{NO_2}{|}}{CH}-CH_3$$

1-Nitropropane

2-Nitropane

All the lower mononitro aliphatics are explosive.

Polynitro Aliphatics

Tetranitromethane is the only polynitro aliphatic compound that has any industrial importance. It is very irritating to all tissue and causes methemoglobin formation. Other toxic effects are not well documented.

Tetranitromethane is extremely explosive when in contact with organic materials but much less so when pure.

$$C(NO_2)_4$$

Tetranitromethane

Halo-Nitro Compounds

The best known member of this series is chloropicrin (trichloronitromethane), which is a lacrimator and an irritant to the skin and respiratory tract. It was used as a war gas in World War I and found to be intermediate in (acute) toxicity between chlorine and phosgene. Warning properties are fairly good, although it is not detectable by odor or irritation at 0.1 ppm, which has been used as the TLV.

$$Cl_3C-NO_2$$

Chloropicrin

One of the interesting commercial uses for chloropicrin is to odorize methyl bromide in a manner similar to the use of mercaptans to odorize natural gas. Chloropicrin is much more irritating than are the natural gas odorants, but it is also much less volatile. When methyl bromide is used as a fumigant, however, pressurized cans are the usual containers. They are used in an all-or-none manner; that is, once opened, cans are emptied completely. Any chloropicrin in the can is thus released, serving to warn of the presence of the much more volatile but odorless methyl bromide.

Other chlorinated nitroparafins that have been studied are also irritants and are less toxic than chloropicrin.

Aliphatic Nitrates

Nitric acid esters of aliphatic alcohols, glycols, etc., are all explosive and this, rather than use as solvents, is the main reason for their production. They all are capable of dilating blood vessels and of causing methemoglobin formation. Signs and symptoms of intoxication are likely to be depression of blood pressure, tremors, ataxia, lethargy, hyperpnea, cyanosis, prostration, and convulsions.

Nitroglycerine (glycerol trinitrate) and ethylene glycol dinitrate (EGDN) are the best known members of this series. The outstanding symptom produced in man by these materials is headache, which may be very severe and which is usually relieved by caffein (from coffee). Following more severe exposures, the headache may be accompanied by flushing, palpitation, and nausea.

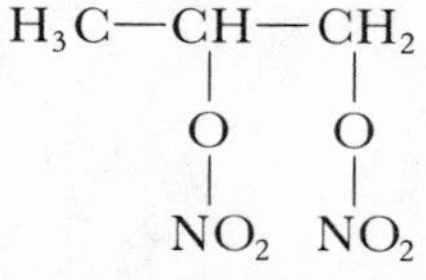

$$H_3C-CH-CH_2$$

Ethylene Glycol Dinitrate

$$H_2C-CH-CH_2$$

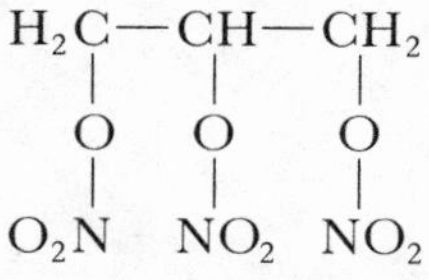

Nitroglycerine

The aliphatic nitrates are very readily absorbed through the intact skin, a factor that greatly complicates a determination of the total absorbed dose by industrial hygiene studies. Because tolerance to the headache develops upon repeated exposure but is lost rapidly, men working with these materials have deliberately exposed themselves to small amounts over a weekend to avoid a Monday morning headache. One way this has been done is to put a little material on one's hatband. Then, the exposure is essentially continuous and tolerance is retained. Reduction of occupational exposures to those that do not cause headache is a much better solution to this problem.

Aromatic Nitro Compounds

This group is characterized by nitrobenzene and 2,4,6-trinitrotoluene (TNT), a well-known explosive. These materials are capable of causing methemoglobin production and cyanosis is the sign of intoxication most frequently observed. With TNT, however, methemoglobin is of little importance. Rather, hepatitis and aplastic anemia are the diseases that have resulted in fatalities after chronic overexposure. Avoidance of cyanosis is not sufficient to assure that chronic overexposure will not occur. People who are deficient in the enzyme glucose-6-phosphate dehydrogenase may be susceptible to acute hemolysis if exposed to TNT [42].

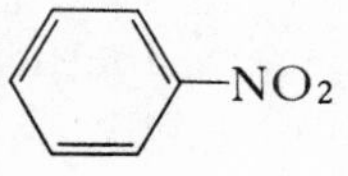

Nitrobenzene

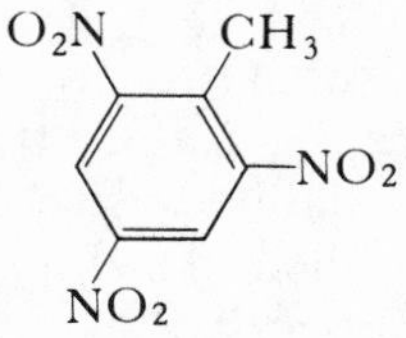

2,4,6-Trinitrotoluene

The nitrophenols, and particularly the dinitrophenols (including those with aliphatic side chains), and cresols have the peculiar ability to increase the basal metabolic rate. In the 1930s, 2,4-dinitrophenol was used extensively in the treatment of obesity. It caused the expected rise in metabolic rate and consequent elevation of body temperature (even hot flashes) and the expected weight loss if dietary intake remained unchanged. Unfortunately, it also caused the formation of cataracts, a side effect that is not shared with the usual experimental animals (rodents, canines), in addition to frank liver and kidney injury leading in some cases to death.

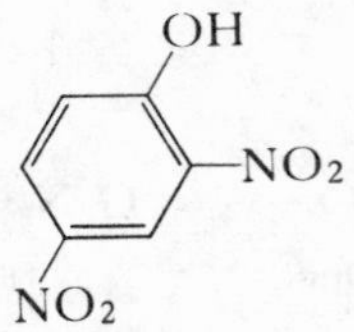

OH

NO$_2$

NO$_2$

2,4-Dinitrophenol

Amines

The word *amine* was derived from *ammonia* and all amines can be considered to be derivatives of ammonia. They share with ammonia an alkaline nature and a sharp, penetrating odor, both of which are most pronounced when ammonia has been substituted the least.

Aliphatic amines are quite basic, have a fishy odor, and are very irritating to all tissue. The lower members of the series (methyl, dimethyl, trimethyl, and ethyl amine) are gases that may be supplied and used in water solution; those of higher molecular weight are liquids. In all cases, the main exposure is to the gas or vapor with attendant severe eye and respiratory tract irritation [27]. Skin absorption may also be important. All the commercially important aliphatic amines have low flash points with consequent high fire/explosion hazards. Ethylamine has a flash point of less than $-15°C$ and di-*n*-butylamine is only $52°C$.

HN $(CH_3)_2$

Dimethylamine

Aromatic amines such as aniline and its derivatives are absorbed through the skin or the respiratory tract to cause methemoglobin formation and/or other changes in blood as their main acute toxic effect. Most of the simple derivatives have no other important acute or chronic systemic effects, but some are skin sensitizers.

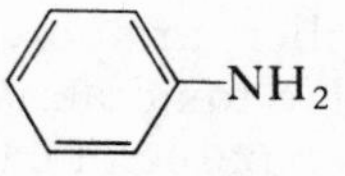

Aniline

When two or more benzene rings appear in the molecule, the picture changes rapidly. Cancer of the bladder is an important consequence of chronic overexposure to β-naphthylamine, benzidine, and (perhaps) α-naphthylamine. In addition, several related materials may be carcinogenic and, therefore, are viewed with suspicion. All of these materials are solids; none is used as a solvent, and most have been used mainly in the dyeing industry.

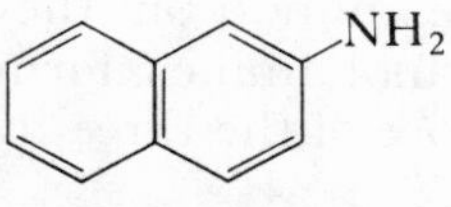

β-Naphthylamine

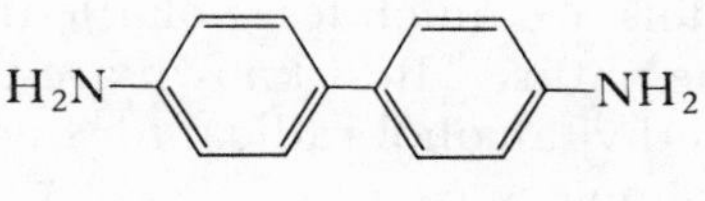

Benzidine

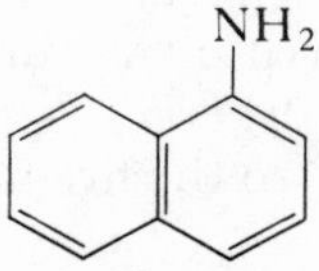

α-Naphthylamine

Long used in the production of rubber and as an isocyanate monomer, 4,4'-diaminodiphenylmethane (methylene dianiline, MDA, or DDM) is capable of causing severe liver injury in susceptible people. In addition, animal experiments have implicated this material as a possible liver carcinogen.

Polyamines

The first member of this homologous series is ethylene diamine (1,2-diaminoethane), which has the structure

$$H_2NCH_2—CH_2NH_2$$

Continuing the series are diethylene triamine, triethylene tetramine, etc.:

$$H_2NCH_2—CH_2—NH—CH_2—CH_2NH_2$$

Diethylene Triamine

In this particular series, all members are polymers of ethylene diamine. All of these homologues have similar toxic actions. They are extremely irritating to all tissue and are also capable of causing lung and skin sensitization. The chronic hazard is essentially the same as the acute hazard, which is great. All have a fishy odor characteristic of amines. Even though the vapors of these materials are quite irritating, the irritation is insufficient to prevent inhalation of acutely toxic amounts.

Flash points for the polyamines range from about 33°C (1,2-propanediamine and tetraethylene pentamine) up, with no obvious relationship to molecular weight or structure.

Alkanol Amines

Amino alcohols are much less volatile than are amines, but they retain the same alkaline nature. Best known are the ethanol amines, formed by substituting an ethyl alcohol radical for one or more of the three hydrogens in the ammonia molecule.

Dimethylamino ethanol is a central nervous system stimulant [43] and other alkanol amines may have effects on blood pressure, urination, and other systems. Because of the irritant nature of these materials, however, such effects are rarely found in workers.

Flash points are all over 90°C so the fire hazard is low.

Amides

Most of the aliphatic amides are solids, are not used as solvents, and are not used in large quantities. *N,N*-Dimethyl formamide (DMF), on the other hand, is a very good solvent for many organic compounds and is widely used. Except for a possible fire/explosion hazard (flash point of 67°C), this material has caused few industrial problems. It is not particularly irritating to the skin or respiratory tract and its rather foul odor appears to limit inhalation exposures adequately although liver injury has been produced in experimental animals [44]. DMF may be absorbed through the intact skin in acutely toxic amounts.

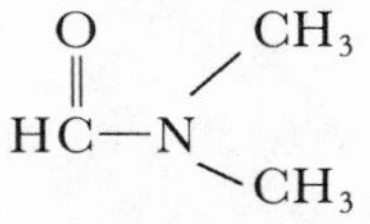

$$\underset{\text{N, N-Dimethyl formamide}}{HC-N\begin{array}{c}O\\ \|\end{array}\begin{array}{c}CH_3\\ \diagup\\ \diagdown\\ CH_3\end{array}}$$

N, N-Dimethyl formamide

Another excellent amide solvent with even greater toxic potential than DMF is hexamethylphosphoric triamide (hexamethylphosphoramide, HMPA) which causes a nasal cancer in rats in concentrations as low as 400 parts vapor per 10^9 parts air by volume. Although chronic exposure appears to be required to induce this effect, the vapor pressure of the compound is high enough so that concentrations orders of magnitude greater than that are easily attainable. Inhalation hazards of this material are severe [45].

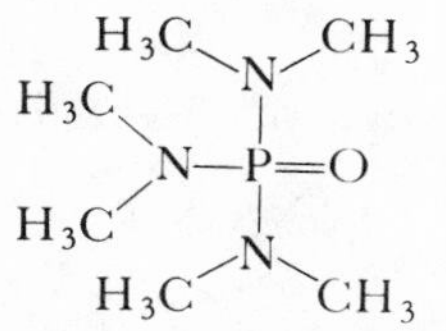

$$
\begin{array}{c}
H_3C \diagdown \quad \diagup CH_3 \\
\quad \quad N \\
H_3C \diagdown \quad \quad | \\
\quad \quad N\!-\!P\!=\!O \\
H_3C \diagup \quad \quad | \\
\quad \quad N \\
H_3C \diagup \quad \diagdown CH_3
\end{array}
$$

Hexamethylphosphoric triamide

Acrylamide is a solid and therefore is not a solvent. It is used as a monomer in the production of polyacrylamides that produce flocs in water to aid in the removal of suspended particulate material. Acrylamide and a few of its simple derivatives are neurotoxic [46], in small doses, usually producing paralysis of the hindquarters in overexposed experimental animals. Exposure may occur industrially by inhalation or skin absorption and warning properties are essentially nil. Acrylamide is a hazardous material to handle but its polymers appear to be physiologically inert.

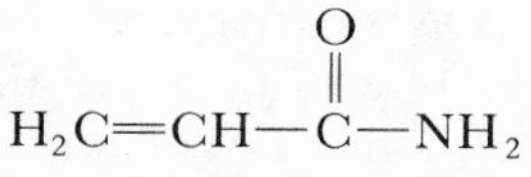

$$
\begin{array}{c}
\quad \quad \quad \quad O \\
\quad \quad \quad \quad \| \\
H_2C\!=\!CH\!-\!C\!-\!NH_2
\end{array}
$$

Acrylamide

Pyridine and Derivatives

As benzene is the first member of the aromatic series of hydrocarbons, pyridine is the first member of the pyridine series. Pyridine behaves like a tertiary amine and a weak base. The pyridines in general, however, closely resemble analogous aromatic compounds.

Pyridine is a narcotic and may produce weakness, ataxia, unconsciousness, and salivation. It is also irritating to the mucous membranes as would be expected from a mildly basic material. Most characteristic is its *strong* fishy, ammoniacal odor.

Pyridine and many of its derivatives can be absorbed by the intact skin in acutely toxic amounts, a hazard that is in addition to the irritation these materials cause. Most of the pyridines are incapable of causing hypersensitization. Pyridine has a flash point of 20°C, which may be compared to that for benzene (−11°C). In general, the pyridines are somewhat less of a fire hazard than aromatics of comparable structure.

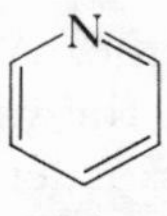

Pyridine

Hydrazines

Chemically, hydrazine can be regarded as *diamine*. Its structure is

$$H_2N\!-\!NH_2$$

Hydrazine

Of course, hydrazine is not organic (and is not usually used as a solvent), but organic radicals can be substituted for one or more hydrogens to produce organic derivatives. Of these, monomethylhydrazine (MMH) and 1,1-dimethylhydrazine (asym-dimethylhydrazine or unsymmetrical dimethylhydrazine, or UDMH) are the most important commercially. Hydrazine, MMH, and UDMH are used as rocket fuels. These materials are *hypergolic* (burst into flame spontaneously) with concentrated nitric acid (red, fuming nitric acid or RFNA).

$$H_3C\!-\!NH\!-\!NH_2 \qquad (CH_3)_2N\!-\!NH_2$$

MMH   UDMH

Hydrazine, MMH, and UDMH are rapidly absorbed through the lungs, GI tract, and skin. The main effects of acute inhalation overexposure are CNS stimulation and consequent convulsions, probably resulting from direct action on the brain. Acute or chronic inhalation of MMH [47] or UDMH [48] can cause hemolytic anemia and kidney injury in addition to convulsions. UDMH is less irritating to the skin, eyes, and mucous membranes than is hydrazine, which is highly irritating.

UDMH, hydrazine, and probably MMH may react exothermically and violently with iron oxides, molybdenum oxides, some stainless steels, and many oxidizing agents.

Not all hydrazine derivatives are so violent in their reactions. Phenylhydrazine has long been used as a reagent in the determination of aldehydes, ketones, and sugars, and its main hazard has been found to be that of skin sensitization. In addition, overexposure can cause hemolysis, some liver injury, and severe kidney damage. Its medicinal use is to combat polycythemia by producing a mild hemolytic anemia.

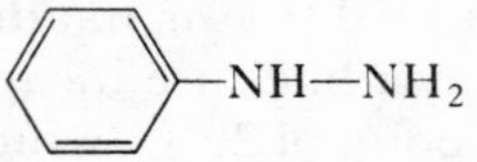

$$\text{—NH}\!-\!NH_2$$

Phenylhydrazine

All commercially important hydrazine derivatives are absorbed rapidly through the intact skin, and percutaneous absorption may be an important route of contact. This is especially important with phenylhydrazine, which is not irritating enough to give adequate warning of its presence on the skin.

Miscellaneous Compounds

Carbon Disulfide

Carbon disulfide is an inorganic compound capable of dissolving many organics and hence is a very useful solvent. Its main uses are as a chemical intermediate in the xanthation of cellulose and as a fumigant. Carbon disulfide is probably encountered most frequently in the laboratory where it is used for its solvent properties. The infrared spectrum of CS_2 is extraordinarily free of absorbance in the region where most organic compounds absorb (from about 7 to 15 μm) and hence its pungent odor typifies an infrared laboratory. Many industrial hygienists have begun to use CS_2 to desorb materials that have been adsorbed onto activated charcoal when obtaining breathing zone air samples.

The main hazards associated with CS_2 are vapor inhalation and fire/explosion. Acutely, CS_2 is a narcotic capable of causing death by inhalation overexposure, but such a serious consequence is very unusual. Instead, persistent alteration of central and peripheral nervous system functions may occur [49], possibly as a result of a single massive overexposure but more commonly following a series of less severe exposures. Symptoms range from headache, fatigue, and listlessness to loss of memory, appetite, and sexual function. In more severe cases the affected person may experience hallucinations and other symptoms of psychosis. Recovery is slow but complete if severe injury is avoided.

One of the more interesting manifestations of chronic CS_2 intoxication is loss of reflexes, another indication of nerve damage. Great strides are being made in the development of tests sensitive enough to use reflex alteration as a biological index of excessive absorption of the material. Such testing is unlikely to supplant environmental monitoring in exposure evaluation but may indeed prove very useful in the diagnosis of problems [50].

Carbon disulfide has one of the lowest autoignition temperatures known, about 120°C. The surfaces of incandescent light bulbs and of some steam pipes are thus hot enough to initiate a fire or explosion if the vapor is in the explosive range (1.25 to 50% in air). The fire hazard is a severe one, compounded by the corrosivity and toxicity of the sulfur dioxide produced by combustion of carbon disulfide.

Triaryl Phosphates

Although of very limited use as solvents, the triaryl phosphates and, particularly, tri-*o*-cresyl phosphate (TOCP), have been the cause of much human misery. Several epidemics have been caused by the accidental or deliberate incorporation of TOCP in human food, resulting in the temporary or permanent paralysis of several tens of thousands of people. The paralytic action of TOCP is delayed, typically appearing from 1 to 3 weeks following ingestion and affecting the legs and feet before the hands and arms. Industrial overexposure is rare.

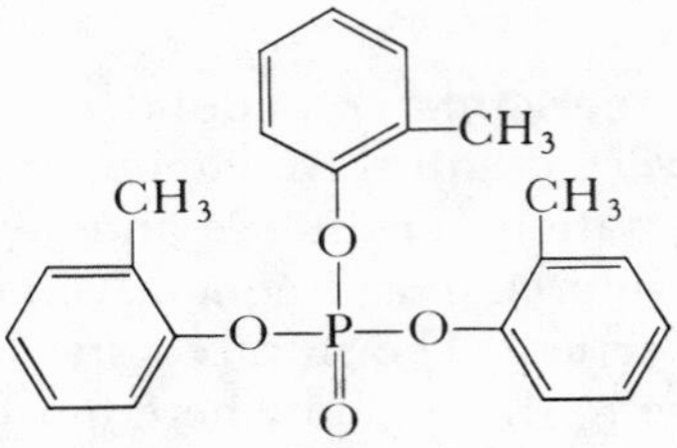

TOCP

Lactones

Lactones are inner esters of acids containing a hydroxyl group on a carbon atom other than that with the carbonyl group:

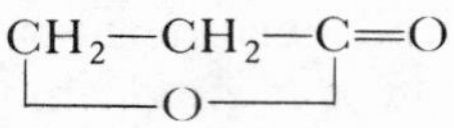

β-Propiolactone

Most of the lactones appear to be low in toxicity, but the first member of the beta series, β-propiolactone, is not. Used occasionally as a solvent but more often in syntheses and as one of the few viricidal agents available, β-propiolactone is a strong primary irritant, a CNS stimulant, and a skin carcinogen. Its toxicity is high by all routes. Exposures, especially to the skin, must be avoided.

REFERENCES

[1] H. F. SMYTHE, JR., ET AL., "Range-Finding Toxicity Data: List VI," *Am. Ind. Hyg. Assn. J.* **23** (1962), 95–107.

[2] H. E. SWANN, ET AL., "Acute Inhalation Toxicity of Volatile Hydrocarbons," *Am. Ind. Hyg. Assn. J.* **35** (1974), 511–18.

[3] T. R. TORKELSON AND V. K. ROWE, "Results of Repeated Inhalation by Laboratory Animals and a Limited Human Sensory Study of a Mixture of Saturated and Unsaturated C$_3$ and C$_4$ Hydrocarbons (MAPP Industrial Gas)," *Am. Ind. Hyg. Assn. J.* **25** (1964), 554–59.

[4] *Reference Data for Hydrocarbons and Petro-Sulfur Compounds*, (Bartlesville, Oklahoma: Phillips Petroleum Company, Special Products Division (1962), pp. 2–4.

[5] J. R. MITCHELL, "Mechanism of Benzene-Induced Aplastic Anemia," Abst. No. 2044 *Fed. Am. Soc. Exp. Biol. Proc.* **30** (1971), 561.

[6] L. J. JENKINS, JR., R. A. JONES, AND J. SIEGEL, "Long-Term Inhalation Screening Studies of Benzene, Toluene, *o*-Xylene, and Cumene on Experimental Animals," *Tox. Appl. Pharm.* **16** (1970), 818–23.

[7] C. A. NAU, J. NEAL, AND M. THORNTON, "C_9–C_{12} Fractions Obtained from Petroleum Distillates: An Evaluation of Their Potential Toxicity," *Arch. Env. Health* **12** (1966) 382–93.

[8] L. D. PAGNOTTO, ET AL., "Industrial Benzene Exposure from Petroleum Naptha: I. Rubber Coating Industry," *Am. Ind. Hyg. Assn. J.* **22** (1961), 417–21.

[9] G. S. PARKINSON, "Benzene in Motor Gasoline—An Investigation into Possible Health Hazards," *Ann. Occup. Hyg.* **14** (1971), 145–57.

[10] H. B. ELKINS, E. M. COMPRONI, AND L. D. PAGNOTTO, "Industrial Benzene Exposure from Petroleum Naptha: II. Pertinent Physical Properties of Hydrocarbon Mixtures," *Am. Ind. Hyg. Assn. J.* **24** (1963), 99–102.

[11] T. M. HELLMAN AND F. H. SMALL, "Characterization of the Odor Properties of 101 Petrochemicals Using Sensory Methods," *J. Am. Poll. Con. Assn.* **24** (1974), 979–82.

[12] W. E. RINEHART, S. A. GENDERNALIK, AND L. F. GILBERT, "Fuel Factors in Automotive Tailpipe Emissions," *Am. Ind. Hyg. Assn. J.* **32** (1971), 179–87.

[13] S. COHEN, "Glue Sniffing," *J. Am. Med. Assn.* **231** (1975) 653–54.

[14] R. D. STEWART, ET AL., "Experimental Human Exposure to Methyl Chloroform Vapor," *Arch. Env. Health* **19** (1969) 467–72.

[15] H. H. CORNISH AND JULITA ADEFUIN, "Ethanol Potentiation of Halogenated Aliphatic Solvent Toxicity," *Am. Ind. Hyg. Assn. J.* **27** (1966), 57–61.

[16] G. J. TRAIGER AND G. L. PLAA, "Chlorinated Hydrocarbon Toxicity—Potentiation by Isopropyl Alcohol and Acetone," *Arch. Env. Health* **28** (1974), 276–78.

[17] H. J. TROCHIMOWICZ, ET AL., "Blood Levels of Fluorocarbon Related to Cardiac Sensitization: Part II," *Am. Ind. Hyg. Assn. J.* **35** (1974), 632–39.

[18] L. C. RINZEMA AND L. G. SILVERSTEIN, "Hazards from Chlorinated Hydrocarbon Decomposition During Welding," *Am. Ind. Hyg. Assn. J.* **33** (1972), 35–40.

[19] M. H. NOWEIR, E. A. PFITZER, AND T. F. HATCH, "Thermal Decomposition of Carbon Tetrachloride Vapors at Its Industrial Threshold Limit Concentration," *Am. Ind. Hyg. Assn. J.* **34** (1973), 25–37.

[20] R. A. SAUNDERS, "A New Hazard in Closed Environmental Atmospheres," *Arch. Env. Health* **14** (1967), 380–84.

[21] T. R. TORKELSON, ET AL., "Toxicity of 1,1,1-Trichloroethane as Determined on Laboratory Animals and Human Subjects," *Am. Ind. Hyg. Assn. J.* **19** (1958), 353–62.

[22] R. D. STEWART, ET AL., "Carboxyhemoglobin Elevation after Exposure to Dichloromethane," *Science* **176** (1972), 295–96.

[23] R. D. STEWART, ET AL., "Experimental Human Exposure to Methylene Chloride," *Arch. Env. Health* **25** (1972), 342–48.

[24] G. D. DIVINCENZO, F. J. YANNO, AND B. D. STILL, "Human and Canine Exposures to Methylene Chloride Vapor," *Am. Ind. Hyg. Assn. J.* **33** (1972), 125–35.

[25] G. D. DIVINCENZO AND M. L. HAMILTON, "Fate and Disposition of [^{14}C] Methylene Chloride in the Rat," *Tox. Appl. Pharm.* **32** (1975), 385–93.

[26] R. A. SCALA AND E. G. BURTIS, "Acute Toxicity of a Homologous Series of Branched-Chain Primary Alcohols," *Am. Ind. Hyg. Assn. J.* **34** (1973), 493–99.

[27] R. A. COON, ET AL., "Animal Inhalation Studies on Ammonia, Ethylene Glycol, Formaldehyde, Dimethylamine, and Ethanol," *Tox. Appl. Pharm.* **16** (1970), 646–55.

[28] N. A. LANGE, *Handbook of Chemistry,* rev. tenth ed. (New York: McGraw-Hill Book Company, 1967), pp. 1798–1807.

[29] J. A. ROBERTS AND H. R. SEIBOLD, "Ethylene Glycol Toxicity in the Monkey," *Tox. Appl. Pharm.* **15** (1969), 624–31.

[30] F. STEINBACK AND P. SHUBIK, "Lack of Toxicity and Carcinogenicity of Some Commonly Used Cutaneous Agents," *Tox. Appl. Pharm.* **30** (1974), 7–13.

[31] HELEN M. ROSENBERGER AND R. H. JOHNSON, "Factors Influencing Peroxide Formation in Diethyl Ether," *Am. Lab* (June, 1970), 8–15.

[32] R. E. HEFNER, ET AL., "Repeated Inhalation Toxicity of Diphenyl Oxide in Experimental Animals," *Tox. Appl. Pharm.* **33** (1975), 78–86.

[33] J. L. GARGUS, W. H. REESE, JR., AND H. A. RUTTER, "Induction of Lung Adenomas in Newborn Mice by Bis(chloromethyl) Ether," *Tox. Appl. Pharm.* **15** (1969), 92–96.

[34] S. LASKIN, ET AL., "Tumors of the Respiratory Tract Induced by Inhalation of Bis(chloromethyl) Ether," *Arch. Env. Health* **23** (1971), 135–36.

[35] S. LASKIN, ET AL., "Inhalation Carcinogenicity of Alpha Halo Ethers—II. Chronic Inhalation Studies with Chloromethyl Methyl Ether," *Arch. Env. Health* **30** (1975), 70–72.

[36] M. KUSCHNER, ET AL., "Inhalation Carcinogenicity of Alpha Halo Ethers—III. Lifetime and Limited Period Inhalation Studies With Bis(chloromethyl) Ether," *Arch. Env. Health* **30** (1975), 73–77.

[37] D. S. ROSS, "Acute Acetone Intoxication Involving Eight Male Workers," *Ann. Occup. Hyg.* **16** (1973), 73–75.

[38] J. R. MCDONOUGH, "Methyl *n*-Butyl Ketone," *J. Occup. Med.* **16** (1974), 412.

[39] M. S. ABDEL-RAHMAN, L. B. HETLAND, AND D. COURI, "Toxicity and Metabolism of Methyl *n*-butyl Ketone," *Am. Ind. Hyg. Assn. J.* **37** (1976), 95–102.

[40] L. LEVINE, G. B. MEYERS, AND LILLIAN LIDDANE, "2-Cyclohexene-1-one: Toxicology and Study of an Occupational Dermal Exposure," *Am. Ind. Hyg. Assn. J.* **33** (1972), 338–42.

[41] V. QUERCI AND D. MASCIA, "Enzymologic and Histologic Observations on the Liver Damage in Experimental Acetate Intoxication," *Med. Lavoro.* **61** (1970), 524–30. Abstracted in *Ind. Hyg. Digest* **35** (April, 1971), 4.

[42] L. S. DJERASSI AND L. VITAHY, "Haemolytic Episode in G6 PD Deficient Workers Exposed to TNT," *Brit. J. Ind. Med.* **32** (1975), 54–58.

[43] H. H. CORNISH, "Oral and Inhalation Toxicity of 2-Diethylaminoethanol," *Amer. Ind. Hyg. Assn. J.* **26** (1965), 479–84.

[44] J. W. CLAYTON, JR., ET AL., "The Inhalation Toxicity of Dimethylformamide (DMF)," *Am. Ind. Hyg. Assn. J.* **24** (1963), 144–54.

[45] J. A. ZAPP, JR., "Inhalation Toxicity of Hexamethylphosphoramide," *Am. Ind. Hyg. Assn. J.* **36** (1975), p 916.

[46] P. M. EDWARDS, "Neurotoxicity of Acrylamide and its Analogues and Effects of These Analogues and Other Agents on Acrylamide Neuropathy," *Brit. J. Ind. Med.* **32** (1975), 31–38.

[47] C. C. Haun, et al., "Acute Inhalation Toxicity of Monomethylhydrazine Vapor," *Am. Ind. Hyg. Assn. J.* **31** (1970), 667–77.

[48] W. E. Rinehard, E. Donati, and E. A. Greene, "The Subacute and Chronic Toxicity of 1,1-Dimethylhydrazine Vapor," *Am. Ind. Hyg. Assn. J.* **21** (1960), 207–9.

[49] J. Teisinger, "1972 Yant Memorial Lecture—New Advances in the Toxicology of Carbon Disulfide," *Am. Ind. Hyg. Assn. J.* **35** (1974), 55–61.

[50] J. Lieben, "III. International Symposium on Toxicology of Carbon Disulphide, Cairo—Alexandria, Egypt, 4–9 May 1974," *J. Occup. Med.* **16** (1974), 483–84.

7

MONOMERS AND POLYMERS

Many materials encountered in everyday life are built on a modular plan, with a few basic units repeated over and over. Examples range from the houses in some subdivisions to the brick walls of those houses to the muscles of the people who live in the houses to the DNA in the nuclei of those muscle cells. In the organic chemistry of resins, a *monomer* is a single chemical unit and a *polymer* is a combination of many monomer units. Polymeric compounds can range from single long chains of monomers to three-dimensional webs or nets of interlocked and interwoven chains and can be formed from a single monomer or from several different monomers in varying proportions. Physically, polymers can range from light, mobile liquids to solids. Solid polymers may be soft and easily deformable (plastic), stretchy (elastomeric), tough, or extremely brittle. Polymers may occur naturally (silk, rubber) or they may be man-made (nylon, polyethylene); they may be amorphous (glass) or crystalline (diamond); they are found everywhere.

Nearly all completely polymerized materials are physiologically inert or closely approach that state regardless of the toxicity of the monomer(s) from which they were formed. The hazards of these materials, then, are associated with the monomers and with the polymerization processes, not with the finished product. Exceptions are those polymeric materials that are allergens (mostly natural polymers), those that are not completely polymerized, or those that have undergone further reaction such as pyrolysis or combustion.

In addition to a plethora of monomers, modern resin, rubber, or fiber production facilities are likely to use many other materials such as catalysts, extenders, plasticizers, blowing agents, dyes and pigments, mold lubricants, emulsifiers, and cleaning solvents that may materially add to the hazards encountered [1]. Processes used range from emulsion polymerization to foam spraying, and the most hazardous job can be cleanup or maintenance rather than normal operation. Hazard evaluation may be a formidable task.

Most of the information in this chapter deals with the toxicity and hazards of monomers but some mention is made of other compounds, especially when they present odd or interesting problems.

Addition Polymerization

Addition polymers are made by processes in which no carbon-to-carbon bond is completely ruptured or broken and are therefore made from monomers containing a carbon-to-carbon double bond. In many cases, a single monomer is reacted with itself to produce the polymer, but many commercially important polymers are formed from a combination of different monomers reacted together; others are made by a grafting process, where a monomer is reacted with an already formed polymer.

Polymerization is an exothermic reaction and very large amounts of heat must be removed from the reacting materials. Heat removal is complicated because, in all cases, the viscosity of the resulting polymer is considerably greater than that of the monomer(s). Steps taken to solve this problem have been ingenious, ranging from the use of reaction vessels containing screws or paddles to stir viscous polymers to the use of solvent, emulsion, or suspension processes in which a second low viscosity phase is present to act as a carrier and heat transfer medium.

Finished resins can be sold as powders, pellets, emulsions, plastisols, sheets, films, or solids with several shapes. In most cases the resin will contain very little residual monomer although many other materials may be incorporated to enhance one or more properties of the product.

Hydrocarbons

Unsaturated hydrocarbons are used in greater tonnage than any other class of monomers. Those most used are ethylene, propylene, isoprene (2-methyl-1,3-butadiene), 1,3-butadiene, isobutylene, and styrene. None presents any unusual toxic hazards; the first five are typical of lower olefins in their effects, and styrene is an acute irritant and narcotic with good-to-excellent warning properties for inhalation.

Ethylene and propylene are used mainly to form polyethylene and polypropylene. The derivatives of butane are used in the production of various rubbers, a use that also accounts for a large tonnage of styrene. Styrene is also used in the production of *homopolymers* (polymers of a single monomer) and in polyesters.

$$H_2C{=}CH_2 \qquad H_2C{=}CH{-}CH_3$$
Ethylene Propylene

$$H_2C{=}CH{-}CH{=}CH_2 \qquad H_2C{=}\underset{\underset{CH_3}{|}}{C}{-}CH_3$$
1,3-Butadiene Isobutylene

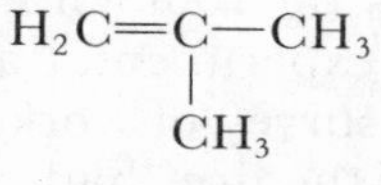

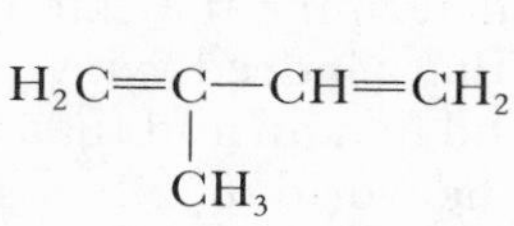

$$H_2C{=}\underset{\underset{CH_3}{|}}{C}{-}CH{=}CH_2 \qquad H_2C{=}CH{-}\bigcirc$$
Isoprene 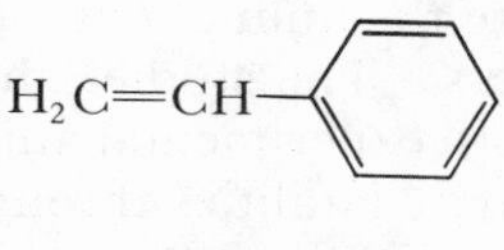 Styrene

Halogenated Hydrocarbons

Vinyl chloride, chloroethylene, is used mainly for the production of its homopolymer, but it is also copolymerized with vinylidene chloride (1,1-dichloroethylene) for the production of Saran.

$$H_2C{=}CHCl \qquad H_2C{=}CCl_2$$
Vinyl Chloride Vinylidene Chloride

In 1961 two papers on the toxicity of vinyl chloride were presented at the annual conference of the American Industrial Hygiene Association, one by a university group [2] and the other by an industrial group [3]. The university paper insisted that the toxicity of vinyl chloride was very low, and the industrial paper (from The Dow Chemical Company, a maker of the material) indicated that the then current TLV of 500 ppm was much too high and that a lower value (50 ppm was subsequently recommended) would be more appropriate. The American Conference of Governmental Industrial Hygienists did not change the TLV of vinyl chloride until much later.

In the late 1960s a peculiar syndrome of bone degeneration in the hands of workers was found. This disease, called *acroosteolysis,* was demonstrated in animals [4] in 1970 and shown by a thorough industrial hygiene survey conducted in 1967 [5] to be associated with the cleaning of polymerization vessels (kettles) in which polyvinylchloride (PVC) had been made. This was the first indication other than the paper previously mentioned that vinyl chloride and/or PVC might be considerably more toxic than had been realized.

A paper published in 1972 reporting results of a very extensive industrial hygiene survey of vinyl chloride workers [6] indicated that time-weighted average exposure to concentrations of 300 ppm or higher may lead to changes in medical laboratory findings related principally to liver function.

Early in 1974 the roof fell in. Chronic vapor exposures were shown to cause cancer in experimental animals [7]. Almost at the same time, an epidemiological survey of workers employed in a rubber plant showed that angiosarcoma of the liver had produced fatalities in men who had worked for a long time with vinyl chloride. Very thorough studies were then under-taken to replicate the animal data and to determine the extent of injury to industrial workers. The studies showed that, indeed, vinyl chloride is a carcinogen in both experimental animals and in man and that it had caused a few tens of cancer fatalities throughout the world [8]. Steps were imme-diately taken by all countries in which this material was handled to reduce exposures of workers to levels hoped to have no ill effects.

One problem that became apparent immediately was that no good esti-mates could be made of concentrations or doses that had led to the human fatalities. What was known, however, was that vinyl chloride had for years been handled as if it were almost completely innocuous and that several of the men who died had, in the past, been subject to doses sufficient to render them unconscious. While possibly helpful, that kind of information was in-sufficient upon which to base a TLV. This illustrates the necessity for nega-tive data, especially where human health is concerned. Thorough industrial hygiene surveys that allow the estimation of worker exposure are a true necessity even in cases where no obvious ill effects have occurred among the exposed group. The survey previously mentioned [6] became extremely valuable when decisions had to be made that could affect the health of thou-sands of people and the economic picture of many more. Other vinyl chlo-rides will very probably be found in the future especially as increasing atten-tion is paid to the effects of chronic exposures.

Chloroprene (2-chloro-1,3-butadiene) is polymerized in the formation of a synthetic rubber (neoprene). By any route it is a lung, liver, and kidney toxin, and it also has the peculiar ability to cause *alopecia* or hair loss follow-ing topical application. Fortunately, the follicles appear to escape injury so that the cosmetic damage is only temporary if further contact is avoided.

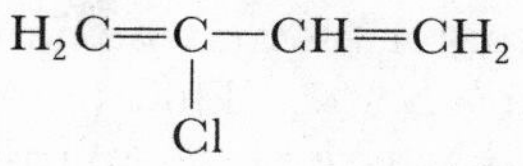

$$H_2C{=}\underset{\underset{\displaystyle Cl}{|}}{C}{-}CH{=}CH_2$$

Chloroprene

Vinyl fluoride, tetrafluoroethylene, and chlorotrifluoroethylene are all used to form polymers that are more chemically inert than most other materials produced by man. The latter two are used mainly to make molding resins, and vinyl fluoride is used in the production of films. All three monomers are gases at room temperature and none presents any particular toxic hazard from inhalation; all are close to inert physiologically.

$$H_2C{=}CHF \qquad F_2C{=}CF_2 \qquad F_2C{=}CFCl$$

Vinyl Fluoride   Tetrafluoroethylene   Chlorotrifluoroethylene

Esters

The commercially important esters used for the production of addition polymers are vinyl acetate and derivatives of acrylic and methacrylic acids. Plexiglass is produced by the homopolymerization of methyl methacrylate. Ethyl acrylate is used as a comonomer with several other materials. Methyl methacrylate and ethyl acrylate are lachrimators with strong odors in very low concentrations. In fact, the odor threshold for ethyl acrylate is about 0.0003 ppm and, even though the odor threshold for methyl methacrylate is about 1000 times higher [9], odor serves as good warning of the presence of both materials.

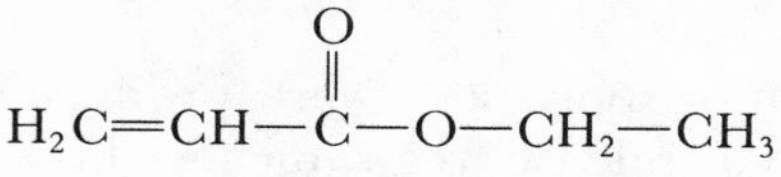

$$H_2C{=}CH{-}\overset{\overset{\displaystyle O}{\|}}{C}{-}O{-}CH_2{-}CH_3 \qquad\qquad H_2C{=}\underset{\underset{\displaystyle CH_3}{|}}{C}{-}\overset{\overset{\displaystyle O}{\|}}{C}{-}O{-}CH_3$$

Ethyl Acrylate 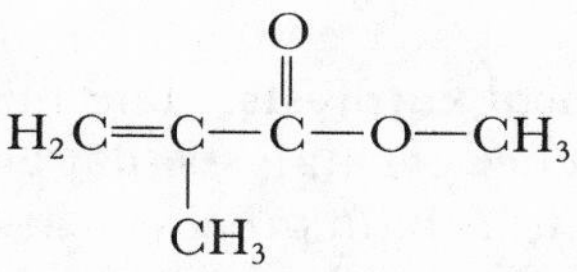   Methyl Methacrylate

Vinyl acetate is usually copolymerized with vinyl chloride although some is used for the production of polyvinyl alcohol, a water-soluble polymer, and some is used in latex paints. Vinyl acetate is more toxic by inhalation than either methyl methacrylate or ethyl acrylate, but its warning properties are also relatively good [10], with a recognition threshold for its rather unpleasant odor of about 0.4 ppm [9].

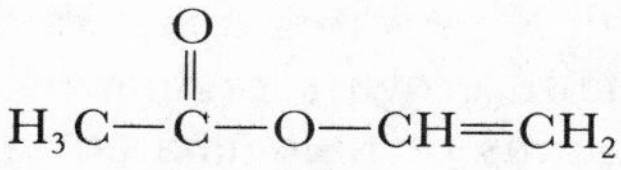

$$H_3C{-}\overset{\overset{\displaystyle O}{\|}}{C}{-}O{-}CH{=}CH_2$$

Vinyl Acetate

Even though all of these materials may be homopolymerized, none of the resulting polymers is called a *polyester*. That name is reserved for some condensation polymers that are discussed later in this chapter.

Nitriles

Acrylonitrile, by far the most important of the nitrile-type addition monomers, is rarely homopolymerized. Instead, it is copolymerized with butadiene to form Buna *N* rubber, with styrene or styrene plus butadiene to form a resin and molding powder, with vinyl chloride to produce *modacrylic* (modified acrylic) fibers, and with one of six or seven other monomers (including vinyl acetate, vinyl chloride, and vinyl pyridine, along with styrene and isobutylene) to produce acrylic fibers.

$$H_2C{=}CHCN$$

Acrylonitrile

Although nitriles are not cyanides, once in the body they act as if they were and, therefore, the toxicity of acrylonitrile is similar to that of hydrogen cyanide qualitatively (but about half as toxic, quantitatively) except for a possible involvement of the liver on chronic exposure [11]. In acutely hazardous concentrations in air, acrylonitrile (vinyl cyanide, VCN) is mildly irritating to the eyes and respiratory tract and therefore it is much less hazardous to handle than is HCN, which is essentially nonirritating.

Catalysts

Free radical catalysts. The first addition polymers were produced under the influence of catalysts that generate or release free radicals. Free radical addition is a chain process where the addition of another unit to the growing chain does not reduce the ability of that chain to add still another unit. For this reason, chain stoppers, or molecular-weight control agents, must be added. Also, agents can be added to the mix to promote branching of the growing chains of monomer units. Rather precise control of average molecular weight and of the viscosity of the resulting polymer is now practical.

Agents that promote free radical formation can be chemicals such as benzoyl peroxide, methyl ethyl ketone peroxide, isopropyl and diisopropyl benzene hydroperoxides, potassium persulfate, lauryl peroxide, some diazo compounds, and even oxygen and ozone. Heat and electromagnetic radiation from ultraviolet through X rays to gamma rays can also be used under some circumstances. Chain growth regulating agents (chain transfer agents) are usually mercaptans or halogenated aliphatic hydrocarbons or an aromatic hydrocarbon such as cumene (isopropylbenzene). Chain stopping is more of an art than a science; agents used may include quinones, oxygen, sulfur, amines, and specific stoppers such as sodium dimethyl-dithiocarbamate or tetramethylthiuram disulfide.

Because of their combination of excess oxygen and combustible carbon and hydrogen all in the same molecule, peroxides can pose severe fire/explosion hazards. Most are irritants to any tissue they contact so that warning is usually good [12], but some may be sensitizers especially of the skin [13]. Hazards of the chain regulators rarely become evident because of the very small amounts needed of such materials.

Ionic catalysts. The most important ionic polymerizations are catalyzed commercially with a combination of an organometallic compound and a metal halide. A typical system might use triisobutyl aluminum and titanium tetrachloride. In general, the organometallic compound is a trialkyl aluminum or dialkyl aluminum halide and the metal halide is usually the trihalide, tetrahalide, or oxyhalide of a transition metal such as titanium, vanadium, or cobalt. Because it is the least expensive, chlorine is the halide normally used; bromine is found only where chlorine will not serve. Ionic polymerization is commercially very important for the production of addition polymers of the lower olefins (ethylene and propylene mainly), but the technique is quite powerful and is also used for other materials.

The aluminum alkyls and alkyl halides are pyrophoric in air and in contact with water. Their effect on the skin or eye is similar to the effect of a flame. When they burn, a smoke consisting mainly of aluminum oxide and unburned carbon is released; the smoke appears to possess no unusual toxicological properties although a condition similar to metal fume fever might result from excessive inhalation. Because of their reactivity, these materials must be handled in closed systems from which air and water are excluded and, therefore, opportunity for contact is very limited, reducing the hazards accordingly.

Condensation Polymerization

Condensation is used in organic chemistry to describe a reaction in which carbon-to-carbon bonds are broken so that

$$A + B \rightleftarrows C + D$$

and one of the products usually has a much higher molecular weight than the other. In polymerizations involving this mechanism, the low-molecular-weight fragment is frequently water (or NH_3, HCl, or $NaCl$), which is removed, helping the reaction go to completion.

The monomers involved in condensation polymerization are usually alcohols or glycols, phenol, aldehydes, acids or acid anhydrides, and amines. Although the first condensation polymers were *thermosetting* (cured to a nonmelting, nonsoftening solid by heat), many of the newer members of this series are *thermoplastic* (soften with heat).

As with addition polymerization, condensation polymerization is used to produce materials with a wide range of physical properties. And, because condensation is a more versatile reaction mechanism than addition, the range of monomers used is broader. Most condensation monomers are considerably more expensive than are addition monomers, however, so that their total tonnage in use is much less [14].

Phenol-Formaldehydes

Phenolic resins (phenolics) have been commercially important since the early 1900s when Leo Baekeland introduced Bakelite and hot molding to the world. These materials are produced in the form of molding powders, usually of the thermoplastic variety. After the addition of plasticizers, colorants, and fillers, the molding powder is mixed with hexamethylenetetramine and heated for final polymerization into the infusible form. There is no free phenol or formaldehyde remaining either in the molding powder or in the finished resin. All the reactants have been discussed except for hexamethylenetetramine, which is handled with no unusual precautions without incident.

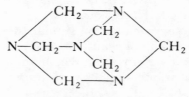

Hexamethylenetetramine

Aminoplasts and Polyamides

The *aminoplasts* are plastics based on the $-NH_2$ group. Nylons are the most important polyamides.

Urea was the first organic compound synthesized from the elements; that synthesis proved life was not necessary for the production of organic materials. Used extensively as a fertilizer as well as in the production of plastics, urea is a solid with a low toxicity by all routes although it is basic enough to be irritating to the eyes.

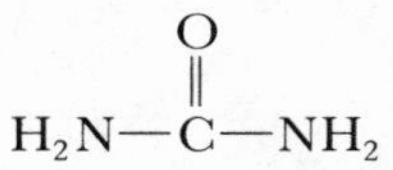

$$\overset{\displaystyle O}{\underset{\displaystyle \|}{H_2N-C-NH_2}}$$

Urea

Melamine has a cyclic structure with conjugated double bonds similar to those in benzene. Reacted with formaldehyde, melamine forms a complex cross-linked polymer that, after the addition of other materials, is cured under conditions similar to those used with the phenolics. Melamine is a solid and poses no unusual handling problems. The few cases of dermatitis that have occurred in this industry have probably been caused by formaldehyde rather than melamine.

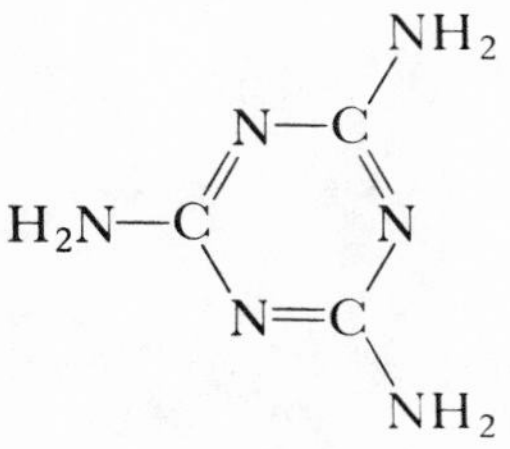

$$NH_2$$
$$N-C$$
$$H_2N-C \quad N$$
$$N=C$$
$$NH_2$$

Melamine

Of the several nylons or polyamides, nylon 66 was the first commercially produced and it is still one of the most important. Chemically it is polyhexamethylene adipamide; its monomers are adipic acid and hexamethylenediamine, which may be made from adipic acid.

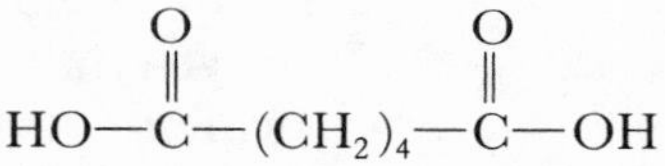

$$HO-\overset{\overset{O}{\|}}{C}-(CH_2)_4-\overset{\overset{O}{\|}}{C}-OH \qquad H_2N-CH_2-(CH_2)_4-CH_2-NH_2$$

Adipic Acid   Hexamethylenediamine

Hexamethylenediamine is a typical aliphatic amine in that it is a fairly strong base and an irritant of all tissue because of that property. In addition, it appears to be a liver toxin in man.

Adipic acid is a solid at room temperature; it poses no unusual handling hazards. No industrial injuries associated with it have been reported.

Nylon 6 is the homopolymer of ε-caprolactam that, along with adipic acid and hexamethylenediamine, is ultimately produced from benzene usually through hydrogenation to cyclohexane followed by another series of steps. Caprolactam is a solid at room temperature but often is handled molten. Even as a solid it has a significant vapor pressure at room temperature. It is irritating to the eyes and to breathe in concentrations above a few ppm and is irritating to the skin even in the vapor form if the concentrations are high enough [15]. Beyond irritation, caprolactam has caused few problems industrially.

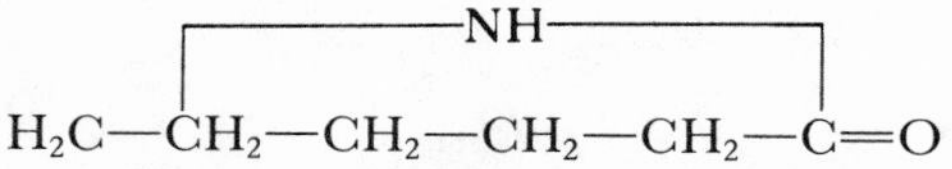

$$
\begin{array}{c}
\overline{}NH\overline{} \\
| \qquad\qquad\qquad | \\
H_2C-CH_2-CH_2-CH_2-CH_2-C=O
\end{array}
$$

ε-Caprolactam

Epoxy Resins

The principal monomer of commercially important epoxy resin systems is epichlorohydrin (1-chloro-2,3-epoxypropane). In addition to its use as a raw material for the production of epoxy resins, epichlorohydrin is used in the manufacture of glycerol and glycidol derivatives and as an intermediate in the production of some other chemicals.

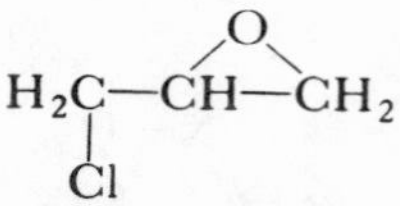

$$H_2C-CH-CH_2$$
$$\underset{Cl}{|}$$
$$\overset{O}{/\backslash}$$

Epichlorohydrin

Epichlorohydrin is rapidly absorbed by any route of contact. It is intensely irritating, although irritation may be delayed, and kills experimental animals either through lung irritation or CNS depression or both. Chronic administration can result in severe kidney injury in addition to the effects experienced upon acute exposure.

Epichlorohydrin is capable of causing skin sensitization although, because of its irritating nature, this is not usually a severe industrial problem. Respiratory tract sensitization is also a possibility.

Allyl glycidyl ether (AGE) is used mainly as a diluent for epoxy resin systems. In this case, AGE not only reduces viscosity of the system but also participates in the polymerization reaction and thus is a true monomer.

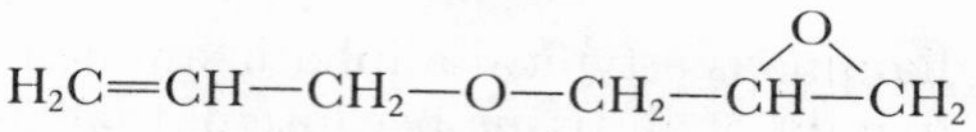

$$H_2C=CH-CH_2-O-CH_2-CH-CH_2$$

Allyl Glycidyl Ether

AGE is much less toxic and much less irritating than is epichlorohydrin, but it is also capable of causing skin sensitization. It is a CNS depressant, and vapor exposure can also cause pulmonary edema. AGE is irritating to the skin upon prolonged contact and can be absorbed through the skin, but its toxicity by that route is low.

The main copolymer in epoxy resin systems is bisphenol A. Most prepolymers are formed by reaction of this material with epichlorohydrin. Bisphenol A is also used to some extent in the formation of other resins.

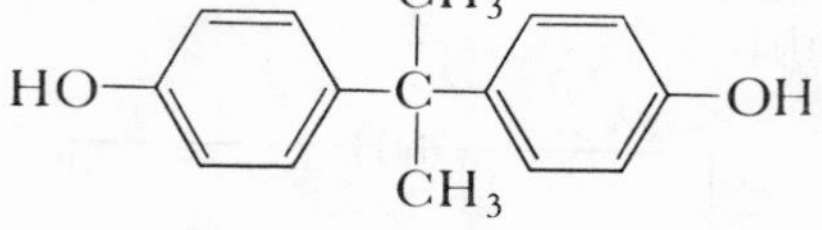

$$HO-\underset{}{\bigcirc}-\overset{CH_3}{\underset{CH_3}{C}}-\underset{}{\bigcirc}-OH$$

Bisphenol A

Bisphenol A is a white-to-tan crystalline material. Its vapor pressure is low enough so that vapor inhalation is not a problem in industry. Its dust is low enough in toxicity so that it is usually classified simply as a nuisance. It is absorbed through the intact skin but not in acutely toxic amounts under normal working conditions. It is not irritating to the skin and does not cause skin sensitization, nor is it irritating to the eyes.

The epoxy resin systems handled by the ordinary user consist of two parts: the resin itself and the catalyst. The resin is the reaction product or prepolymer of epichlorohydrin and bisphenol A, which may be in a solution of a reactive diluent such as allyl glycidyl ether. The catalyst may be one of the polyamines, acid anhydrides, or organic acids, perhaps also in solution. In use, the resin and catalyst are mixed thoroughly, applied, and allowed to cure at room or elevated temperature. Once the two components of the system have been mixed, the curing will take place in a time that depends on their nature and relative proportions.

Occasionally an extender may be used, especially when the epoxy resin is not being used for its strength or adhesive properties. For the potting of electrical components, several extenders have been advised, ranging from nonconductive powders to coal tar. These materials have in common their relatively low cost; their toxic properties may vary considerably.

In past practice, the catalyst component of epoxy resin systems has caused more industrial problems than has the resin itself. With the newer anhydride and acid catalysts, the balance has swung the other way. Almost invariably the main problem associated with handling epoxy resin systems is skin sensitization, resulting in a rash similar to that produced by poison ivy. Once a person is sensitized, further rashes are usually produced by even extremely minor contact. Animal experiments on more than 60 epoxides concerning carcinogenicity and inhalation toxicity in addition to sensitization potential indicated little relationship of toxic properties to structure [16]; each material must be evaluated separately.

Polyurethanes

Almost any material with a functional hydroxyl or amine group can participate in the reaction that produces a polyurethane. Most common polyols are the polyglycols and polyethers; the most common amine is 4,4'-methylene bis(2-chloroaniline). Variation in this part of the finished molecule is used to control the character of the finished resin, which may range from a very hard, tough solid to a soft, flexible foam. None of these materials poses much of a handling problem except possibly for 4,4'-methylene bis(2-chloroaniline), which on the basis of animal testing was declared to be a potential carcinogen in man by the U.S. Occupational Safety and Health Administration in 1973 [17].

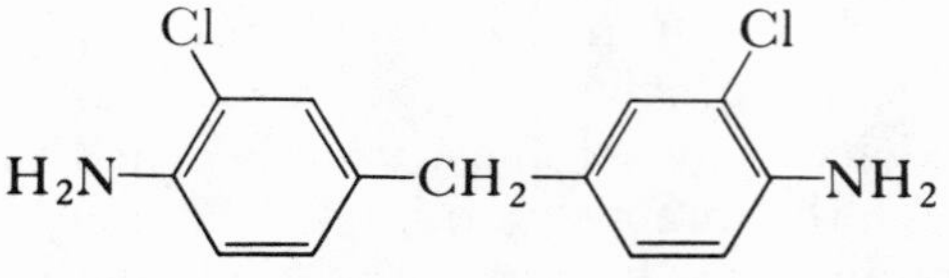

$$H_2N-\underset{}{\bigcirc}-CH_2-\underset{}{\bigcirc}-NH_2$$

Methylene bis(2-chloroaniline)

The other half of the polyurethane system consists of a diisocyanate, usually toluene-2,4- (and -2,6-) diisocyanate (TDI) or methylene diphenyl-diisocyanate (MDI), which is also used extensively. Several other diiso-cyanates have limited use for special applications.

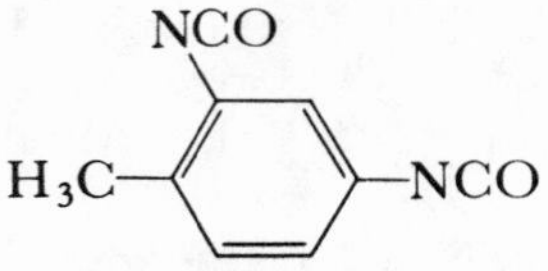

$$H_3C-\underset{}{\bigcirc}-NCO \qquad OCN-\underset{}{\bigcirc}-CH_2-\underset{}{\bigcirc}-NCO$$

TDI 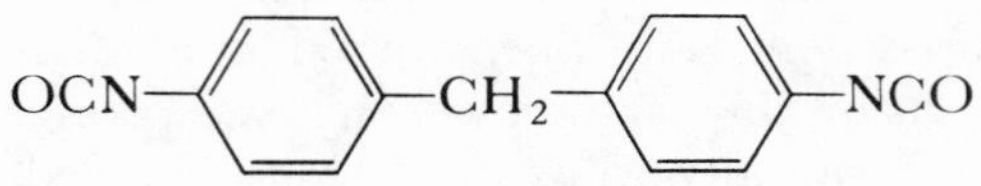   MDI

Diisocyanates as a group are strong irritants and are capable of causing severe liver and kidney injury [18]. Nevertheless, almost all the industrial problems associated with the manufacture and handling of the isocyanates or the polyurethanes (either solids or foams) can be attributed to respiratory tract sensitization to the diisocyanates [19].

In very low doses (low concentrations and short durations) these materials are capable of sensitizing man's respiratory tract, giving rise to asthma-like symptoms. Once sensitization has occurred, the exposure required to initiate an attack of asthma may consist of a few molecules in a breath or two of air.

Diisocyanates react with polyols or polyamines to produce resins that may be used as films (varnishes) or molded products, depending on the composition of the reacting materials. Furthermore, diisocyanates react exothermically with water, producing a urea and CO_2. The first polyure-thane foams were produced by mixing a polyol, a diisocyanate, and water. As the diisocyanate reacted with the polyol to form the polymer, it also reacted with water to produce CO_2. With the proper balance of reactants, foams with widely varying properties could be made. This type of reaction, however, wastes the diisocyanate, which is the most expensive of the reac-tants. Present practice, therefore, is to dissolve one of the fluorocarbons (typically dichlorodifluoromethane) in the polyol. Heat of the reaction causes the fluorocarbon to vaporize, resulting in an action similar to that caused by the release of CO_2. These materials can be used for foam-in-place operations (production of matresses, for instance) or can be sprayed onto surfaces with attendant greater hazards [20, 21].

Polyesters

The term *polyester* is not used in the plastics industry in its generic sense. Instead, it refers to the reaction products of a more or less specific polyester with a vinyl type monomer, typically styrene. The polyester part of the formulation is usually based on the reaction product of maleic anhydride with ethylene glycol, diethylene glycol, and/or propylene glycol; phthalic anhydride is usually added to the reaction to reduce price and to promote mixing of the polyester with styrene later. In addition to phthalic anhydride, adipic acid is used where more flexibility in the finished product is desired.

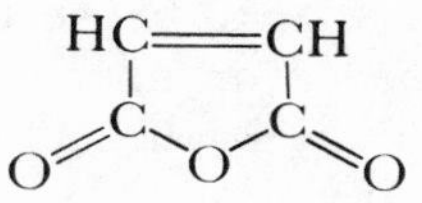

Maleic Anhydride 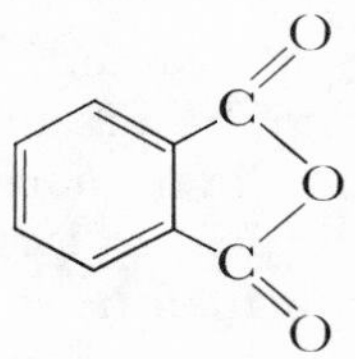   Phthalic Anhydride

Finished products are made by mixing any additives (in particular, fibrous glass) with the polyester, styrene, and a free radical catalyst. Depending on the catalyst used, curing may take place at room or at an elevated temperature. Excess styrene is always present; the double bonds of the maleic anhydride molecule in the polyester will react only with the styrene (not with themselves), but styrene will react with itself; in this manner the finished resin contains no unreacted monomer and no vulnerable double bonds.

The free radical catalysts used are almost all peroxides, typically benzoyl peroxide or methyl ethyl ketone peroxide, or ozonides, azones, or azo compounds for special purposes. Toxic properties of the latter materials are not well known; peroxides were discussed on p. 110.

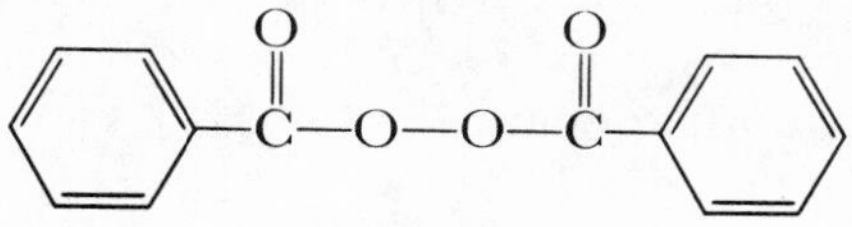

Benzoyl Peroxide

Polyester resins are quite often encountered by the man-in-the-street as fix-it kits for automobile bodies or other structures where a strong, permanent, but easily applied and shaped patch is required. For this kind of use as well as the manufacture of fiberglass (fibrous glass is responsible for much of the strength of the finished material, but the resin is usually a poly-ester) plastics, the chief hazard arises from styrene inhalation. As styrene has reasonably good warning properties (irritation), the overall hazards associated with this use are low. The problem of skin sensitization, for instance, is almost vanishingly small, especially when the polyesters are compared with epoxy resin systems.

Polymers

Nearly all completely polymerized materials are, or come close to being physiologically inert [21, 22]. No matter how toxic the monomer(s) may be, the finished polymers bear little resemblance to the individual units that form them. On the other hand, if the polymerization is not complete, un-reacted monomer may be present and able to cause its typical reactions. For instance, several different resins based on formaldehyde have caused contact dermatitis typical of formaldehyde itself. This is so whether the resin is used in bulk form or used to modify a facial tissue or a textile fabric. Similar problems have been encountered with most plastics, including even nylon.

Decomposition of a finished polymer by application of heat or by any other means can give rise to materials that are considerably more toxic and/ or more hazardous than the polymer or its monomer(s). Only in a very few cases has the thermal decomposition of a polymer been investigated in much detail [24, 25, 26, 27]. Almost notorious for its ability to cause a syndrome similar to metal fume fever is polytetrafluoroethylene and/or its thermal decomposition products [28]. This resin is very stable to heat and also to chemical attack by most other materials. When small particles have been vaporized to form a fume (in a contaminated cigarette), however, polymer fume fever has resulted. This syndrome may be particle size-related and may be caused by a polymer rather than its decomposition products.

Another syndrome associated with complete polymers has been termed *thesaurosis* and may be caused by the inhalation of aerosolized hair sprays, especially those containing polyvinylpyrrolidone. While hair spray pneumonconiosis or thesaurosis may possibly be a hazard to beauticians, its incidence and severity both appear to be low [29].

References

[1] R. E. ECKARDT AND R. HINDIN, "The Health Hazards of Plastics," *J. Occup. Med.* **15** (1973), 808–19.

[2] D. LESTER, L. A. GREENBERG, AND W. R. ADAMS, "Effects of Single and Repeated Exposures of Humans and Rats to Vinyl Chloride," *Am. Ind. Hyg. Assn. J.* **24** (1963), 265–75.

[3]   T. R. TORKELSON, F. OYEN, AND V. K. ROWE, "The Toxicity of Vinyl Chloride as Determined by Repeated Exposure of Laboratory Animals," *Am. Ind. Hyg. Assn. J.* **22** (1961), 354–61.

[4]   P. L. VIOLA, "Pathology of Vinyl Chloride," *Med. Lavoro.* **61** (1970), 174–80. Abstracted in *Ind. Hyg. Digest* **34** (Oct., 1970), 10.

[5]   W. A. COOK, ET AL., "Occupational Acroosteolysis: An Industrial Hygiene Study," *Arch. Env. Health* **22** (1971), 74–82.

[6]   E. D. BARETTA, R. D. STEWART, AND J. E. MUTCHLER, "Monitoring Exposures to Vinyl Chloride Vapor: Breath Analysis and Continuous Air Sampling," *Am. Ind. Hyg. Assn. J.* **30** (1969), 537–44.

[7]   C. MALTONI AND G. LEFEMINE, "Carcinogenicity Bioassays of Vinyl Chloride," *Env. Res.* **7** (1974), 387–405.

[8]   L. B. THOMAS, ET AL., "Vinyl-Chloride-Induced Liver Disease," *N. E. J. Med.* **292** (1975), 17–22.

[9]   T. M. HELLMAN AND F. H. SMALL, "Characterization of the Odor Properties of 101 Petrochemicals Using Sensory Methods," *J. Air Poll. Con. Assn.* **24** (1974), 979–82.

[10] D. E. DEESE AND R. E. JOYNER, "Vinyl Acetate: A Study of Chronic Human Exposure," *Am. Ind. Hyg. Assn. J.* **30** (1969), 449–57.

[11] H. SAKURAI AND M. KUSUMOTO, "Epidemiological Study of Health Impairment among Acrylonitrile Workers," *J. Sci. Labour* **48** (1972), 273–82. Abstracted in *Ind. Hyg. Digest* **36** (Oct., 1972), 10.

[12] E. P. FLOYD AND H. E. STOKINGER, "Toxicity Studies of Certain Organic Peroxides and Hydroperoxides," *Am. Ind. Hyg. Assn. J.* **19** (1958), 205–12.

[13] R. L. POOLE, J. F. GRIFFITH, AND F. S. K. MACMILLAN, "Experimental Contact Sensitization with Benzoyl Peroxide," *Arch. Derm.* **102** (1970), 400–464.

[14] J. A. KENT, ED., *Riegel's Handbook of Industrial Chemistry,* seventh ed. (New York: Van Nostrand Reinhold Company, 1974) p. 258.

[15] W. S. FERGUSON AND D. D. WHEELER, "Caprolactam Vapor Exposure," *Am. Ind. Hyg. Assn. J.* **34** (1973), 384–89.

[16] C. S. WEIL, ET AL., "Experimental Carcinogenicity and Acute Toxicity of Representative Epoxides," *Am. Ind. Hyg. Assn. J.* **24** (1963), 304–25.

[17] "Emergency Temporary Standard on Certain Carcinogens," *Fed. Reg.* **38** (1973), p. 10929.

[18] R. NIEWENHUIS, ET AL., "Toxicity of Chronic Low Level Exposures to Toluene Diisocyanate in Animals," *Am. Ind. Hyg. Assn. J.* **26** (1965), 143–49.

[19] N. L. LAPP, "Physiological Changes as Diagnostic Aids in Isocyanate Exposure," *Am. Ind. Hyg. Assn. J.* **32** (1971), 378–82.

[20] J. E. PETERSON, R. A. COPELAND, AND H. R. HOYLE, "Health Hazards of Spraying Polyurethane Foam Out-of-doors," *Am. Ind. Hyg. Assn. J.* **23** (1962), 345–52.

[21] K. E. GRIM AND R. E. KNOX, "Factors Influencing Hazards in Isocyanate Foam-Spraying," *Am. Ind. Hyg. Assn. J.* **27** (1966), 62–67.

[22] G. W. H. SCHEPERS, "Pulmonary Reaction to Sheet Fiberglas—Plastic Dust," *Am. Ind. Hyg. Assn. J.* **20** (1959), 73–81.

[23] O. E. FANCHER, ET AL., "Toxicity of a Butylene Glycol Adipic Acid Polyester," *Tox. Appl. Pharm.* **25** (1973), 58–62.

[24]   D. R. PAULSON AND G. F. MORAN, "Analysis of Some Toxic Combustion Products of Low-Density Flexible Polyurethane-Ether Foams," *Env. Sci. Tech.* **8** (1974), 1116–18.

[25]   J. H. PETAJAN, ET AL., "Extreme Toxicity from Combustion Products of a Fire-Retarded Polyurethane Foam," *Science* **187** (1975), 742–44.

[26]   K. L. PACLOREK, ET AL., "Oxidative Thermal Degredation of Selected Polymeric Compositions," *Am. Ind. Hyg. Assn. J.* **35** (1974), 175–80.

[27]   R. I. THRUNE, "Gases Released When Fire Resistant Epoxy Resins Are Burned," *Am. Ind. Hyg. Assn. J.* **26** (1968), 475–80.

[28]   "Pyrolysis of Polytetrafluoroethylene," *Am. Ind. Hyg. Assn. J.* **29** (1968), 19–60.

[29]   A. G. McLAUGHLIN AND P. L. BIDSTRUP, "The Effects of Hair Lacquer Sprays on the Lungs," *Food Cosmetics Tox.* **1** (1963), 171–88.

8

PESTICIDES

The suffix *-icide* means *killing*. A person who kills another is said to have committed homicide, but a person who steps on an ant is not accused of committing insecticide. Instead, in the industry of economic toxicology, the *-icide* suffix is used only to refer to materials that kill, and the prefix indicates what is killed. Therefore, the generic term for such a material is *pesticide* because it kills man's pests.

In almost all cases where pesticides are used, a more specific term will apply, depending on the pest. That is, *insecticide, herbicide,* and *rodenticide* are the terms, respectively, for materials that are used to kill insects, vegetation, and rodents. Just because their use may be obvious from the descriptive term does not indicate that only one specific pest will be killed from an application of the material. For instance, the insecticide chlordane can be used to kill Japanese beetles in a lawn; when so-used it will often kill crabgrass, in which case it also functions as a herbicide. Nevertheless, the science of economic toxicology depends on a phenomenon called *selective toxicity,* which takes advantage of the fact that toxicity of a material probably

121

varies widely among the species of living things. By application of knowledge, insight, and experiment, materials are found which are highly lethal for the pest of concern but which are far less toxic to man and/or to useful species. In a sense, this phase of toxicology is similar to the science of pharmacology, which seeks remedies which are not particularly harmful to man for man's diseases.

In seeking useful pesticides, the practitioner is most troubled when the species is closest to man on the tree of evolution. Trying to find something lethal to great apes in small doses but harmless to man in those same doses would probably be a fruitless task. On the other hand, many materials have been found which are extremely toxic to vegetation but which have no discernible effect on man or other animals in the dosages usually applied.

In addition to the laborious research required to find better pesticides, the whole science of economic toxicology is plagued with two problems that appear to be growing in their importance. Historically, the first to develop was that of immunity or increased resistance while the second, bioconcentration, was found much later.

Resistance

In Chapter 2 the concept of variability of response to a toxin was developed. If the same dose of a material is given to a group of animals (or insects or grasses or fungi), the response of each individual in that group will probably be qualitatively similar but probably quantitatively different from that of others in the group. For instance, if the amount of material administered is equal to the LD_{50}, all in the group probably will have similar signs of the resulting illness but only about half will die. Those that die will obviously be those with least resistance to the toxic effect; those that live will be the ones best able to withstand the effect—those with the greatest immunity or resistance.

Reasons for greater resistance can be either hereditary or environmental, or a combination of both. An individual just recovering from an illness, for instance, might well be more susceptible than others in the group to the action of *any* stress, no matter what his inherited background. Nutritional state, age, sex, and other factors obviously play a role but if the exposed group is large enough, all factors but those related to heredity will cancel out. The individuals that survive the onslaught of the toxin will, in general, be those whose genes have best equipped them to handle the problem. When those hardy individuals mate, their offspring are likely to be resistant to the toxin. And, as man's pests have much shorter reproductive cycles than he does, the result is a relatively rapid development of resistance to any weapon employed, including pesticides, but not excluding even traps and guns. (Any person who has hunted or trapped for a long time can talk about the increased wariness of his prey if the hunting pressure on that prey has been chronic. Despite animated cartoons, animals and birds do not talk or write and use genetics as the most effective means of communicating behavior patterns to their offspring.)

Ordinarily, resistance or immunity does not build up in a single individual; it is the result of several generations where only the fit survived. This phenomenon, then, is one that cannot possibly be avoided, but its problems can be minimized.

One of the best ways of negating the problem of immunity is to change the method of attack. If the pest has developed resistance to a certain chemical, then attack with a substance that kills by a different mechanism may well succeed. Techniques other than direct lethality can also be used on occasion or mixed with pesticide use. The other methods range from the release of sterile males to the use of sex attractants to the cultivation of one of the pest's natural enemies or diseases. No one method is likely to be completely successful in eradicating the pest and, therefore, the chances are excellent that there will always be a need for new and better ways of ridding man of such organisms.

Bioconcentration

Bioconcentration in a single step is exemplified by the use of DDT to kill the tree bark beetles that carry Dutch elm disease. When the beetles die, they are easily found by birds such as the robin. If a robin eats a few dead (poisoned) beetles, the effect will probably be small, but robins have been found who ate literally scores of dead or dying beetles. Each beetle has only a small amount of DDT either in or on its body, but the sum of the DDT in or on scores of beetles may be sufficient to in turn kill the robin even though DDT is much less toxic to birds than it is to beetles.

The problem of bioconcentration is compounded when the pesticide is persistent in the environment because then several generations of poisoned prey may be able to, in turn, poison their predators. Of course, the problem may not be one of direct lethality but may be much more subtle. Such a subtle effect has been observed on some birds whose egg shells become thinner and therefore weaker if the mother has absorbed more than a certain minimal amount of DDT.

Bioconcentration is not necessarily confined to a single step nor need it culminate with lethal or other effects on birds alone. Several of the more persistent pesticides are found in man even when there has been no direct exposure of any kind to the material. For the most part, man avoids serious problems associated with bioconcentration, possibly because he almost always eats a varied diet. However, the experience in Japan, Finland, and other countries with fish contaminated with methyl mercury (*not* a pesticide) has caused a reappraisal of this problem with the result that in the United States the use of persistent pesticides has been severely restricted.

In the following discussion of specific pesticide groups, many have been totally omitted. For instance, no mention will be found of slimicides, miticides, or even of avicides. Instead, attention is focused on those categories that come closest to being ubiquitous in man's environment: rodenticides, insecticides, herbicides, and fungicides.

Rodenticides

Rodenticides are used mainly to kill rats and, to a lesser extent, mice. All other rodents, and indeed other animals as a class, together cause much less trouble to man than the rat. Rodenticides therefore have a very long history troubled by the fact that the rat is a warm-blooded animal. Chemicals that kill rats are quite likely to be very toxic to man. All man's infamous poisons of the past from strychnine to arsenic trioxide to cyanides have been used in attempts to rid his quarters of disease-bearing hungry rats. All of those discussed in this section are still used, but more specific toxins (warfarin and norbormide) are tending to displace the others.

Red Squill

Red squill or simply squill is a natural mixture of materials extracted from the sea onion, *Urginea maritima,* which grows in the Mediterranean Sea. Although more toxic to rodents than to most other animals, red squill has similar actions on all species, namely gastric upset followed by respiratory and circulatory disturbances that may result in death. This substance is a potent emetic and, unless the dose is overpowering, most animals (including man) will vomit soon after ingestion, thus reducing the dose to a tolerable level. Rats, however, cannot vomit and therefore red squill remains in their bodies until it paralyzes the heart. It is fast-acting and kills quite close to the bait location.

Sodium Fluoroacetate

Sodium fluoroacetate, compound 1080, is an extremely potent enzyme inhibitor with a high toxicity for all animals. Absorbed through the GI tract or the skin, it causes hyperexcitability, convulsions, and finally ventricullar fibrillation. The lethal dose for a man is estimated to be about 50 mg; it is far too hazardous to be used indiscriminately.

Thallium Salts

Another rather general toxin used much more in Europe than in the United States is thallous sulfate (Tl_2SO_4). Baits have been eaten by children who subsequently experience gastroenteritis, respiratory difficulties, and then tremors, convulsions, paralysis, and death. If the victim survives, alopecia is likely and the nails may be lost as well. In addition, thallous sulfate can cause mental aberrations (psychoses), blindness, kidney damage, and general peripheral neuritis. The toxicity and lack of warning properties of thallium salts make them much too hazardous for anything except very careful use.

ANTU

α-Naphthylthiourea [ANTU or 1-(1-naphthyl)-2-thiourea] has a relatively low toxicity for man and his pets but causes massive lung edema in rats and probably other rodents. An indication of its utility as a rodenticide is the fact that its acute oral LD_{50} for rats is about 6 mg/kg and that for guinea pigs is about 140 mg/kg. Nevertheless, if the toxicity for man is similar to that for a guinea pig, a gram or two could be fatal to a child.

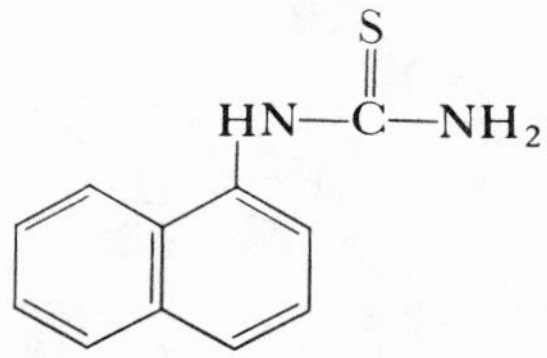

ANTU

Norbormide

Norbormide [5-(α-hydroxy-α-2-pyridylbenzyl)-7-(α-2-pyridylbenzylidene)-5-norbornene-2,3-dicarboximide] was discovered in the 1960s to be extremely toxic to rats and much less toxic even to other rodents. Its acute oral LD_{50} for rats is about 5 mg/kg, for guinea pigs it is over 600 mg/kg, and for mice it is about 2,000 mg/kg. In the rat, norbormide causes peripheral vasoconstriction leading to death. Rather than *rodenticide, raticide* is more appropriate for this substance.

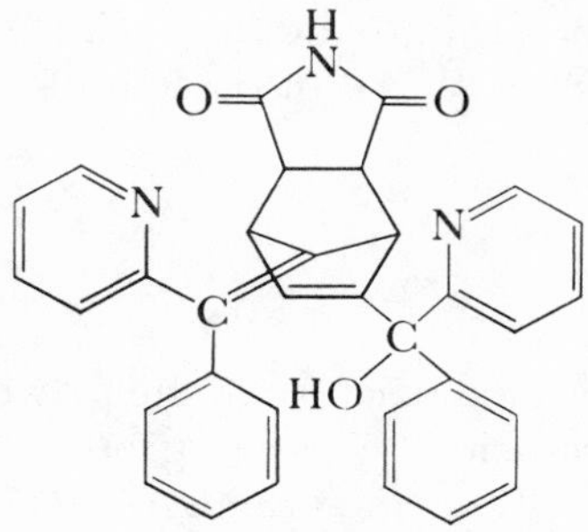

Norbormide

Warfarin

All the rodenticides mentioned to this point have high acute oral toxicities for rats. Warfarin [3-(α-acetonylbenzyl)-4-hydroxycoumarin] does not. In fact, its acute oral toxicity for rats is only about 300 mg/kg (for the mouse, however, it is about 2 mg/kg), but warfarin is one of the most successful rat poisons. Rather than killing directly, warfarin and other materials of a similar nature interfere with the clotting mechanism of blood. A scratch can then be fatal. In the rat, warfarin also increases capillary permeability so that even the first scratch is not necessary. Furthermore, because of its relatively low acute oral toxicity and slow action, adult rats are able to carry bait back to the nest so the young are killed.

Even if a child or other human should accidentally ingest a large amount of warfarin, administration of vitamin K is antidotal. Warfarin is still one of the safest of all rodenticides to use.

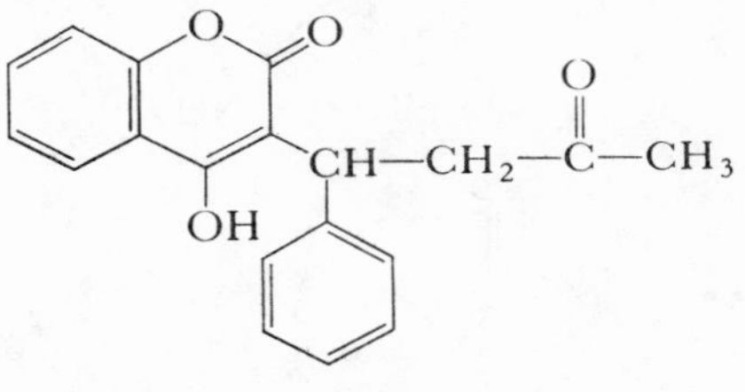

Warfarin

Fumigants

Other than those materials that are dispensed with or on baits and enter the pest from ingestion, several gases and vapors have been used to kill by fumigation. None of these materials is at all selective and, therefore, fumigants tend to exterminate insects, spiders, birds, and other living things along with rats and mice. Most popular are hydrogen cyanide (HCN), methyl bromide (H_3CBr), and phosphine (PH_3), although most of the highly toxic vapors and gases have been used for this purpose. In addition, because of its ready availability, carbon monoxide (from the tail pipe of an automobile, for instance) is sometimes used. All of these materials have been discussed previously.

Insecticides

Almost all the insecticides used because of their direct action fall into one of two categories. First, historically, is the group that could be called *CNS toxins* but instead it is usually termed *chlorinated hydrocarbons*. The materials comprising this group are, in fact, all chlorinated hydrocarbons but many chlorinated hydrocarbons are not good insecticides. The second group is usually called *cholinesterase inhibitors* because of the common mode of action. The two groups will be discussed separately.

Chlorinated Hydrocarbons

Chlorinated hydrocarbon insecticides have in common an organic base, chlorine in the molecule, a largely unknown mechanism of action on insects, moderate-to-long persistence in the environment, accumulation and storage in the lipids of birds and mammals, and toxic action on the CNS (through a largely unknown mechanism) of birds and mammals. Some of these materials are strongly suspected of being potential carcinogens in man.

The first organochlorine insecticide was DDT [1,1,1-trichloro-2,2-bis (*p*-chlorophenyl) ethane], synthesized in 1874, but first used to kill insects in 1942. DDT was found to be cheaper and much more effective against a wide range of insect species than the natural insecticides nicotine, pyrethrum, and rotenone or the synthetic dinitrophenols and cresols. In addition, its toxicity to man and other mammals was relatively low [1] and because of its lack of reactivity with oxygen, its low water solubility, and its low vapor pressure, it persisted on sprayed or dusted surfaces for long periods of time in insecticidal concentrations. DDT revolutionized agriculture as well as public health practices and caused the development of many other chlorinated hydrocarbon insecticides in efforts to increase effectiveness and decrease cost. Because of its bioconcentration and experimental data indicating possible carcinogenic effects in man, its use in the United States was essentially banned as of 1974 except for a few isolated well-controlled uses. It is still used widely in other countries, primarily for malaria mosquito control.

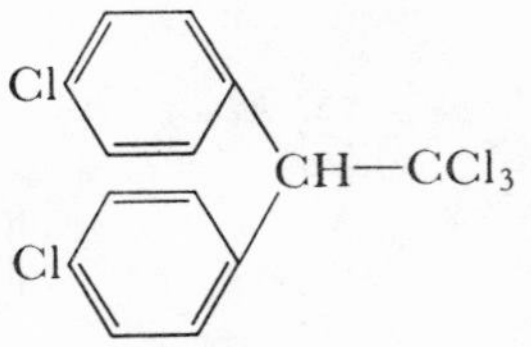

DDT

Once DDT had been banned, suspicious eyes were turned toward several other more or less persistent organochlorine insecticides. More thorough animal experiments than had been done were undertaken in order to determine whether or not, at any dose level, these materials were capable of causing either cancer or precancerous tumors in one or more species. Included with DDT on the suspect list at one time or another were aldrin, dieldrin, chlordane, heptachlor, mirex, toxaphene (a chlorinated camphene), and hexachlorocyclohexane (also called benzene hexachloride, not to be confused with hexachlorobenzene, which actually is a benzene derivative). Even though none of these materials (including DDT) had been shown to cause cancer in man, the fear was that persistence in lipid would cause extremely prolonged exposure even though exogenous exposure had ceased and that perhaps the endogenous exposure would cause cancer.

On the one hand, regulatory agencies were under rather great political pressure to adopt a conservative attitude (*use nothing until it has been proved safe*); on the other hand, manufacturers and users of the materials complained that a negative theory (*this substance does not cause cancer*) can never be proved. The ban of DDT occurred without a great deal of fuss mainly because in the United States many insect species had developed almost complete resistance to its toxic action and use of the material had been declining for that reason. This was not true for the other chlorinated hydrocarbon insecticides and therefore attempts to ban were contested vigorously.

Proponents of banning said, in essence, "we cannot take the chance of human cancer so long as any other alternatives are available;" opponents insisted that there was no evidence of human cancer from the materials and that alternate insecticides (where they existed) were more acutely toxic to man and thus more hazardous along with being more expensive and less persistent. At stake were not only political careers and manufacturer's profits but the real possibility (but usually admitted small probability) that a wrong decision would cause either untold misery because of cancer or untold misery because of starvation and/or disease.

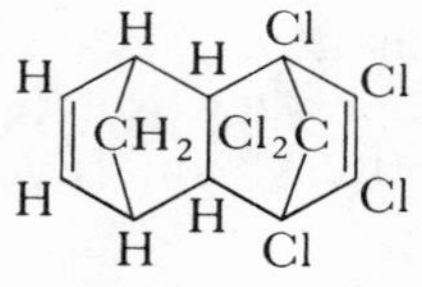

Aldrin

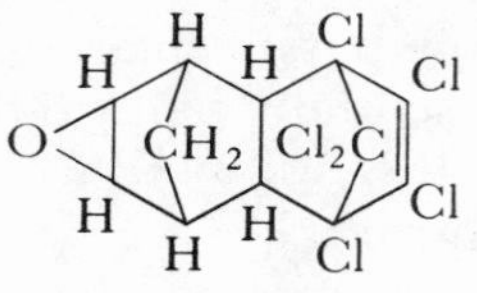

Dieldrin

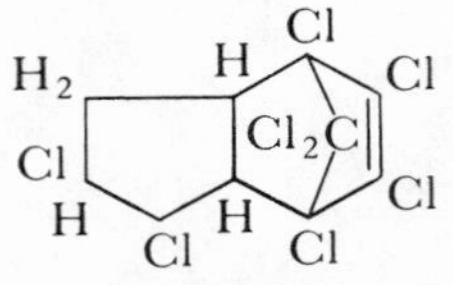

Chlordane

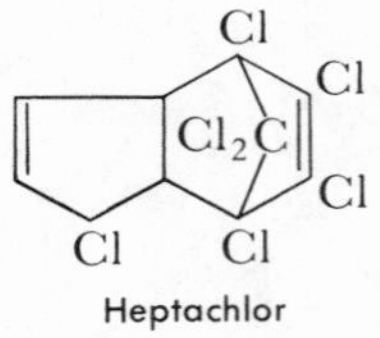

Heptachlor

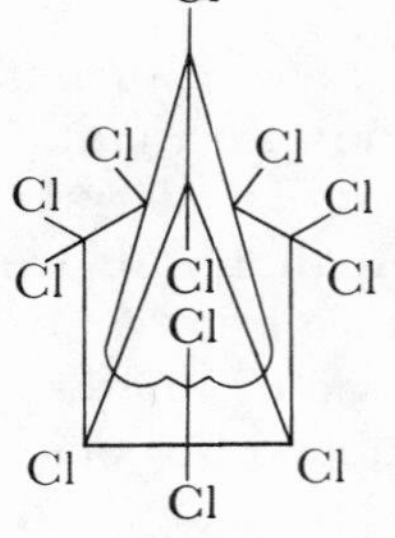

Mirex

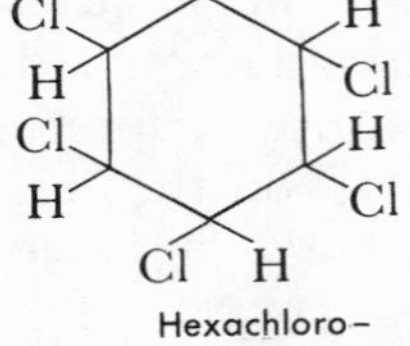

Hexachloro-
cyclohexane

At no time was the issue philosophically simple. In addition to the alternate materials (almost exclusively cholinesterase inhibitors), alternate methods of controlling insects were available, at least in theory. Bacterial or viral insecticides, chitin inhibitors, growth regulators, pheromone (sex) attractants, use of sterile males, and even the development of insect-resistant crops were in the laboratory or early trial stages and one of these or even one of the newer insecticides such as the pyrethrin derivatives might obviate the question given enough research money and effort. Finally, some people insisted that as there was no evidence that insecticides had caused even a minor drop in total insect population on earth, the use of all insecticides should be totally banned. Controversy will probably continue into the forseeable future.

All the chlorinated hydrocarbon insecticides are stored in fat once they enter the body, usually by ingestion but occasionally (during application) by inhalation. Fat (lipid) acts as a storage depot [2, 3] accumulating these materials when their concentration in blood is high and releasing them when the blood concentration is low [2, 4, 5, 6]. These insecticides have little or no effect on the lipid; instead, all affect the central nervous system, and some affect the liver.

Central nervous system effects are manifested in hyperirritability and tremors. If the exposure is great enough, there may be convulsions and death, probably from respiratory failure. Nevertheless, human experience with most of these materials has been very good; a few incidents of acute intoxication caused by mishandling or from suicidal intent have appeared in the literature but there have been very few fatalities. The problem with these materials is not overt toxicity [7] but that of the potential for causing cancer in man [7, 8, 9] and for causing major disturbances all along a food chain.

Cholinesterase Inhibitors

The part of a living organism's nervous system that lies physically outside the central nervous system (brain and spinal cord in vertebrate animals such as man) is called the *peripheral nervous system*. Very briefly, this system consists of *afferent* nerves (going toward) and *efferent* nerves (going away) that carry impulses to and from the CNS. Efferent nerves can be classed as *somatic* (terminating on skeletal muscle) or *autonomic* (terminating on cardiac or smooth muscle or on glands). Finally, autonomic nerves are classified as either *sympathetic* (having to do with the "fight or flight" response) or *parasympathetic* (having to do with maintenance of homeostasis or the normal state of the body).

Transmission of impulses along the body of a nerve takes place by means of electrical phenomena but transmission from the termination of a *neuron* (nerve body) to a muscle cell, gland, or another nerve is by means of chemical mediators. That is, the nervous impulse acts to release a chemical that diffuses across to receptor sites on the other (receptor) cell where changes in cell wall permeability to ions occur, leading to a propagation of the impulse. In the sympathetic nervous system, the chemical mediator for all but terminal receptor sites is acetylcholine.

A somatic nerve is a single-cell system that starts in the CNS and terminates at a muscle cell where a release of acetylcholine transmits the impulse to the muscle. In the autonomic nervous system, each nerve (with one exception) consists of at least two neurons, the first leading from the CNS to a junction called a *synapse* and the second beginning at the synapse and terminating at the heart, smooth muscle, or a gland. (The one exception is that the adrenal medulla is reached by a single neuron with no intervening synapse.) A synapse can have several afferent and/or several efferent neurons. Synapses outside the CNS (where they abound) are called *ganglia*. The chemical mediator at all ganglia is acetylcholine. Finally, acetylcholine also mediates transmission of nervous impulses at all terminal receptor sites of the parasympathetic nervous system. (At some other locations another mediator, possibly norepinephrine, is used instead of acetylcholine but acetylcholine is of most interest here.)

$$(H_3C)_3-N^+-CH_2-CH_2-O-\overset{\overset{\displaystyle O}{\|}}{C}-CH_3$$

Acetylcholine

When acetylcholine reaches a receptor site, it causes a large increase in the permeability of that cell wall to certain ions. Entrance of these ions into the receptor cell alters the electric charge balance across the cell membrane in such a way that an impulse (called an *action potential*) is started. Then the acetylcholine must be inactivated rapidly so that the cell permeability can be restored to its former value or ions will continue to pour in, triggering a constant stream of action potentials. Inactivation of the acetylcholine is accomplished by an enzyme called *acetylcholinesterase,* found everywhere in the body.

A molecule of acetylcholine fits into an active site on the much larger molecule of acetylcholinesterase in a manner analogous to the fitting of a key into a lock. The enzyme splits acetylcholine into choline and acetic acid, which are then released. (Another enzyme within the neuron uses choline and acetate to make acetylcholine.)

Acetylcholinesterase inhibitors act by competing with acetylcholine for active sites on the enzyme molecule [10]. Once one of the active sites is occupied by an alien molecule, that site can no longer be used for the hydrolysis of acetylcholine, which then builds up in concentration at ganglia and nerve endings, resulting in overstimulation first of the parasympathetic nervous system and then of the remainder of the peripheral and central nervous systems. Overstimulation carried to the extreme results in paralysis because the nerves can no longer carry information, and the paralysis can result in death.

In man, the initial symptoms of cholinesterase inhibition are those of sympathetic nervous system stimulation; intestinal cramps, tightness in the chest, blurred vision, headache, diarrhea, decrease in blood pressure, and salivation [11]. This first stage is followed minutes or hours later by the second stage where skeletal muscle and the CNS are involved, leading to respiratory failure, convulsions, paralysis, and death.

Atropine, a drug obtained from the belladonna plant, is able to completely block the action of acetylcholine at the terminal receptor sites of the postganglionic parasympathetic nervous system. It does this by rendering the effector structure insensitive to acetylcholine so that even though a great excess of acetylcholine may be present, permeability of cell walls remains unchanged.

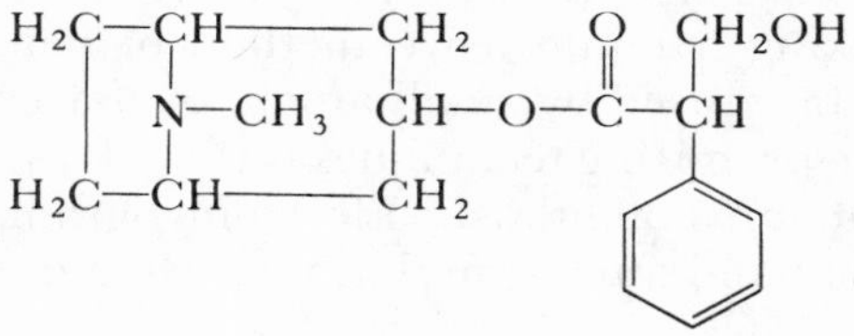

Atropine

Atropine has no particular effect on acetylcholinesterase, so that after an overdose of a cholinesterase inhibitor either time [12] or another drug can be used to reactivate the enzyme. Usually when medical aid is available, 2-PAM chloride (or iodide) (the chloride is pyridine-2-aldoxime methochloride, also called Pralidoxime) is used to reactivate cholinesterase. The use of atropine and 2-PAM chloride (or iodide) is another of the few illustrations available in toxicology where the mechanism of action is known and antidotal materials are available and effective [11, 13, 14].

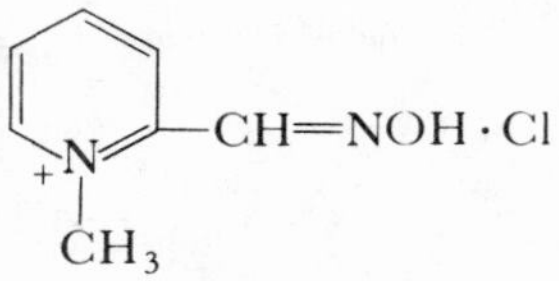

2-PAM Chloride

So far in this discussion, *cholinesterase* and *acetylcholinesterase* have been used interchangeably. They are not the same as there are at least two cholinesterases in the body but only one acetylcholinesterase. A material that inactivates acetylcholinesterase almost always also inactivates other cholinesterases.

Industrial exposures by inhalation or, more often, by skin absorption [15, 16] to cholinesterase inhibitors can be monitored easily and simply by means of blood sampling. Usually determined are total cholinesterase in plasma (largely pseudocholinesterase) and in red blood cells (largely acetylcholinesterase). Plasma cholinesterase responds rapidly to inhibitors and on occasion will even fall from 100% of normal to zero before any effect is seen on red cell cholinesterase. Red cell cholinesterase values are more likely to be indicative of chronic than of acute exposures so that by monitoring both plasma and red cell values the industrial hygienist has available information on short- and long-term effects of exposure. If weekly blood samples are obtained, any statistically significant decrease in the plasma cholinesterase value should trigger an investigation to find and eliminate the cause; any significant decrease in the red cell value should cause the worker to be removed from the exposure. With this regimen, there will almost certainly be no other effects of the exposure [17].

In many cases, workers are reluctant to be "stuck" with a needle for a blood sample even once, not to mention weekly. For that reason, several researchers are seeking a noninvasive method of determining very early changes that might be caused by small amounts of a cholinesterase inhibitor. One of the more promising techniques is that of *electromyography* (EMG), recording muscle action potentials. Electromyograms may be obtained from a muscle in the hand after stimulating a nerve in the wrist, for example.

Cholinesterase-inhibiting insecticides in widespread use are either organophosphorus compounds or derivatives of carbamic acid. The organophosphorus compounds have been used longer than the carbamates and lead in total consumption by a factor of about 2.5 in the United States. Most popular among the first group are methyl parathion, malathion, parathion, and diazinon; the carbamates used most are carbaryl and carbofuran.

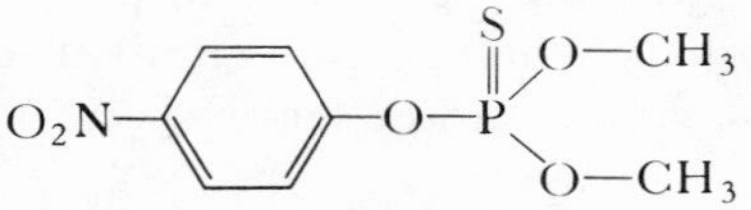

Methyl Parathion

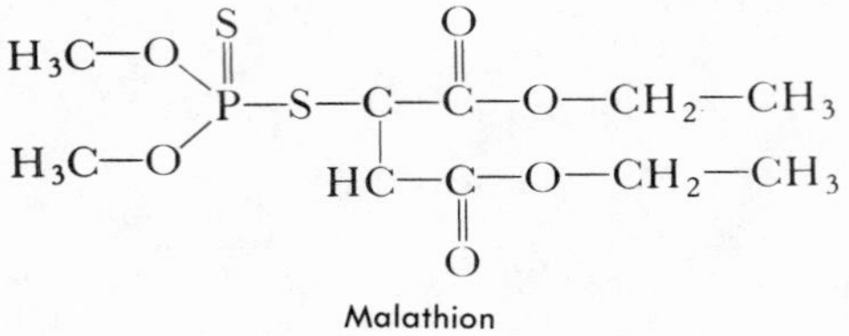

Malathion

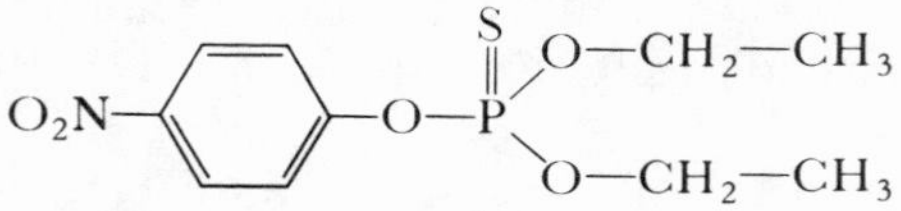

Parathion

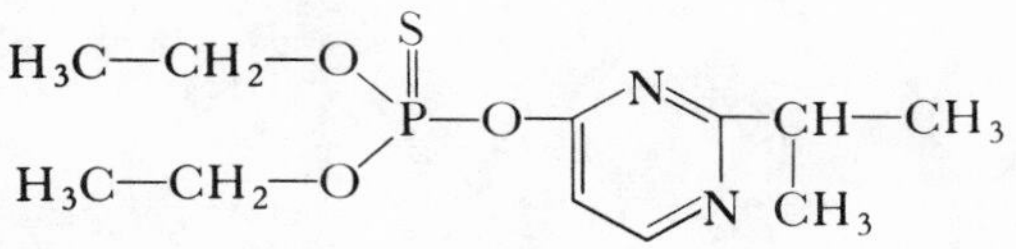

$$H_3C-CH_2-O \quad \overset{S}{\underset{||}{P}}-O-\text{(pyrimidine ring)}-CH-CH_3, CH_3$$

H_3C-CH_2-O / P with =S / -O- / N=, pyrimidine ring, -CH-CH_3 / N, CH_3

Diazinon

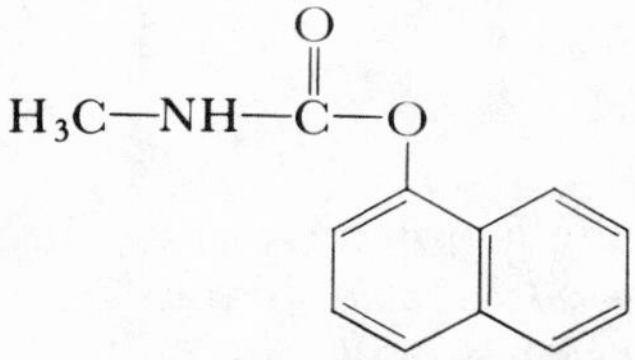

$$H_3C-NH-\overset{O}{\underset{||}{C}}-O-\text{(naphthyl)}$$

Carbaryl

$$H_3C-NH-\overset{O}{\underset{||}{C}}-O-\text{(benzofuran ring)}-(CH_3)_2$$

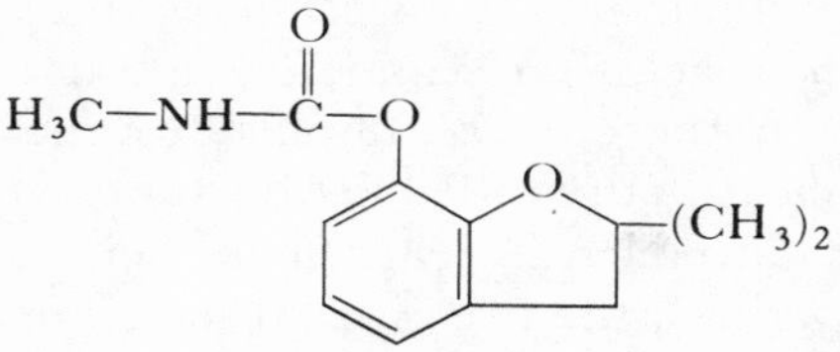

Carbofuran

Both methyl parathion and parathion are highly toxic to warm-blooded animals, with acute oral LD_{50} values ranging from about 3.5 to 25 mg/kg, depending on species. All the others are much less toxic, with LD_{50} values ranging upward from 500 mg/kg.

One of the most interesting organophosphates is DDVP, or dichlorvos. This material has a much higher vapor pressure than most other insecticides and an extremely high toxicity for many insects but a low hazard to man from inhalation. DDVP is applied as a film to a plastic or semisorbent strip, which is suspended in the air of the space to be cleared of flying insects. Without causing any discernible harm to human or animal inhabitants of the space [18], flies and mosquitos as well as several crawling bugs and insects are killed by the exposure.

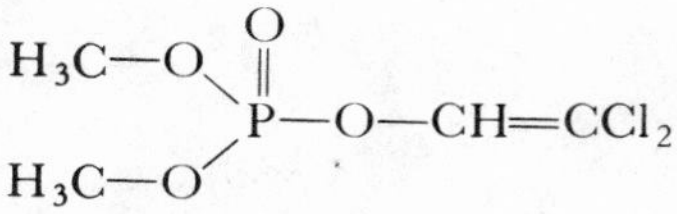

$$H_3C-O \quad \overset{O}{\underset{||}{P}}-O-CH=CCl_2$$

H_3C-O / P / H_3C-O

Dichlorvos

Ronnel is a cross between chlorinated hydrocarbon and organophosphorus compounds. It has an LD_{50} for the rat of about 3,000 mg/kg and therefore is one of the least toxic of all insecticides to that species (and probably to man). It is a weak cholinesterase inhibitor that has been fed to animals to kill external as well as internal pests.

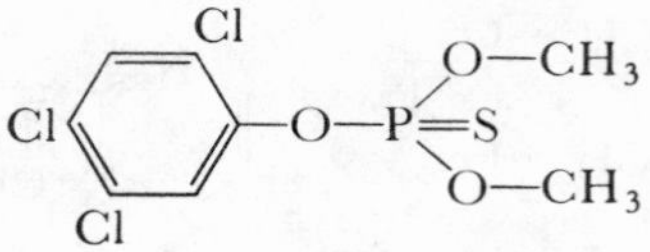

Ronnel

Herbicides

The most common herbicide problem is to kill broad-leafed or *dicotyledenous* plants amid grasses (*monocotyledenous* plants) or vice versa. In addition, broad-spectrum herbicides that will kill or at least *defoliate* (remove the leaves from) many kinds of vegetation have extensive uses. To kill or control one kind of vegetation amid another is a real problem in selective toxicity; to kill all or most of the vegetation in an area without seriously harming animals in the vicinity is another; and both must be done at the lowest cost.

Most of the herbicides in use today act by simulating plant hormones or *auxins*. Auxins are produced by the leaf; when their concentration decreases because of *senescence* (aging), the leaf can no longer remain attached to the stem. Application of several herbicides causes auxin levels to drop, resulting in defoliation and some other changes. Increasing auxin levels causes the leaf to remain attached to the plant and even stimulates growth. Too much stimulation causes the plant to increase respiration beyond its ability to manufacture food; it literally grows to death.

Both 2,4-D (2,4-dichlorophenoxyacetic acid) and 2,4,5-T (2,4,5-trichlorophenoxyacetic acid) simulate auxins. In their usual application, they cause the plant to try to grow too rapidly. Both materials are much more toxic to dicotyledonous than to monocotyledonous plants so that they can be used, for instance, to kill dandelions in a lawn. The slight change in structure makes 2,4,5-T a good brush killer, which 2,4-D is not. Toxicity to animals and man for both is quite low [19, 20].

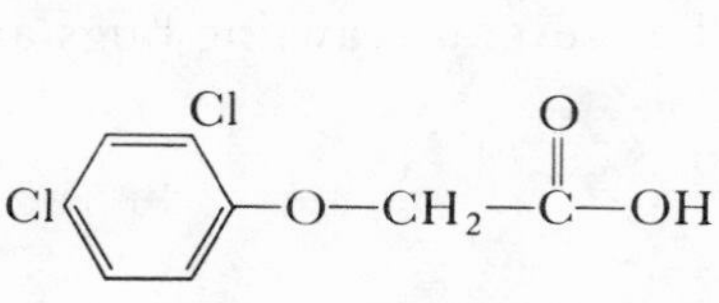

2,4-D 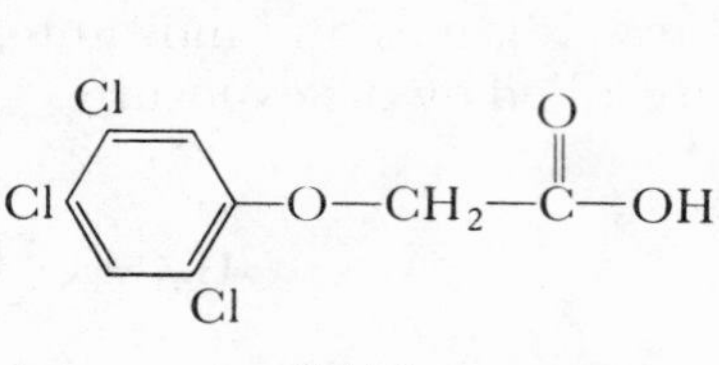   2,4,5-T

2,4,5-T was used extensively in the Viet Nam conflict to defoliate forest trees, reducing their ability to conceal enemy troops. During the 1960s, it was produced in vast amounts from the intermediate, 2,4,5-trichlorophenol. Unfortunately, the phenol can undergo a rearrangement, reacting with itself to produce 2,3,7,8-tetrachlorodibenzo-*p*-dioxin, or TCDD.

TCDD is one of the most toxic synthetic organic compounds ever produced; the LD_{50} of TCDD for mice is about 0.1 mg/kg and a dose of 0.001 mg/kg weekly was found to produce significantly increased liver weights [21, 22]. In addition, this material is an *acnegen* (causes acneform dermatitis complete with blackheads) in man and *hyperplasia* (thickening) of rabbit ears on topical application. TCDD is teratogenic to experimental animals in low doses. Because of that, several studies were made of people in Viet Nam to determine if there had been an increase in mutation rate (birth defects) but there were essentially negative results. Apparently TCDD contamination of the 2,4,5-T used was low enough so that the doses people received were insufficient to cause this kind of injury.

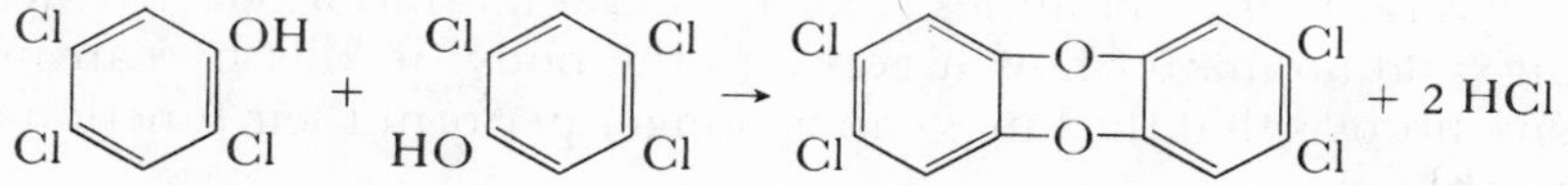

TCDD over the years has caused 2,4,5-trichlorophenol production facilities in Germany and the United States to be abandoned and/or burned to the ground because the contamination on equipment and structures was too difficult to remove and so hazardous to workers. Fortunately, a process for the phenol production that does not cause TCDD formation was discovered (in Germany) so that present production is free of that hazard.

Herbicides so far discussed have been effective on broad-leafed plants. Dalapon (2,2-dichloropropionic acid) is a very effective killer of monocotyledonous plants of all kinds. It, too, has a very low toxicity for man and animals.

Dalapon

Herbicides that are acids are not usually used as such for several reasons, one being volatility. Low molecular weights are associated with high vapor pressures and short persistence in the environment. Therefore, these materials are usually applied as salts or esters, not acids, to increase effectiveness. All are readily destroyed by bacterial action, however, so that after a week or so in the usual environment little trace remains.

Paraquat (1,1'-dimethyl-4,4'-bipyridinum dichloride) has been used since the mid 1960s as a very versatile herbicide. It will kill both mono- and dicotyledenous plants if sprayed on leaves but causes little injury otherwise. One of its main uses is to assist in the harvest of cotton. Just before harvest, the cotton plants are sprayed with paraquat, which has two main effects.

First, it desiccates the leaves and defoliates the plants so that leaves will not be picked along with the cotton bolls (in concentrations used, it does not kill the plant). Second, it causes the mature bolls to open so that they are easily picked. Mechanical harvest of cotton would be greatly hindered without use of a material such as paraquat.

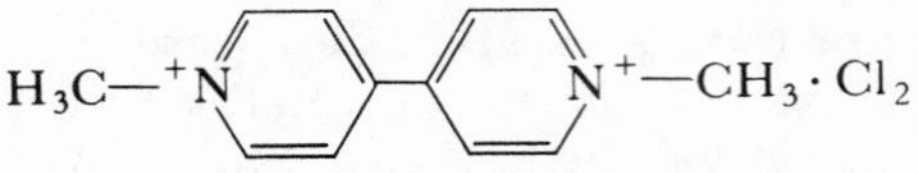

$$H_3C-^+N\text{<=>}N^+-CH_3 \cdot Cl_2$$

Paraquat

Unfortunately, paraquat is quite toxic to warm-blooded animals and has caused several human fatalities [23, 24]. Paraquat has a selective action on the lung, and no matter how it gets into the body, it kills by causing such extensive fibrosis that the lungs can no longer perform their function of oxygenating blood.

Inorganic salts such as sodium chloride and sodium chlorate have been used to kill plants in a very nonselective manner. Sodium chloride (Na_2ClO_4) is much more toxic to vegetation than is sodium chloride (NaCl). In addition, it is an active skin and eye irritant, a methemoglobin-former, and a potent oxidizer of many materials.

Several materials have been used as volatile soil sterilants. These materials can kill all fungi, nematodes, plants, and most seeds on application and then, being volatile, evaporate, leaving the soil sterile. Methyl bromide has been used, for instance, to sterilize soil of football/baseball stadia prior to the installation of artificial turf. For this purpose the ground is first covered with impervious plastic film and then the gas is injected beneath the film and allowed to remain in contact with the soil for a day or two. The plastic is then removed and artificial turf is applied. Other soil sterilants in more or less common use are chloropicrin, sodium *N*-methyldithiocarbamate, ethylene dibromide, and mixed dichloropropenes, all liquids which are usually applied in a trench which is immediately covered.

$$H_3CBr \qquad\qquad Cl_3C-NO_2$$

Methyl Bromide   Chloropicrin

$$H_3C-NH-\overset{\overset{\textstyle S}{\|}}{C}-S-Na \qquad\qquad BrCH_2-CH_2Br$$

Sodium N-Methyldithiocarbamate   Ethylene Dibromide

Fungicides

Fungi are very difficult to kill and therefore in many instances materials sold and used for that purpose should probably be called *fungistats* (preventing growth of fungi) rather than fungicides. Perhaps because fungi are so difficult to kill, there is no true economical broad-spectrum fungicide with a low toxicity to other forms of life. The mercurials (largely mercuric chloride and phenyl and alkyl mercury compounds) probably come as close to being broad-spectrum as any, but they are expensive and either they or their bio-concentration products are quite toxic to many forms of life [25]. For the most part, mercurials are used to protect seeds and bulbs from fungal at-tack and to eradicate many fungi already present.

Dithiocarbamates

The largest group of commercially important fungicides is the dithiocarba-mates, represented by disulfiram, ferbam, maneb, nabam, ziram and zineb. These compounds are mainly protective (fungistatic) in nature but will kill some fungi [26]. None is so toxic to man that minor contact will have seri-ous consequences, but all may be skin sensitizers especially after prolonged repeated contact. They may be somewhat more toxic to birds [27].

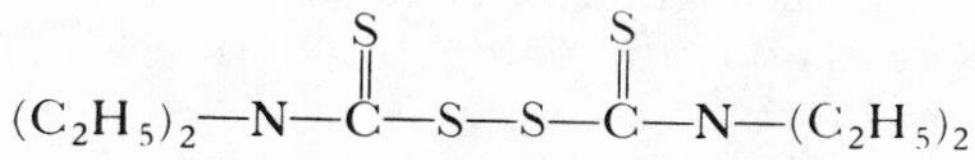

$$(C_2H_5)_2-N-\overset{\overset{\displaystyle S}{\|}}{C}-S-S-\overset{\overset{\displaystyle S}{\|}}{C}-N-(C_2H_5)_2$$

Disulfiram

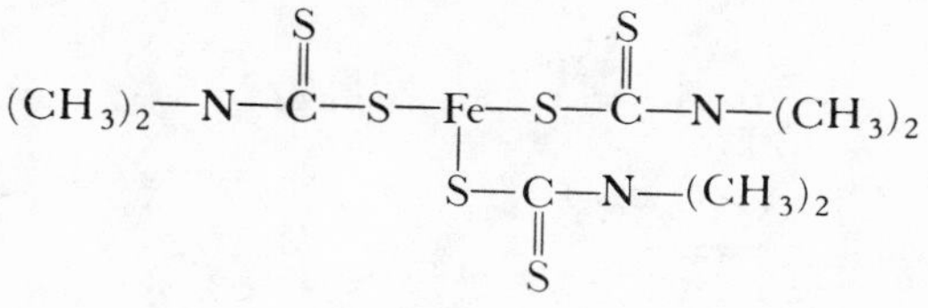

$$(CH_3)_2-N-\overset{\overset{\displaystyle S}{\|}}{C}-S-Fe-S-\overset{\overset{\displaystyle S}{\|}}{C}-N-(CH_3)_2$$
$$S-\overset{\overset{\displaystyle}{}}{\underset{\underset{\displaystyle S}{\|}}{C}}-N-(CH_3)_2$$

Ferbam

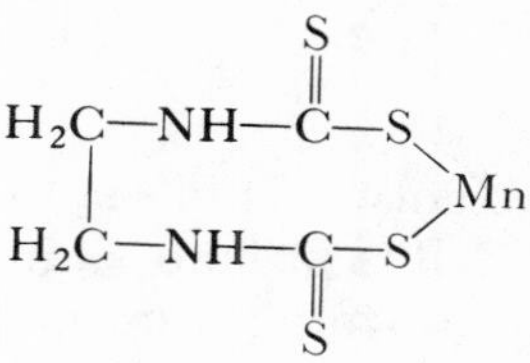

Maneb

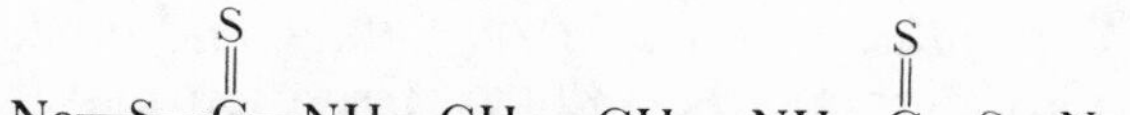

Nabam

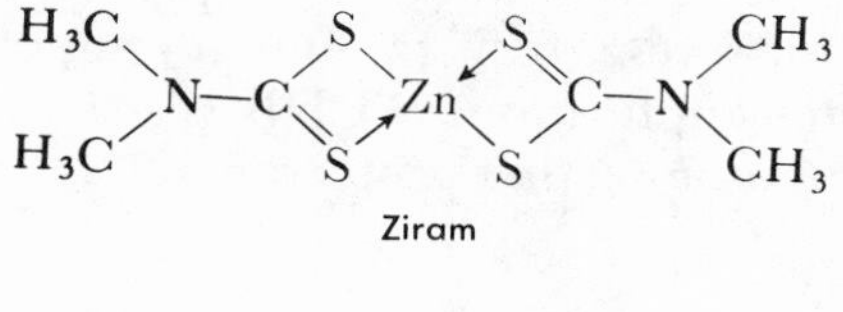

$$Na-S-\overset{\overset{\displaystyle S}{\|}}{C}-NH-CH_2-CH_2-NH-\overset{\overset{\displaystyle S}{\|}}{C}-S-Na$$

Ziram

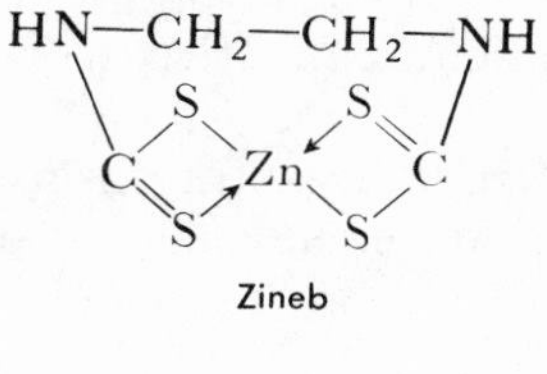

Zineb

Phthalimids

The phthalimids are represented by only two important materials, folpet and captan, both of which are fairly broad-spectrum fungistats and fungicides with a low toxicity to man and animals.

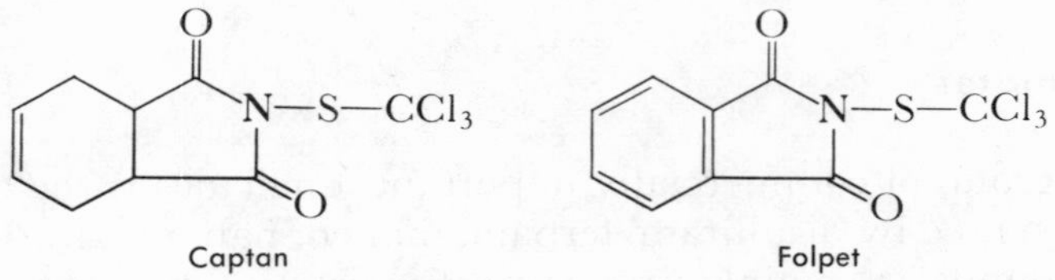

Captan   Folpet

Quinones

Two quinones, chloranil and diclone, are commercially important. Both are quite irritating to the skin and to mucous membranes and both are CNS depressants. Neither is a good fungicide, but both are good fungistats, especially in cool weather [26].

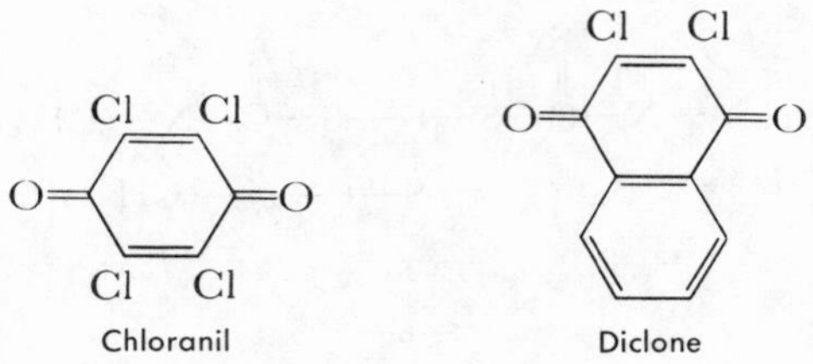

Chloranil   Diclone

Miscellaneous Fungicides

Fungicides that seem not to fall into any general category include PCNB (pentachloronitrobenzene), dodine, and dinocap. PCNB is probably quite irritating to breathe and that irritation may give good warning of hazardous concentrations. Both PCNB and dinocap may have effects on the metabolic rate of man and animals from overexposure. Dodine is irritating to the skin, eyes, and GI tract; it causes vomiting if ingested. Dinocap is one of the few fungicides effective in eradicating powdery mildew; it is specific for that fungus. Dodine is a good erradicant for several fruit tree fungi and, in addition, gives long-lasting protection after application. PCNB is used mainly on soils rather than vegetation; it gives long-lasting protection [26].

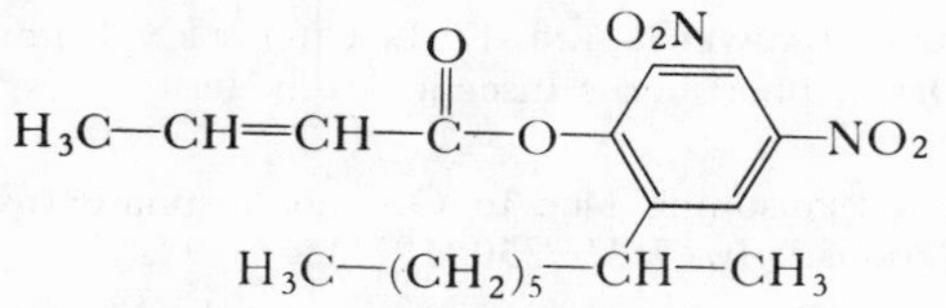

H3C—CH=CH—C(=O)—O—[benzene ring with O2N and NO2 substituents]

H3C—(CH2)5—CH—CH3

Dinocap

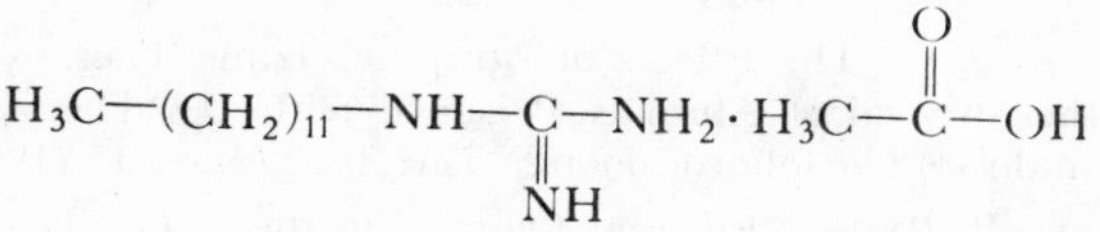

H3C—(CH2)11—NH—C(=NH)—NH2 · H3C—C(=O)—OH

Dodine

[benzene ring with Cl, Cl, Cl, Cl, Cl substituents and NO2]

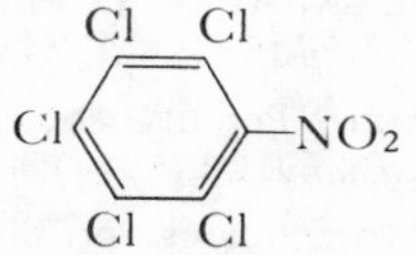

PCNB

The first fungicide ever used, sulfur, is still popular, mainly because of its low cost. Its toxicity is low by all routes of exposure; it is not so effective as the newer fungicides and fungistats whether used as the wettable powder or in chemical combination as in lime sulfur.

REFERENCES

[1] W. J. HAYES, JR., W. E. DALE, AND C. I. PIRKLE, "Evidence of Safety of Long-Term, High, Oral Doses of DDT for Man," *Arch. Env. Health* **22** (1971), 119–35.

[2] M. E. ZABIK AND R. SCHEMMEL, "Dieldrin Storage of Obese, Normal, and Semistarved Rats," *Arch. Env. Health* **27** (1973), 25–30.

[3] V. M. ADAMOVIC AND B. SOKIC, "Lower Level Phenomena of DDT Cumulation in Female Abdominal Fatty Tissue," *Arh. Hig. Rada i Toksikol.* **24** (1973), 303–06. Abstracted in *Ind. Hyg. Digest* **38** (1974), Abst. No. 1011/74.

[4] T. H. MILBY, A. J. SAMUELS, AND F. OTTOBONI, "Human Exposure to Lindane: Blood Lindane Levels as a Function of Exposure," *J. Occup. Med.* **10** (1968), 584–87.

[5] D. L. MICK, K. R. LONG, AND D. P. BONDERMAN, "Aldrin and Dieldrin in the Blood of Pesticide Formulators," *Am. Ind. Hyg. Assn. J.* **33** (1972), 94–99.

[6] C. G. HUNTER, J. ROBINSON, AND M. ROBERTS, "Pharmacodynamics of Dieldrin (HEOD)," *Arch. Env. Health* **18** (1969), 12–21.

[7] L. LYKKEN, "Chemical Control of Pests," *Chemistry* **44** (1971), 18–21.

[8] K. KAY, "Toxicology of Pesticides: Recent Advances," *Env. Res.* **7** (1974), 202–43.

[9] K. KAY, "Occupational Cancer Risks for Pesticide Workers," *Env. Res.* **7** (1974), 243–71.

[10] S. D. MURPHY, R. R. LAUWERYS, AND K. L. CHEEVER, "Comparative Anticholinesterase Action of Organophosphorus Insecticides in Vertebrates," *Tox. Appl. Pharm.* **12** (1968), 22–35.

[11] T. NAMBA, ET AL., "Poisoning Due to Organophosphate Insecticides: Acute and Chronic Manifestations," *Am. J. Med.* **50** (1971), 475–92.

[12] L. W. REITER, ET AL., "Parathion Administration in the Monkey: Time Course of Inhibition and Recovery of Blood Cholinesterases and Visual Discrimination Performance," *Tox. Appl. Pharm.* **33** (1975), 1–13.

[13] L. W. HARRIS, ET AL., "The Effects of Atropine-Oxime Therapy on Cholinesterase Activity and the Survival of Animals Poisoned with *O,O*-Diethyl-*O*-(2-isopropyl-6-methyl-4-pyrimidinyl) Phosphorothioate," *Tox. Appl. Pharm.* **15** (1969), 216–24.

[14] I. L. NATOFF AND B. REIFF, "Effect of Oximes on the Acute Toxicity of Anticholinesterase Carbamates," *Tox. Appl. Pharm.* **25** (1973), 569–75.

[15] H. R. WOLFE, J. F. ARMSTRONG, AND W. F. DURHAM, "Pesticide Exposure from Concentrate Spraying," *Arch. Env. Health* **13** (1966), 345–50.

[16] R. J. FELDMAN AND H. I. MAIBACH, "Percutaneous Penetration of Some Pesticides and Herbicides in Man," *Tox. Appl. Pharm.* **28** (1974), 126–32.

[17] G. CLARK, "Organophosphate Insecticides and Behavior, a Review," *Aerosp. Med.* **42** (1971), 735–40.

[18] J. S. LEARY, ET AL., "Safety Evaluation in the Home of Polyvinyl Chloride Resin Strip Containing Dichlorvos (DDVP)," *Arch. Env. Health* **29** (1974), 308–14.

[19] A. W. DUDLEY, JR. AND N. T. THAPAR, "Fatal Human Ingestion of 2,4-D, a Common Herbicide," *Arch. Path.* **94** (1972), 270–75.

[20] J. A. BURTON, ET AL., "Absorption of Herbicides from the Rat Lung," *Arch. Env. Health* **29** (1974), 31–33.

[21] J. G. VOS, ET AL., "Toxicity of 2,3,7,8-Tetrachlorodibenzo-*p*-dioxin (TCDD) in C57B1/6 Mice," *Tox. Appl. Pharm.* **29** (1974), 229–41.

[22] R. D. KIMBROUGH, "Toxicity of Chlorinated Hydrocarbons and Related Compounds: A Review Including Chlorinated Dibenzodioxins and Chlorinated Dibenzofurans," *Arch. Env. Health* **25** (1972), 125–31.

[23] G. M. COPLAND, ET AL., "Fatal Pulmonary Intra-Alveolar Fibrosis After Paraquat Ingestion," *N. E. J. Med.* **291** (1974), 290–92.

[24] "The Paraquat Puzzle," *Med. J. Australia* **2** (1974), 800–801.

[25] L. G. DALES, "The Neurotoxicity of Alkyl Mercury Compounds," *Am. J. Med.* **53** (1972), 219–32.

[26] M. C. SHURTLEFF, *How to Control Plant Diseases . . . in Home and Garden,* 2nd ed. (Ames, Iowa: Iowa State University Press, 1966), pp. 100–106.

[27] A. R. RASUL AND J. M. HOWELL, "The Toxicity of Some Dithiocarbamate Compounds in Young and Adult Domestic Fowl," *Tox. Appl. Pharm.* **30** (1974), 63–78.

9

SENSITIZATION AND DERMATITIS

Materials that contact the skin may have little or no deleterious effect, or they may cause dermatitis in two quite different ways. First and most familiar is that called *primary skin irritation,* typical of a solvent (defatting) rash or the burn produced by hydrofluoric acid. Second is the phenomenon of *sensitization* in which two outwardly similar people can have different reactions from contact with a material. There may be no effect at all, or a weeping, blistery rash may cover great areas of the body; there may also be a full-blown attack of asthma, which is not a skin reaction at all. Because it is the most perplexing problem, sensitization will be discussed first.

Hypersensitization

Nearly everyone knows someone or has heard of someone who can treat poison ivy with impunity—who can pull the vines out by the roots, who perhaps can stand in the smoke of a fire burning the plants, or who can put the

141

leaves in his mouth and chew them, all without perceptible effect. Unfortunately, the body's immune response system does not respect bravado, and many of the people who boasted of being too tough for poison ivy found one day by breaking out in a rash that they, too, were susceptible. This is one aspect of sensitization and one aspect of dermatitis.

Sensitizers (also called *allergens*) are chemical agents that are normally harmless to the majority of people but when they are brought into contact with some individuals, they produce a disturbance of the normal physiological state. People who are so affected by an agent are generally referred to as *hypersensitive* or *allergic* to that agent. Poison ivy, or rather the chemical mixture called *urushiol*, is a sensitizer, and the reaction it produces is one kind of *dermatitis* (skin irritation). Not all materials are sensitizers, and not all reactions to sensitizers occur on (or in) the skin, but poison ivy is a common potent sensitizer and dermatitis is a common reaction to such materials.

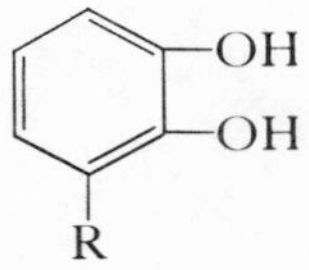

Urushiol

R may be any of five radicals (one of which is *n*-pentadecyl) that, when hydrogenated, completely yield *n*-pentadecyl $[-(CH_2)_{14}-CH_3]$.

The individual who found to his dismay that he could no longer handle poison ivy with impunity illustrates another very important fact about the phenomenon of sensitization. That is, in order for an agent to produce an allergic reaction, a prior sensitizing exposure is required, either with the same agent or with one closely related chemically. One who has become sensitive to poison ivy is probably also sensitive to poison oak and poison sumac although the mixture of R side chains in urushiol differs for the three materials.

Mechanisms of Allergic Reaction

There is no one simple explanation for the phenomenon of hypersensitization, and even the multiple, complicated explanations are still in the process of development by people working in that field of science called *immunology* (which is concerned with the operation of the systems that protect against a second attack of an infective agent). Hypersensitization, however, is known to be just one unwanted side effect of ways in which the body handles invasions of any foreign material.

According to at least one theory, the body has two immune systems, one concerned mainly with the production of antibodies to combat bacterial infections and the other concerned mainly with cell-mediated responses more closely associated with the phenomenon of inflammation than with antibodies [1]. Skin reactions such as the dermatitis of poison ivy and the respiratory tract reactions such as asthma associated with ragweed pollen are probably caused by operation of the second immune system, that associated with cells whose ancestors came from the thymus gland. Because of its importance to medicine and because of the relative accessibility of antibodies, however, the first (thymus independent, TI) system has been studied much more thoroughly than the second. Nevertheless, the systems are not independent of one another and an outline of the operation of the TI system may be sufficient to allow some feel for the complexities involved.

First, there must be an exposure to the allergen. If that material happens to be a protein (virus, fungus, bacterium, etc.), it in itself may be sufficient to elicit the next stage of response. In the occupational environment, however, very few allergens are proteins; some are as simple as ions (hexavalent chromium, for instance) or very uncomplicated organic molecules (formaldehyde, for instance) and these alone will not act as *antigens* (materials that cause or allow the formation of antibodies). Instead, the nonproteinaceous molecules (called *haptens*) probably combine with proteins already present in the body, altering their surface structure sufficiently so that the immune system can recognize the resultant as foreign—as *not self*. A combination, then, of a hapten and a body protein becomes an antigen [2]. (If the small foreign molecule does not combine with protein to form an antigen, it is not a hapten.) The body protein can come from any of the layers of the skin or mucosa, from the blood, or from any organ the hapten can reach.

If the amount of hapten is small, or if the hapten is not a very active allergen, the body may not respond at all to its presence. This may be because the right kind of body protein was not found to produce an antigen or simply because any antigen formed was in too small amounts or was too local to produce any further generalized response. On the other hand, if the amount of hapten is larger or if the exposure is frequently repeated (*frequently* in this case is relative to the disappearance of an antibody population and may be measured in weeks or months rather than hours or days), sufficient antigen may be formed to trigger the second stage of the system. Details of what is called the *second stage* here have not been worked out completely, but much more than a single stage of response is brought into play. One end effect of the second stage is that antibodies specific to the antigen begin to circulate throughout the body. Another result is that once this stage is complete, the body then has the capacity to mobilize great numbers of antibodies in a short period of time long after that second stage was triggered.

If the thymus-dependent system is the main one triggered, the number of circulating antibodies may be relatively low, but the capacity for large and rapid response will still be present [1]. With either of the immune systems, some time is required for the second stage to be completed after the process has begun. The necessary latent period varies somewhat but is always at least a few days and rarely more than 2 weeks. At the end of that time there will probably be no outward appearance that anything at all has changed, but a blood sample may find an increased concentration of certain cells (eosinophils—white cells that stain easily with the dye eosin—for instance) typical of this reaction.

Mediator Actions

Sometime after the latent period has elapsed, there may be a subsequent exposure to the hapten that initiated the whole process. If the subsequent exposure takes place while there is a plentiful supply of the specific antibodies, the hapten may react directly with them, forming hapten-antibody complexes. In addition, the hapten will probably combine with much the same kind of protein as it did in the sensitizing exposure, forming the antigen again. The antigen will react with antibodies, forming antigen-antibody complexes, and may also react directly with immunocytes (white cells involved in the formation of antibodies and perhaps in other phases of the immune process) and other cells, especially those at or close to the site of contact. Through a process that still is obscure, the hapten-antibody or antigen-antibody complexes or both cause the disruption of several kinds of cells to release into the tissue histamine and similar materials that cause many of the manifestations of the reaction known as inflammation. Histamine and other mediators of this response increase capillary permeability, dilate arterial vessels, constrict venous vessels and smooth muscles, increase the number of eosinophils, activate immunocytes, and convert lymphocytes (a type of white cell) to other types of cells.

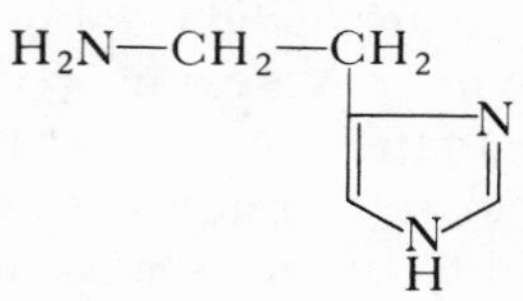

$$H_2N—CH_2—CH_2$$

Histamine

Because of the dilation of blood vessels and increased permeability of capillary walls, tissue at and near the inflamed site becomes swollen and flushed; if the site is the skin, it begins to itch. If the site of contact is the upper respiratory tract, constriction of smooth muscle causes a narrowing of bronchi, because bronchi and *bronchioles* (small bronchi) are encased in smooth muscle. This happens in addition to the normal swelling associated with inflammation with the result that there may be a sudden and dramatic closure of air passages typical of an asthmatic attack. If the supply of hapten and/or antigen is large, the subsequent major release of mediators may trigger a whole-body response, called *anaphylactic shock,* that can be fatal. Typically, death is associated with respiratory failure.

One of the most important facts associated with the concept of hypersensitization is that the response is caused by histamine and related substances produced by the body. The response is triggered by the allergen, but its manifestations are always those associated with the histamine-like mediators. Therefore, a physician cannot tell by looking whether a rash was caused by poison ivy or by formaldehyde. In each case the rash is simply typical of a sensitization response.

Contrary to popular belief, fluid from the blisters that can be associated with a dermatitis of this type will *not* spread the rash. If the rash appears at a point distant from the known contact, then there was probably a secondary contact with the allergen itself; the delay in appearance may be simply because the supply of allergen at that point was much less than that associated with the primary target area. Even in hypersensitivity, response is proportional to dose.

Increased and Reduced Sensitivity

Further contacts with the allergen beyond that which elicited the response are very likely either to put the immune system into a heightened state of readiness or, if the contacts are very minor and frequently repeated, to reduce the state of readiness. In the first case, the amount of allergen required to trigger the response becomes smaller and smaller and/or the response to a given amount of allergen becomes ever greater, involving more tissue with more severe inflammation. In the second case, the phenomenon of *desensitization* (or *hyposensitization*) occurs in which the whole process is gradually turned off.

Reactions of the immune system change with age or probably with turnover of the cells responsible for the response to a specific hapten or antigen. If the sensitizing dose was encountered early in life, and the exposure was never repeated, then the chances are good that the person can or will outgrow his reactions to the substance. If the sensitization takes place well after puberty, however, the chances are good that the hypersensitivity will persist, perhaps decreasing in potential intensity of response with age.

Antihistamines

One of the more successful concepts of *chemotherapy* (the use of drugs for treatment) was that since the response of the immune system is mediated through histamine and histamine-like substances, it might be possible to alter the response by interfering with the action of histamine.

Antihistamines are drugs that were developed to compete with histamine for sites on enzymes and other receptors. If those sites are occupied by a drug that induces no response, then even though histamine has been released and is present, it cannot occupy the receptor sites to cause its typical responses. The antigen-antibody complex still causes the release of histamine but if an antihistamine is present simultaneously or beforehand, at least some of the responses associated with histamine will be blocked.

Use of antihistamines does not confer unmixed blessings. The inflammatory response was selected by nature over the long process of evolution and (in some diseases) inflammation is necessary if a cure is to ensue. Even in cases of allergy, use of antihistamines only blocks some of the manifestations of histamine, and then perhaps partially. Antihistamines, for instance, have essentially no effect on the progress of a sensitization; they do not desensitize nor do they prevent a sensitization from becoming more and more severe.

Cross-Sensitization

A person who has developed an allergy to a particular material may have also developed an allergy to several substances usually chemically similar to the initial sensitizing agent. The extent of cross-sensitization that develops is a function not only of the allergen but also of the state of the person's immune system.

Some people develop many more cross-sensitizations than others, just as some people become hypersensitive to more materials than others do. Whether a person will become sensitive to a particular allergen is not predictable with the present state of knowledge nor is the extent of cross-sensitizations that will be experienced by a sensitized person.

Factors Influencing Allergic Disease

One of the perplexing problems encountered by the industrial physician or hygienist is why some people appear to react much more vigorously to materials that contact the skin than others do. The other side of that coin is exemplified by the query, "How can I select a work force that will be naturally resistent to skin problems?" Although the question is easy to ask and even relatively easy to answer, implementation of the answer may call for techniques well beyond those usually employed in personnel departments.

A number of factors are known to affect the development of dermatitis; some pertain mainly to allergic contact dermatitis and others pertain even to primary skin irritation.

A. Psychosomatic factors. Emotional upset will not cause an allergic disease of the skin or of the respiratory tract or of any other organ or organ system. Nevertheless, emotional upset may be the actual trigger that precipitates an attack of asthma or the development of a dermatitis. Furthermore, an emotional upset may aggravate an allergic disease, and an allergic disease may cause emotional upset, setting up the proverbial vicious circle that can be very difficult to break. Moreover, emotions may aggravate not only sensitization dermatitis but also just about any skin condition ranging from acne to psoriasis.

B. Fatigue. Akin to emotion, both chronic and acute fatigue can be involved in or trigger skin problems.

C. Other toxins. Several drugs routinely used for disease therapy can cause dermatitis and/or increase the intensity especially of a hypersensitization response. Both alcohol ingestion and tobacco smoking can increase the severity of some reactions.

D. Sex. Women have far fewer skin problems in the usual industrial environments than do their male counterparts who may perform the very same jobs. One reason for this is undoubtedly that women are more fastidious than are men and therefore keep their skin much cleaner. Others may relate to sex hormones and/or to the generally thicker subcutaneous fat layer in women.

E. Age. In animals as well as in man, the young are more suceptible to dermatitis-causing agents than the old or even middle-aged. This phenomenon is probably found most often with sensitizers but may appear with primary irritants as well.

F. Extraoccupational contacts. The occupational environment is not the only one in which people may contact skin irritants. Consider the man who patches his car with an epoxy resin formulation or the woman who wears nickel-plated earrings or other jewelry, for instance. Not all dermatoses are of occupational origin.

G. Weather. People tend to have more skin problems from an occupational source in the summer than in the winter. This probably is related to the fact that in summer they wear less clothing and thus more skin area is available for easy contact. In addition, perspiration itself can cause or aggravate frictional irritation and can give microorganisms several of their needed nutrients. Copious perspiration, on the other hand, tends to wash the skin of foreign matter and thus can be protective.

H. Skin color. People with dark skin have fewer dermatoses from any source than people with fair skins. This relationship refers to natural skin color and not that associated with tanning, which does give protection against sunburn and may confer some benefit even against other agents. The benefits of tanning, however, seem small in comparison with the possibility of skin cancer caused by excessive ultraviolet radiation (see Chapter 10). Protection from dermatitis associated with increasing darkness of skin color appears to be a continuous function and is not a property only of black or nearly black skins.

In sum, to best avoid occupational dermatitis, all workers should be chosen from among the group of black females who work only in the fall and winter and then not hard enough to perspire; who avoid alcohol, tobacco, and other drugs; who are emotionally stable; who are very neat and fastidious; who are over the age of 50; who wear no jewelry or fabrics containing wool; and who have no hobbies or other possible extraoccupational contacts with materials capable of causing dermatitis.

Common Allergic Disorders

By far the most commonly encountered allergic disorder is hay fever, which has no association with occupation but which is caused by pollens and molds airborne in the fall of the year. Along with the typical hay fever signs of running nose (*rhinitis*), red eyes, and puffy face, there also may be the dyspnea associated with asthma.

The occupational disease that has the closest relationship to hay fever is *farmer's lung,* an allergy manifested in the respiratory tract and caused by handling moldy fodder or hay. There is some evidence that hay dust is the true causative factor, perhaps augmented by mold (especially *Micropolyspor faeni*) in the early inflammatory stages [3]. Symptomology of farmer's lung is similar to that of metal fume fever with dyspnea followed by chills and fever [4]. It is usually diagnosed from symptoms and by demonstrating the presence of circulating antibodies specific to the mold.

Contact dermatitis (skin irritation resulting from direct action of the causative agent on the skin) is by far the most prevalent health problem associated with occupations. Although much dermatitis is the result of contact with primary irritants, a large proportion is still the result of true sensitization. In most contacts with sensitizers there will be no demonstrable changes in the skin after the first or even the first few. Several days after the contact or series of contacts that actually trigger the reacting mechanism, however, a further contact will cause the development of a dermatitis. Severity of the problem is extremely variable, depending on both the agent and the responder, but the pattern is usually that of

    A. Reddening of the skin (*erythema*)
    B. Formation of *vesicles* (tiny blisters)
    C. Oozing or weeping when the vesicles break
    D. Swelling of the affected tissue
    E. Formation of scales
    F. Itching

This pattern holds true no matter what allergen is contacted because the response is mediated through the immune system. Some materials that are sensitizers are also primary irritants, however, thus altering the pattern somewhat. Typical materials in this latter catagory are formaldehyde, some derivatives of phenol, and some of the compounds of hexavalent chromium and divalent mercury.

Once a person has been sensitized, his response to the allergen may become more severe with each subsequent contact or it may alter in severity only very slowly. The only sure way to prevent the allergic response is to avoid contacting the offending substance.

Diagnostic Criteria

In the late 1800s, Robert Koch in Germany formulated a series of postulates or criteria that must be met to prove that a particular microorganism is the cause of a certain disease. He was concerned with tuberculosis, but since that time his postulates have been modified to pertain to everything from air pollution to dermatitis. His original statements were [5]

> A. The micro-organism should be present in all cases of the disease, its distribution within the body being in general conformity with the nature and site of the pathological lesions.
> B. The micro-organism should be isolated from the diseased tissues and grown in pure culture outside the body for a number of generations.
> C. Administration of the micro-organism in pure culture to a susceptible animal should give rise to the disease in question, and the micro-organism should be recovered from the tissues of the diseased animal.

Modified for use in dermatology, these three simple statements become

A. An occupational dermatosis is one in which the role of an occupational causal factor has (at some previous time) been established beyond reasonable doubt.
B. The person has been working in contact with an agent known to be capable of producing similar changes in the skin.
C. The time relationship between exposure to the agent and the onset of the dermatosis is correct for that particular agent and that particular abnormality of the skin.
D. The site of the onset of the cutaneous disease and the site of maximum involvement are consistent with the site of maximum exposure.
E. The lesions present are consistent with those known to have followed the reputed exposure or trauma.
F. The person is employed in an occupation in which similar cases have previously occurred.
G. Some of the person's fellow workers using the same agent are known by the examiner to have similar manifestations due to the same cause.

H. As far as the examiner can ascertain, there has been no exposure outside of occupation that can be implicated.
I. If the diagnosis is dermatitis, the following items are important:
   a. Attacks after exposure to an agent followed by improvement and clearing after cessation of exposure constitute most convincing evidence of the occupational factor as a cause.
   b. The results of patch tests performed and interpreted by competent medidcal personnel corroborate the history and examination in the majority of cases.

These criteria, with the exception of a few words, were first presented by a committee of dermatologists in 1942 [6]. They further stated that it is not necessary or even likely that all the criteria would be met in each individual case of occupational dermatitis. Nevertheless, it is quite apparent that establishing the cause of an isolated rash or other skin disease sign may be a difficult undertaking. The most potent aids the dermatologist has at his disposal are allergy tests.

Allergy Tests

For the most part, allergy tests depend on either finding evidence of circulating antibodies or finding tissue reactions to the offending substance. Circulating antibody precipitin tests are well developed [2] but quite often do not apply to cases of allergy manifested in the skin simply because very few, if any, circulating antibodies are produced in reaction to the allergen. Cases of hypersensitization to occupational agents are usually diagnosed with the help of one of the tissue-reaction tests.

Several tissue tests have been used in which the skin is scratched or the material is injected just under the skin surface, but by far the most widely used is the patch test [7]. Here the physician simply applies a small portion of the possible allergen to the skin and then covers the area with a patch of gauze for a predetermined period of time (up to about 24 hr). The area under the patch is then examined and any responses are graded, usually on a numerical scale. From the extent of reaction, the dermatologist can determine his patient's degree of sensitivity to that material.

Of course, in the usual case the dermatologist does not work with one material and one patch at a time. In fact, he may well have a whole series of materials at the correct test concentrations available and may apply many of those at the same time, being careful to note the location of each. Furthermore, patch testing can be used not just to diagnose hypersensitivity to particular agents but also to determine just what potential any particular agent has for sensitizing a population of people.

Every now and then an employer working with materials capable of causing sensitization reactions has an inspiration. Having heard of patch testing, he would like to use it to determine the degree of sensitivity to the allergens prior to hiring new members of his work force. To this end he may persuade a physician to patch test all potential employees with most or all the substances they will be in contact with in the plant. He or the physician then will use results of the testing to decide whom to hire and whom to avoid. Logical as this approach appears to be, it nevertheless is fraught with problems, not the least of which is the fact that the patch test itself may in some cases be the sensitizing exposure [8]. That is, the person would not be sensitive to the agent if he had not been patch-tested. Because of the latent period required between the sensitizing exposure and the exposure that shows a reaction, the person sensitized by the test will probably show no reaction to the substance at all because 24 hr is not sufficient time for the typical tissue reactions to develop. Preemployment patch testing is more of a trap than an aid; it should be avoided.

Diagnostic patch testing, on the other hand, if performed skillfully can provide large amounts of information with little added risk. This is so because diagnostic testing is not done unless there is an apparent problem that must be solved; the people involved are already exposed to a greater possible hazard than that posed by the test itself. All such testing must, of course, be done by a physician skilled in the art—and it is an art—if only because of the possible hazards of the testing procedure.

Allergic Potential

Not all materials capable of sensitizing man have the same ability to sensitize; there is a spectrum of allergic potential ranging from near zero to near certainty. Furthermore, sensitization is a very individual response and no two people are likely to react in identical ways to all sensitizers and no animal species reacts at all like man. A fairly large amount of allergic potential testing is done with guinea pigs because that species appears to have an immune system that functions in a manner similar to man's. Lack of a guinea pig reaction, however, does *not* mean that man will not react to the substance nor does a positive reaction imply that all men will also react. Instead, if guinea pigs do react, the substance is usually assumed able to sensitize man; if guinea pigs do not react, the toxicologist may suggest that cautious human trials are warranted.

Urushiol from poison ivy, oak, and sumac; oil of wintergreen (methyl salicylate); and *p*-phenylenediamine are all substances that have high allergic potential. With sufficient contact, most of a human population could probably be sensitized to all of them and *sufficient contact* means very small amounts in most cases. On the other hand, boric acid, cotton fabric, and rice rarely cause sensitizations even after extremely long contact. Most materials encountered occupationally fall between these extremes.

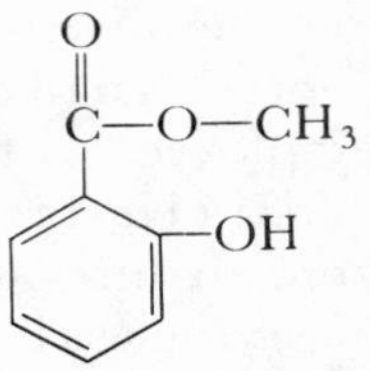

Methyl Salicylate

p-Phenylenediamine

The incidence of positive reaction to patch testing was the subject of two papers in 1973 [9, 10]. Materials found to induce reactions most often among the patient populations and that appeared among the top ten on both lists were: nickel sulfate, potassium dichromate, p-phenylenediamine, ethylene diamine, and turpentine oil. Other materials among the top offenders on one of the lists were mercuric chloride, mercaptobenzothiazole, thimersal (Merthiolate), formaldehyde solution, poison ivy, neomycin sulfate, and bismarck brown. Materials on these lists are not necessarily those that are chief offenders in industry as several drugs are included.

A literature survey does not suffice to reveal which among all materials cause the worst problems occupationally. People (mainly physicians) who write papers on this subject are prone to emphasize findings that differ from the usual experience rather than the norm and therefore the list in Table 9-1 has little relationship to what is usually found. Nevertheless, it does give

Table 9-1   Some of Man's Sensitizers

Benzalkonium chloride	Hydroxylamine hydrochloride
Benzophenone	Isopropanol
Benzoyl peroxide	Kerosine
Benzoyl salicylate	Lanolin
p-*tert*-Butyl phenol	Leather
Cashew nuts	Lichens
Castor seed dust	Lily
Cedar (wood)	Liverwort
Chromium (3^+ and 6^+)	Methylene diamine
Chrysanthemum	Moldy malt
Cobalt	Piperazine
Copper	Ragweed pollen
Corkwood leaves	Resorcinol monobenzoate
p-Diamino benzene	Hyacinth
Diethylthiourea	Sisal
Ethyl acetate	Subtilisins
Ethylene glycol methyl	Tea dust
ether acetate	Tulip
Fumaronitrile	Vitamine E
Gold	

some indication of the variety of materials which can sensitize man and which can be found in a rather casual survey of the relevant literature.

Primary Skin Irritation

The most prevalent industrial health problem appears to be primary skin irritation caused by cutting and lubricating oils. These materials are rather poor irritants and prolonged, repeated contact is usually necessary to cause any kind of a reaction. The rash that may then appear is usually a red, roughened area of skin that may be painful or itchy. Usually the hands are affected most seriously although any skin area that is uncovered or in contact with wetted clothing may also be involved.

Most of the oils are mixtures of hydrocarbons ranging from the naphthas (hexanes and highers) to the heavy lubricating oils. When the oils are used for cutting, they are normally emulsified with water using various soaps or detergents for the emulsification. Additives for various purposes may be present and may be responsible for the dermatitis. Very occasionally, the oils may prove to be reasonably good culture media, providing for the growth of bacteria or fungi that then may cause effects on the skin.

Strong acids and bases cause burns within a few seconds or minutes of contact unless they are very rapidly removed. Because rapid removal is so important, the only good decontaminating agent usually available is running water, which should be used copiously with rubbing and/or scrubbing for 10 or 15 min to assure complete removal.

Among the most difficult burns to treat are those caused by hydrofluoric acid. The fluoride ion adheres tenaciously to protein and in addition is quite toxic to cells. Treatment of HF burns has been investigated and prompt application of a calcium gluconate solution appears to be rational [11, 12]. Burns from other materials do not usually require any material-specific treatment.

Protective Measures

Two methods of protecting people against skin allergens are specific to that problem, namely desensitization and barrier creams. Other measures include the use of protective clothing and respirators.

Rarely used in industry, desensitization involves frequent exposure to the offending substance in extremely minute concentration at first, with the concentration gradually being increased over a period that may range up to a few years. At the end of that time the person will be hyposensitive to the material if the treatment has been successful. Unfortunately, this procedure often fails so that at some point in the regimen the person reacts violently to the material and the whole process must be restarted after the reaction has been controlled. Nevertheless, hyposensitization treatment has had some success against poison ivy and other potent allergens and therefore remains at least a possibility [13].

Barrier creams or ointments have had a long and stormy history. They have been developed for use by people who either cannot or will not wear impervious clothing such as gloves, sleeves, and the like. In theory, the cream is applied to the skin prior to exposure and then acts as a barrier against the offending material. Oily or greasy materials that are not wet by water are used as barriers against water solutions of various chemicals; substances that are not wet by oily materials are used as barriers against them. In some cases, creams have been developed that are said to be barriers against water and oily materials simultaneously.

One way in which barrier creams protect is that they may well cause the user to wash his skin much more frequently than he would otherwise. In this manner, any contact with offending agents will be limited in time. All types of creams and lotions have this side effect that may, indeed, be the main effect in some cases.

No cream, lotion, or ointment is as effective in guarding the skin from contact as an impervious glove, and some of the creams on the market offer no protection at all; most give some protection if they are used properly. In general, those barrier materials claimed to offer the widest range of protection are the least effective for any particular type of offending agent. Those that have been developed for a specific purpose are likely to be most effective for that purpose. That kind of a generalization is also valid when protective clothing is considered. There is no universal impervious material that can be worn to protect against all hazards but materials have been developed to protect against almost any conceivable hazard (consider a space suit, for instance).

Quite often the best way to protect people is to arrange the equipment and/or process so that there will be essentially no exposure, thus obviating the need for personal protective equipment of any kind.

Measurement of the amount of solvent or other material in the breath after exposing a known area of skin to the material is one of the best ways available of determining the rate of penetration of the material through the skin [14]. This technique has been used to evaluate the efficacy of barrier creams under use conditions [15]. A body of knowledge could be built up that would allow not only the selection of efficacious barrier creams but also the development of better barriers for specific (groups of) materials by widespread application of this biological test method.

REFERENCES

[1] *Biology Today* (Del Mar, California: Communications Research Machines, Inc., 1972), 437–55.

[2] T. D. BROCK, *Biology of Microorganisms*, 2nd ed. (Englewood Cliffs, New Jersey: Prentice-Hall, Inc., 1974), 517–36.

[3] S. H. ZAIDI, ET AL., "Experimental Farmer's Lung in Guinea Pigs," *J. Path.* **105** (1971), 41–48.

[4] D. C. MORGAN, ET AL., "Chest Symptoms and Farmer's Lung: A Community Survey," *Brit. J. Ind. Med.* **30** (1973), 259–65.

[5] "Bacterial and Infectious Diseases," *Encyclopaedia Britannica* **2** (1957), 893–99.

[6] "Industrial Dermatoses, A Report by the Committee on Industrial Dermatoses of the Section on Dermatology and Syphilology of the American Medical Association," *J. Am. Med. Assn.* **118** (1942), 613–15.

[7] B. C. KORBITZ, "Possible Improved Technique of Skin Patch Testing," *Arch. Env. Health* **27** (1973), 409–11.

[8] G. AGRUP, "Sensitization Induced by Patch Testing," *Brit. J. Derm.* **80** (1968), 631–34.

[9] R. L. BAER, D. L. RAMSEY, AND E. BIONDI, "The Most Common Contact Allergens," *Arch. Derm.* **108** (1973), 74–78.

[10] N. AMERICAN CONTACT DERMATITIS GROUP, "Epidemiology of Contact Dermatitis in North America," *Arch. Derm.* **108** (1973), 537–40.

[11] T. D. BROWNE, "The Treatment of Hydrofluoric Acid Burns," *J. Soc. Occup. Med.* **24** (1974), 80–89.

[12] S. A. CARNEY, ET AL., "Rationale of the Treatment of Hydrofluoric Acid Burns," *Brit. J. Ind. Med.* **31** (1974), 317–21.

[13] W. L. EPSTEIN, ET AL., "Poison Oak Hyposensitization. Evaluation of Purified Urushiol," *Arch. Derm.* **109** (1974), 356–60.

[14] R. D. STEWART AND H. C. DODD, "Absorption of Carbon Tetrachloride, Trichloroethylene, Tetrachloroethylene, Methylene Chloride, and 1,1,1-Trichloroethane Through the Human Skin," *Am. Ind. Hyg. Assn. J.* **25** (1964), 439–46.

[15] M. GUILLEMIN, ET AL., "Simple Method to Determine the Efficiency of a Cream for Skin Protection against Solvents," *Brit. J. Ind. Med.* **31** (1974), 310–16.

10

CARCINOGENESIS

Carcinogenesis is the production of cancer. It is a process that operates at the subcellular level and therefore a sketchy review of microbiology is appropriate.

The cell (Fig. 10-1) is the basic biological unit of structure and function in the living body. The *eucaryotic* (having a nucleus) cell is composed of a darkly staining compact nucleus, a *cytoplasm* (literally, cell tissue or cell protoplasm), which surrounds the nucleus, and a membrane, which encloses the whole structure. Many other components of the cell exist but are ignored here.

The nucleus is the genetic control center of the cell; processes within the nucleus regulate the structure, function, and reproduction of the cell. Within the nucleus are *chromosomes* (literally, colored bodies, referring to the fact that they stain darkly) which are made up of discrete strands of deoxyribonucleic acid (DNA) forming *genes* (biologic units of heredity) which carry the genetic orders. Human cells each have 46 chromosomes; the number per cell varies quite widely among living things.

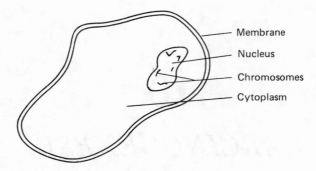

Membrane
Nucleus
Chromosomes
Cytoplasm

Figure 10-1 The cell.

Genetic DNA of the cell controls the manufacture of special *enzymes* (proteins that catalyze biochemical reactions) that are necessary for the cell's function. Another nucleic acid, ribonucleic acid (RNA), is used by the cell in the synthesis of protein and in the transfer of genetic information. Genes exert a strict control over the process of development; they discipline and restrain growth of individual cells within the body.

Definition of Cancer

Cancer is the *malignant* (harmful), unrestrained, and undisciplined growth of cells. Once cancer cells are established, they differ from their tissue of origin in that they have become independent of the control mechanism that ordinarily limits cell growth and division (*mitosis*) in normal differentiated tissue. In general, cancer cells continue to carry on functions of the tissue from which they came; their unrestrained growth due to a lack of normal *contact inhibition* (contact with other cells somehow keeps normal cells from growing without limit) is their common characteristic.

Cancer is derived from the Greek word *karkinos,* which means *crab*. Among the many synonyms are *malignant tumor* and *malignant neoplasm* (literally, harmful new tissue). Subgroups of cancer include carcinoma, lymphoma, sarcoma, melanoma, and many others. The prefix usually indicates the type of tissue in which the cancer arose, while the suffix *-oma* means swelling or tumor.

Cancer cells can continue their cycles of growth inexorably, disrupting the normal relationships between adjacent cells, invading capillaries and lymph channels, and hence being carried throughout the body, eventually killing the host. The process of malignant spreading is called *metastasis*. Some types of cancer are much more prone to metastasize than others; skin cancer, for instance, rarely involves other body organs but liver cancer almost always does.

Benign tumors are clumps of cells that grow in a limited area and do not spread throughout the body; they stay in one place and just grow. Cells at the surface of benign tumors appear to be restrained in their growth by contact with other kinds of normal cells. Not contacting normal cells, the ones inside the tumor have no such restraint and continue to multiply. Benign tumors are harmful only when they press against other organs and disturb the normal functions of those organs. Benign tumors are not cancer because they do not spread out and invade other organs and tissues; they do not metastasize.

Environmental Carcinogenesis

A *carcinogen* is any agent that has been shown to cause cancer. Carcinogenesis is the production of cancer in living cells by any (unspecified) mechanism. *Environmental carcinogenesis* is the production of cancer by agents from the environment external to the host. The field of environmental carcinogenesis is concerned with identification of carcinogenic hazards, elucidation of their modes of entry into the body, the mechanism of their actions within the host, and the characteristics of populations affected by them.

Both the *incidence* (rate of occurrence of new cases) and the *prevalence* (the number of cases at any one time) of many types of cancer vary considerably among subgroups of the human population. Variations are found as a function of occupation [1], site of residence, genetic history (race), socioeconomic class, and many other variables.

Factors Influencing the Development of Cancer

Occupation. The first occupation-related cancer identified as such was scrotal cancer among a large percentage of chimney sweeps in London as early as 1775. This cancer was related to the soot from coal burned in fireplaces, usually for home heating.

In the 1870s, miners in Germany were found to have lung cancer caused by inhaling radon from radioactive ore. This pattern has continued, with lung cancer being found in excess among uranium miners [2] and hematite miners [3] but not coal miners [4], again probably caused by the inhalation of radon.

During the early production of dyes based on aniline, workers in this industry were found to have a much greater than normal risk of bladder cancer, so much so that aniline itself was thought by many to be a cause. Aniline, however, is not a carcinogen; some related materials such as benzidene [5] (not to be confused with the drug Benzidrene) and 2-naphthylamine [6] are definitely carcinogens and were (and are) the cause of bladder cancer among those overexposed.

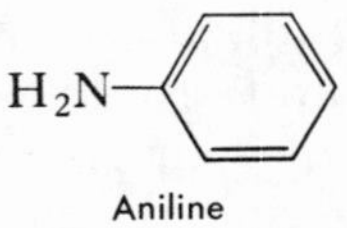

Aniline

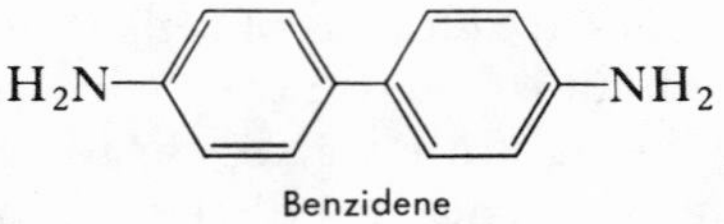

Benzidene

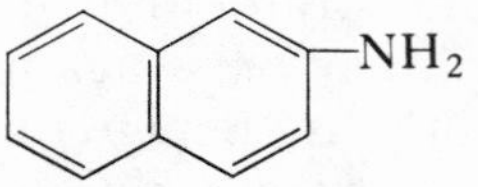

2-Naphthylamine

Radium ingestion and radon inhalation among radium dial painters was found to cause bone cancer. Lung cancer has been found among chromate workers [7], coke plant workers [8], people exposed to asbestos (especially when they smoke cigarettes) [9], bis(chloromethyl) ether workers, and those who produced mustard gas [bis(2-chlorethyl) sulfide] [10], during World Wars I and II. Vinyl chloride was found to be a carcinogen in the mid 1970s, [11, 12].

$$ClH_2C—CH_2—S—CH_2—CH_2Cl$$

Mustard gas

Site of residence. *Medical demography* (medical study of various populations) has shown that skin cancer (of the neck, especially) is more prevalent in western and southwestern areas of the United States than in other parts of the country. This cancer is probably caused by the high exposures to ultraviolet radiation experienced there because of the sunshine and the general outdoor living and working customs of the residents.

Respiratory cancer rates are higher in urban than in rural areas, especially in the United States [13]. The cause of this phenomenon has been attributed to many possible carcinogens [14] or to generalized air pollution as measured by one index or another [15]. Lung cancer prevalence rates around the world appear to be highest in England and Wales and lowest in Japan and Norway. In the United States, Green Bay, Wisconsin, has one of the lowest prevalence rates of lung cancer and Birmingham, Alabama, has one of the highest.

Part of the reason for the rather large site-of-residence effects may well be one or more air pollutants [16], but several studies have shown that air pollution as presently measured is not the whole answer. Consideration must also be given to regional diets, drinking water, ethnic mixture, and other factors that may influence the development of cancer in various body organs, including even the kind of clothing worn.

Ethnic origin. The difficulties of comparing cancer (or other disease) rates between countries are enormous, beginning with varying degrees of statistical validity of data. Notwithstanding, some comparisons can be made.

For instance, the prevalence rate of breast cancer in Japanese women is about one-fourth that among women in the United States. Furthermore, cancer of the stomach is much more frequent among people of the USSR, Poland, and the Scandinavian countries than elsewhere in the world. Cancer of the liver is very prevalent in some areas of Africa. Skin cancer rates among American blacks are less than one-fourth those of whites. Some of these differences may relate to inherited tendencies.

Socioeconomic class. Socioeconomic class is a variable that becomes *confounded* (mixed in) with several others such as site of residence, ethnic origin, occupational history, diet, smoking habits, and air pollution exposure. Perhaps because of the operation of the "other" variables, those in the lowest economic classes in the United States appear to have the highest cancer prevalence rates. Cancer rates also appear to be higher in countries with relatively poor public health practices and personal hygiene than in others, and this may reflect the class.

Smoking. The largest single factor related to cancer, in the United States at least, appears to be smoking history. Those who smoke cigarettes are much more likely to have and die of lung cancer than any other group. Smoking history is therefore one of the factors that tends to obscure the effect of other variables when the epidemiology of cancer is studied. No one knows for sure what weight to assign the kind of smoking material used (pipe tobacco, cigars, cigarettes, filtered cigarettes, long cigarettes, short cigarettes, short cigars, etc.), the frequency of smoking, whether or not the smoke is inhaled, or even such elementary data as how long a person has smoked or the age at which the habit was started or stopped. So many people smoke or have smoked that to ignore them is to ignore much possibly useful data. The problem is even more perplexing because smoking is related not only to the development of lung cancer but to many other diseases as well, so that this is a problem appearing in most statistical treatments of morbidity or mortality data, not only those related to cancer or diseases of the respiratory tract.

Carcinogens

There are three basic catagories of carcinogens, namely chemical, physical, and biological. Because so little is known about cancer and its origin, no list of carcinogens can hope to be exhaustive but must, rather, be illustrative.

Chemical Carcinogens

Almost all chemical carcinogens can be grouped into a few general classes of materials and this is done for the sake of convenience, not to imply that actions within a class are identical or even similar. Furthermore, classifications tend to give the impression that carcinogenesis is an all-or-none phe-nomenon at least with respect to materials that are its cause. Typically, however, when pathways have been partially elucidated, chemical agents act in concert with others, not alone, to produce cancer. Thus, any particu-lar chemical may not be at all active in this regard unless a cocarcinogen is present.

Polynuclear aromatic hydrocarbons. Polynuclear (or polycyclic) aromatic hydrocarbons (PAH or PNA) are present in crude petroleum, coal (and other) tars, and in the products of combustion of most organic materials. They are almost ubiquitous and are found, when sought, in samples of the ambient atmosphere for suspended particulates and even in the charred por-tion of a grilled steak. Not all PAH are carcinogenic, but several are so that this class of materials is always suspect. Several PAH known to have car-cinogenic activity at least by animal experiment are 3,4-benzpyrene (benzo[a]pyrene), most or all of the methyl-substituted benz[a]anthracenes (but perhaps not benz[a]anthracene, itself), the methyl-substituted benzo [c]phenanthrenes, and even the fluorobenzanthrenes and phenanthrenes [17]. The two lowest (or first) members of this (PAH) series, naphthalene and anthracene, have shown no carcinogenic activity, however.

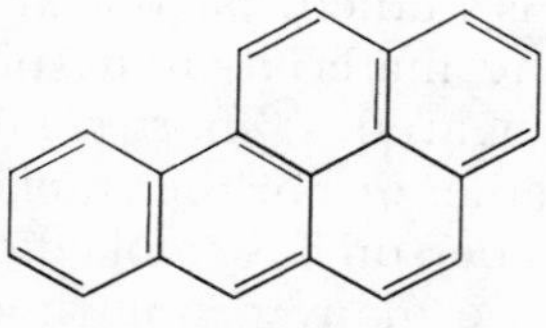

Benz[a]pyrene

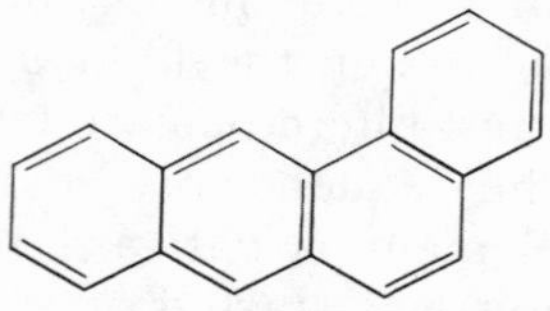

Benz[a]anthracene

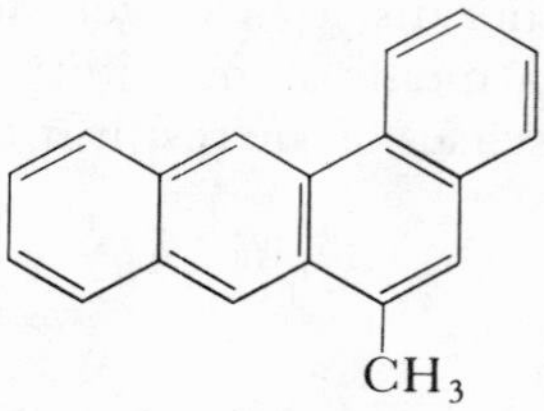

6-Methylbenz[a]anthracene

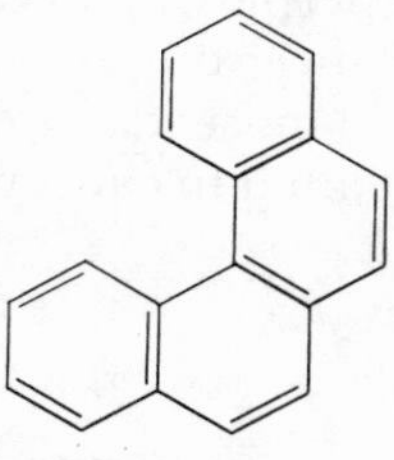

Benzo[c]phenanthrene

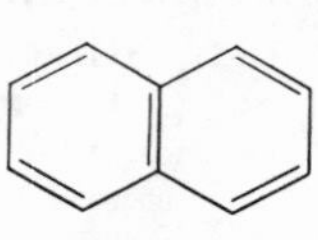

Naphthalene

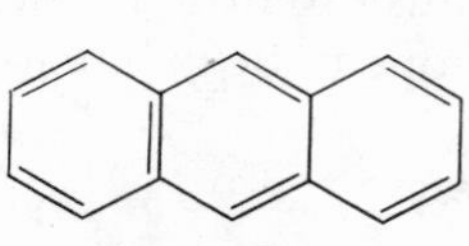

Anthracene

Other PAH strongly suspected of having the ability to cause cancer in man include dibenzanthracene and methyl cholanthrene. For the most part, the cancer suspicion comes from experiments done by painting the shaved backs of cancer-prone mice with the substance once or twice a week for several months and then determining whether or not there has been an increased incidence of tumors. As these materials are found in smoke and in suspended particulate air pollution [18], the suspicion is that they may cause lung or skin cancer in man, and possibly stomach cancer as well. Cases of human cancer caused by this class of compounds are rare in the literature, however.

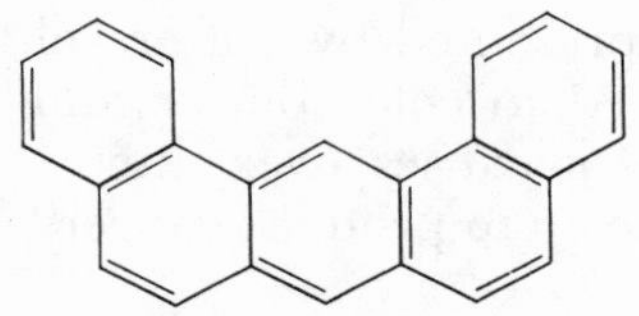

Dibenz[*a,j*]anthracene 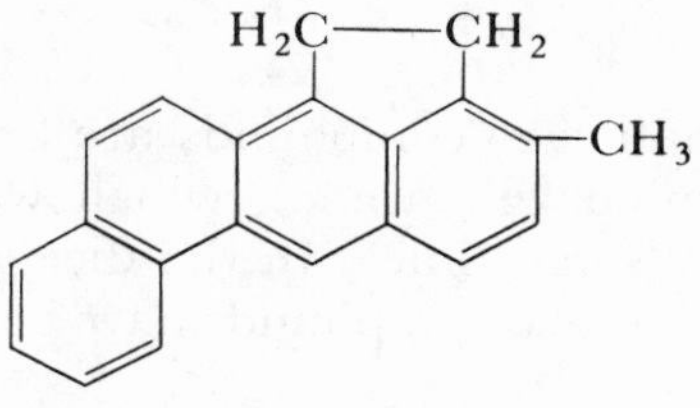   3-Methylcholanthrene

Mineral oil [19] and *n*-dodecane [20] have been shown to be cocarcinogens with several of the PAH, potentiating the action of these materials by orders of magnitude. As many experiments require that the active compound be in solution, and as mineral oil and the heavier aliphatic hydrocarbons are good solvents for the PAH, cocarcinogenicity may have been demonstrated unwittingly by some of the earlier experiments.

Nitrogen-containing compounds. Nitrogen is very active biochemically, and many of its organic compounds have been found to be carcinogenic in one or more species of animal and/or man. Subgrouping may tend to indict materials with little or no activity but may also prove useful. In this case, most of the known carcinogens are aromatic amines or amides, aromatic azo compounds, nitrosoamines, and carbamates.

Aromatic amines (or amides) that have caused cancer in man or are strongly suspected of having the potential include benzidene [5], 2-naphthylamine [21], 2-acetylaminofluorene (*N*-2-fluorenylacetamine), 4-aminobiphenyl, 4-aminostilbene, 4,4'-methylene bis(2-chloroaniline), and 3,3'-dichlorobenzidene. After the first three in this list, the evidence for human carcinogenicity becomes quite sparse, and the last two may not be carcinogens in man in doses that do not cause frank acute toxic manifestations [22, 23, 24].

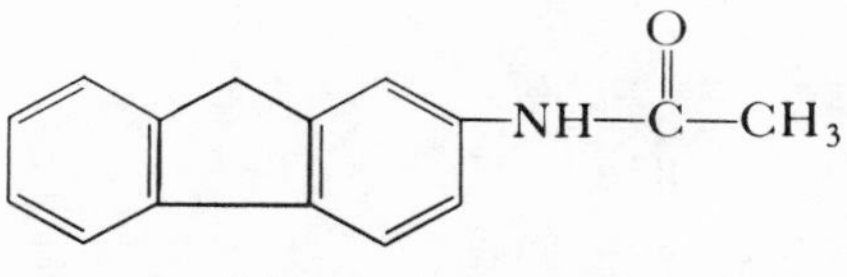

2-Acetylaminofluorene 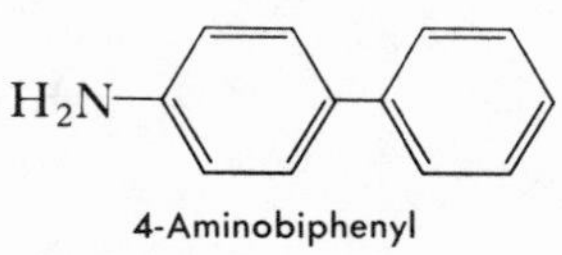   4-Aminobiphenyl

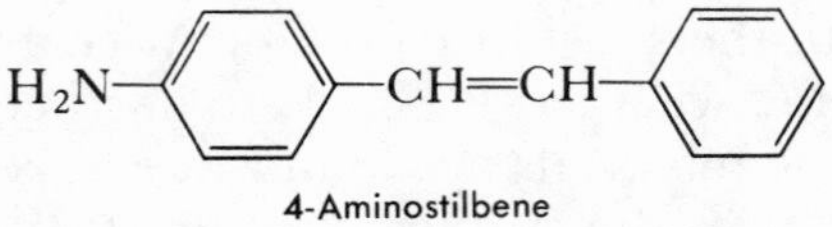

$$H_2N- \bigcirc -CH{=}CH- \bigcirc$$

4-Aminostilbene

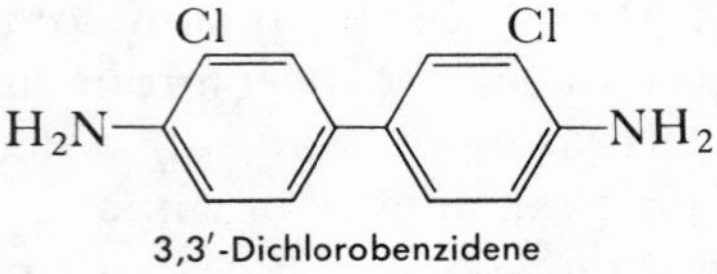

$$H_2N- \bigcirc - \bigcirc -NH_2$$

3,3'-Dichlorobenzidene

$$H_2N- \bigcirc -CH_2- \bigcirc -NH_2$$

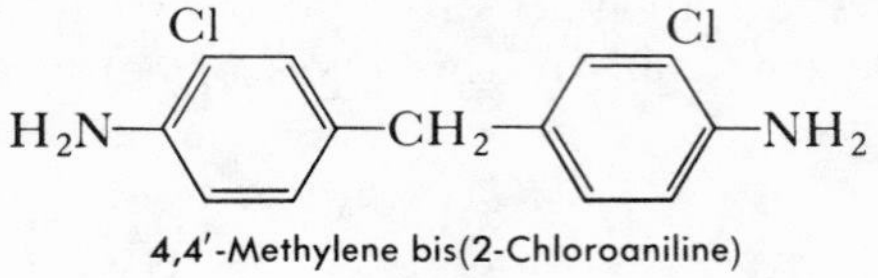

4,4'-Methylene bis(2-Chloroaniline)

Aromatic azo compounds are best represented by butter yellow (*p*-dimethylaminoazobenzene), which was used to color butter and oleomargarine for some time. In the diet of rats it causes liver tumors. Several other amino-azo compounds have been shown to produce cancers in experimental animals.

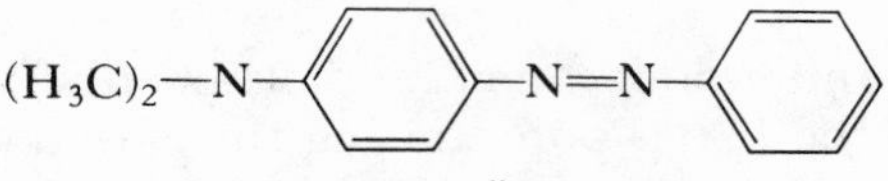

$$(H_3C)_2-N- \bigcirc -N{=}N- \bigcirc$$

Butter Yellow

Nitrosamines (in particular, *N*-nitrosodimethylamine) are methemoglobin-formers and active carcinogens in animals.

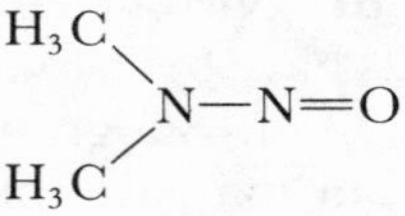

$$\begin{matrix} H_3C \\ \quad\quad\quad N-N{=}O \\ H_3C \end{matrix}$$

N-Nitrosodimethylamine

In the case of carbamates, ethyl carbamate (urethane) is the only member so far to have shown much carcinogenic activity. It can produce lung cancers in mice but it does not produce skin cancer when fed to or painted on mice. Neither does croton oil (a powerful vesicant found in the seeds of a certain plant). If urethane is fed to mice that then have their skins painted with croton oil, however, skin cancers are produced. This was one of the first confirmed cases of cocarcinogenesis. [Urethane (also called urethan) is a widely used anesthetic for laboratory animals.]

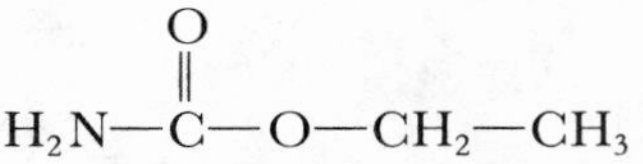

$$\overset{\displaystyle O}{\overset{\displaystyle \|}{H_2N-C}}-O-CH_2-CH_3$$

Ethyl Carbamate

Hexamethylphosphoramide has been found to be an extremely potent carcinogen in rats following chronic inhalation exposure to concentrations as low as 400 parts in 10^9 parts of air (see p. 96).

Reactive organic compounds. This category of materials has also been called *alkylating agents* but it contains several non-alkyl members so that the more general title seems appropriate. Included are the nitrogen mustards (from mustard gas), the diepoxides, the halo ethers, and vinyl chloride with perhaps other halogenated alkenes such as trichloroethylene as members.

The nitrogen mustards have formulas similar to that of mustard gas but are derivatives of nitrogen rather than sulfur. These materials can cause cancer in experimental animals, but their main reason for being is to treat cancers such as lymphomas. Two members of this group are mechloreth-amine hydrochloride (2,2'-dichloro-*N*-methyldiethylamine hydrochloride) and mechlorethamine oxide hydrochloride (2,2'-dichloro-*N*-methyldiethyl-amine *N*-oxide hydrochloride). Several other derivatives have been made, mainly by substituting other radicals for the methyl group, and tested in hopes of developing a good cancer chemotherapeutic agent.

Mechlorethamine Hydrochloride   Mechlorethamine Oxide Hydrochloride

Mouse skin painting tests of a large number of epoxy- and diepoxy-materials revealed four to have high carcinogenic activity [25]. They were butadiene dioxide, vinyl cyclohexene dioxide, 3,4-epoxy-6-methylcyclo-hexylmethyl-3,4-epoxy-6-methyl cyclohexane carboxylate, and "modified bisphenol diglycidyl ethers." None of the monoepoxides tested was found to produce cancer.

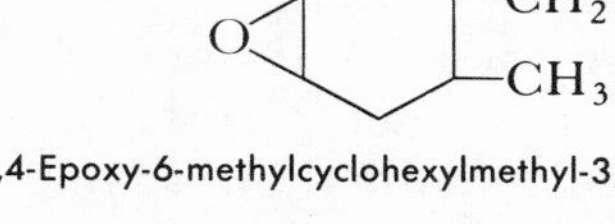

Butadiene Dioxide 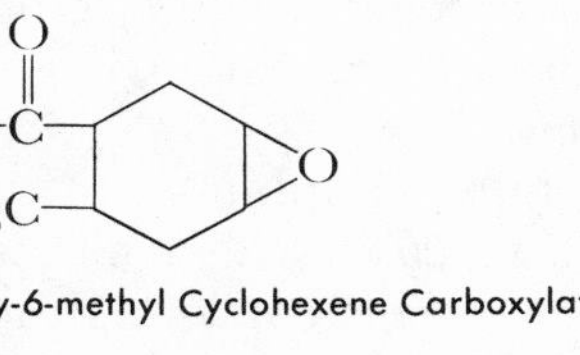   Vinylcyclohexene Dioxide

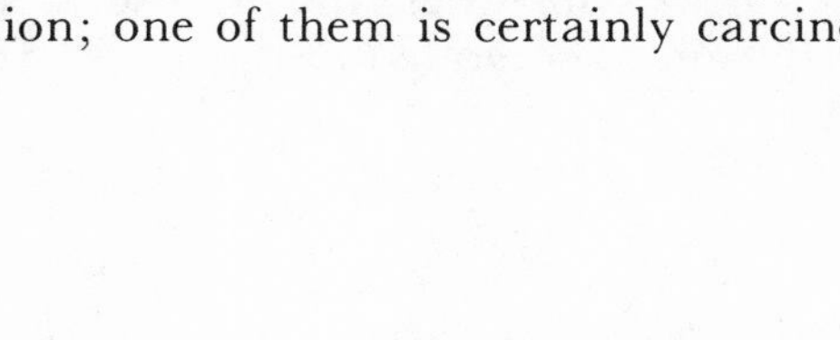

3,4-Epoxy-6-methylcyclohexylmethyl-3,4-epoxy-6-methyl Cyclohexene Carboxylate

Two halo-ethers have been found to be carcinogenic in animals by in-halation and injection; one of them is certainly carcinogenic in man [26]

and the other probably is [27]. The two are bis(chloromethyl) ether [28] (the more carcinogenic) and chloromethyl methyl ether [29]. Brominated and iodinated analogs are probably also carcinogenic and thus the term *halo-* rather than *chloro-* ethers. Both substances are used industrially in the modification of polymeric materials.

$$ClH_2C—O—CH_2Cl \qquad ClH_2C—O—CH_3$$

bis(Chloromethyl) Ether   Chloromethyl Methyl Ether

Discovery of the carcinogenicity of vinyl chloride [30, 31] is detailed in Chapter 7. Since this substance was found to cause cancer in man, several of its near relatives have been examined very closely. The first indication that other similar substances might also cause cancer was an announcement that trichloroethylene had been found to cause cancer when fed at high levels to experimental rats. As with the others, brominated and iodinated analogs of the chlorinated materials are strongly suspect.

At least one lactone, β-propiolactone, has been found to be carconogenic in experimental animals [32]. Suspicion was thus thrown on all lactones and the compound with a contained lactone structure with the greatest contact with people was easily determined to be penicillin. Penicillin was tested at extremely high doses and was, indeed, found to be carcinogenic but it is not carcinogenic at therapeutic levels in man or animals. Penicillin is not a single material but a mixture of several. All have the lactone linkage in tautomeric equilibrium with the acid form as illustrated below for penicillic acid.

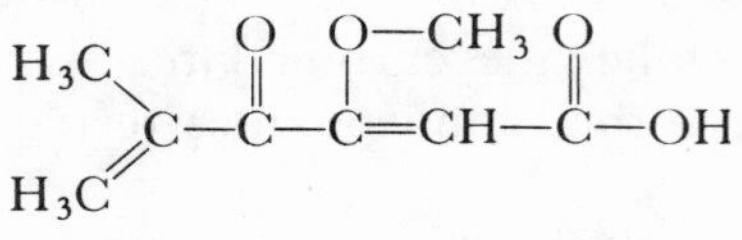

Acid Form 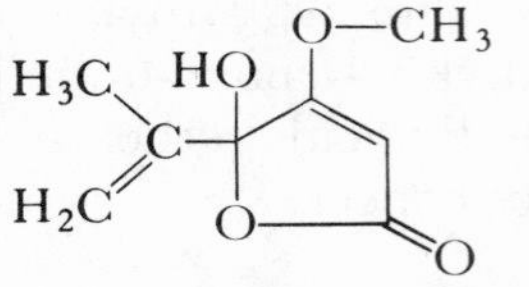   Lactone Form

Estrogens. The female sex hormones, including artificial compounds with estrogenic activity such as diethylstilbestrol (DES), produce a wide variety of cancers in rodents at sites distant to the site of application but in tissues that respond physiologically to these materials. Recent evidence appears to indicate that estrogens may also be involved in uterine and/or cervical cancer in women.

DES has been implanted in cattle and poultry to increase protein production and to cause other changes such as increased fat in chickens but decreased fat in beef cattle. Although this synthetic hormone is destroyed by cooking temperatures, its use in poultry was banned (in the United States) because of control difficulties and its use in beef and mutton is looked upon with suspicion by several groups.

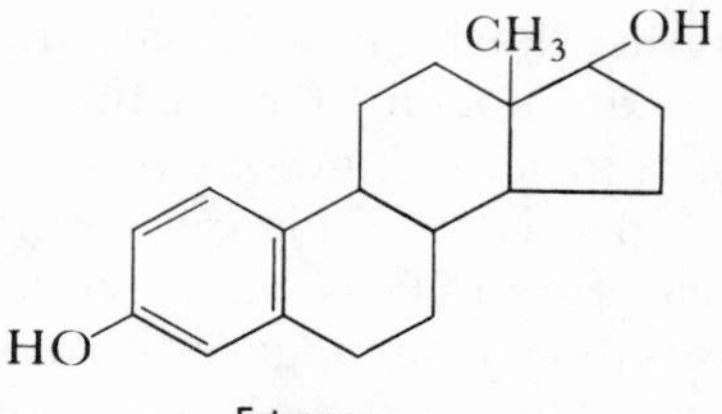

Estrogen

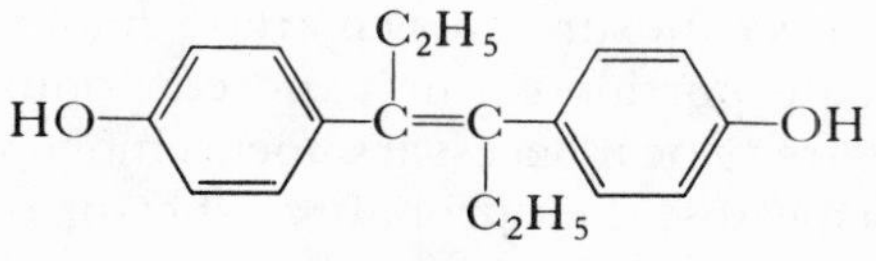

Diethylstilbestrol

Metals. The subject of metal carcinogens is discussed in Chapter 4. In summary, both chromium (in the hexavalent form and perhaps in the trivalent as well) and nickel are known to be carcinogens in man and animals. It appears that arsenic [33] and beryllium [34, 35] cause cancer in man; whether they can do so in the absence of other signs of disease is not so clearcut. At one time or another, cadmium [36], cobalt, lead, and even titanium have been said to be carcinogens but so far there has been little evidence to support the statements.

Asbestos. Carcinogenicity of asbestos is discussed in Chapter 3. All forms of asbestos appear able to cause mesothelioma and other kinds of cancer, especially of the lung and larynx [37] in people, particularly those who smoke cigarettes. Cancer may appear in the absence of other signs or symptoms of asbestos-related illness.

Cigarette smoke. In 1964 the Surgeon General of the United States reported on lung cancer and smoking: "The risk of developing lung cancer increases with duration of smoking and the number of cigarettes smoked per day and is diminished by discontinued smoking. In comparison with non-smokers, average male smokers of cigarettes have approximately a ninefold to tenfold risk of developing lung cancer and heavy smokers at least a twentyfold risk." Later, a law required every package of cigarettes and every advertisement of cigarettes to carry a warning about the health hazard of using that product. Cigarette advertising was barred from television. All this happened before anyone had demonstrated that cigarette smoke could cause lung (or other) cancer in experimental animals. In fact, multiple experiments prior to and since the Surgeon General's warning have failed to demonstrate the animal carcinogenicity of cigarette smoke.

If cigarette smoke is a carcinogen in man, then certainly some of the animal experiments—in fact, probably most of them—should result in cancer. Perhaps one of the reasons for failure of the animal experiments is that a cocarinogen is necessary. Almost invariably animals are exposed to air in which tobacco (usually cigarette) smoke is the only (albeit heavy) contaminant. On the other hand, man does not usually expose himself to as much smoke as the experimental animals must breathe but man also breathes air much more contaminated than that usually found in today's air-conditioned toxicology laboratories. Epidemiological studies have shown that, in all probability, asbestos alone is at best a weak carcinogen in man but that asbestos exposure plus cigarette smoking results in some kind of cancer in a large proportion of the exposed population. In this case, cigarette smoke appears to be a necessary cocarcinogen. If cigarette smoke is a cocarcinogen for asbestos, then probably asbestos is a cocarcinogen for cigarette smoke. If asbestos is a cocarcinogen for cigarette smoke, then other materials, possibly present to a greater extent in the ambient atmosphere of some urban environments than elsewhere in the world, may also act as cocarcinogens for cigarette smoke. Cigarette smoke alone may be no more of a carcinogen for man than it is for experimental animals, but cigarette smoke plus one or more contaminants commonly found in polluted air may be a reasonably efficient carcinogen in both man and animals.

Cigarette smoke contains carbon dioxide, water vapor, carbon monoxide, particles of unburned carbon, and nicotine in addition to several materials known to be able to induce cancer in experimental animals. The carcinogens include dibenzpyrene, benzo[a]pyrene, 1,2,5,6-dibenzanthracene, nitrosonornicotine, and ^{210}Po (an α emitter). With the exception of radioactive polonium, the known carcinogens may require cocarcinogens to be very active [19, 20]. Smoking cigarettes may not be good for one's health; smoking cigarettes while breathing polluted air may be fatal.

Physical Carcinogens

Nonionizing and ionizing radiations have been shown to cause cancer in man and in experimental animals. Other physical agents such as trauma (bruising, cutting, etc.), noise, heat, magnetic flux, and pressure do not appear to have the ability to cause cancer.

Nonionizing radiation. Of all the nonionizing radiations, only the most energetic, ultraviolet, appears to be capable of causing cancer in man. This subject is discussed in Chapter 14.

Ionizing radiation. All ionizing electromagnetic radiation and all ionizing particulate radiations appear able to cause cancer. The carcinogenic aspects of this subject are *not* discussed in detail in Chapter 15, but quite a bit of necessary background is explained there.

In general, the scientific community accepts the carcinogenicity of ionizing radiation. All concerned agree that such radiations are carcinogenic, but some take the stand that to the basic statement must be added "...if the dose is large enough." Others maintain that there is no threshold for this effect, and that *any* ionizing radiation administered to a population will result in some increase in the incidence of cancer in that population.

Ionizing radiation causes several different kinds of injury at the cellular level including chromosome breaks, inhibition of mitosis, modification of chromosomes, and alterations of molecules of DNA and RNA. When one of the kinds of injury results in "turning off" the factor(s) in the cell that inhibit mitosis, the result is cancer.

Experimentation at high dose levels of external or internal radiation shows that the incidence of cancer in experimental animals is more or less proportional to the total dose given, but that the dose may be somewhat less effective if given over a long period of time rather than acutely. When dose levels are decreased, however, the cancer incidence rate even from acute exposure decreases faster than does the dose, so that a linear extrapolation to zero dose is not possible. The problem is further complicated by the fact that the cancer need not appear at once but may become manifest only after a long delay and therefore may never appear in short-lived experimental animals.

Cancer induced by radiation is indistinguishable from cancers that arise spontaneously in a population, especially in the human population. *Spontaneously* may mean simply that the source of the cancer is unknown, not that man or any other living thing has an innate spontaneous cancer rate; part of man's spontaneous cancer incidence is probably caused by cosmic radiation and background ionizing radiation from the earth, man's structures, and so forth, if there is no threshold for this effect. If there is a threshold for radiogenic cancer, then the causes of much of the spontaneous rate may be environmental contaminants because lack of a threshold for one effect does not imply lack (or presence) of another. Because the cancer incidence at low radiation doses is so low in experimental animals, it may be impossible ever to determine whether there is a true threshold for this effect.

Biological Carcinogens

In 1911, Rous succeeded in transmitting several sarcomas of chickens from one chicken to another by means of cell-free, bacteria-free filtrates. Because the agent obviously had the characteristics of a virus, it was called the Rous sarcoma virus, or RSV. Since then the presence of Rous associated virus (RAV) has been found to be necessary before cancer will ensue—another example of cocarcinogenicity. No other viral cancer was discovered until 1933 when Shope successfully caused invasive cancer in rabbits with cell-free, bacteria-free filtrates from a skin wart found on wild rabbits.

Because cancer had been shown to be caused by viruses in animals, the suspicion grew that perhaps one or more of man's cancers had a viral origin. Perhaps surprisingly, many experimental attempts have been made to transfer a cancer from one person to another. This can sometimes be done, but it has never been done with a cell-free, bacteria-free filtrate. Despite this lack of success, some people believe that a liver cancer found in the peoples of some African areas may be caused by a virus because its pattern of attack is similar to the patterns of some infectious agents. The riddle of the many diseases grouped under the label *cancer* will probably not be unraveled all at once; there may be almost as many solutions as problems.

Modes of Entry and Sites of Action

Carcinogenic agents other than external radiations contact and enter the body in the same manner as do other chemicals. The main sites of contact and portals of entry are thus the skin, mouth, and lungs. As with other toxic agents, carcinogens can cause their main effect either at the site of contact or at some site remote from that. This is particularly true if the carcinogen is not the material with which contact is made but instead is a metabolite of that material.

Skin contact with a carcinogen usually results in a skin cancer rather than a cancer at another site. Ultraviolet and ionizing electromagnetic radiations alone or acting in concert with cocarcinogens such as certain unrefined petroleum products (tars) cause skin tumors as well as other manifestations of toxicity.

Inhalation of carcinogens is usually associated with lung cancer rather than cancer at a remote site. This is true of nickel and chromium and perhaps with PAH. However, vinyl chloride appears able to cause cancer in many remote areas of the body, particularly the liver; asbestos inhalation is associated with mesothelioma not only of the pleura but also of the peritoneum as well as cancer of the lung and larynx.

Ingestion can be associated with kidney and bladder cancers as well as with cancers of the stomach. As with inhalation, when the cancer appears at a site remote from the site of contact, a suspicion always arises that the real carcinogen may not be the material contacted at all but a metabolite of the material. The liver is the main organ of detoxification and the kidneys and bladder are associated with excretion; these organs appear ideally suited as sites for cancer when metabolism is a necessary adjunct. This certainly seems to be the case with several of the aromatic amines such as benzidene and its derivatives and perhaps with several other substances.

Animal Testing

The test most often used for carcinogenesis is to paint the material on the shaved backs of a group of mice, either once or over a period of time. Most often, the mice are of a strain prone to develop skin or other cancer and what is noted is whether the substance increases the incidence of cancer. Other tests involve injecting the material, usually subcutaneously [38] or, far less often, having the animals inhale it over prolonged periods of time. Especially if the tested material has a potential use that will bring it into man's food, chronic ingestion by one or more species will be used.

The main problem with all of these tests other than the general one of extrapolating from animal to man is that they require long periods of time. Cancer does not develop rapidly in most cases, and chronic exposure is usually required to hasten its appearance. Furthermore, tests that take large amounts of time are proportionately expensive, not only in terms of cash outlay but also in providing space for the animals, people to take care of them, and so forth. For these reasons, a good deal of effort has been expended searching for short-term carcinogenicity tests. Many such tests have been developed, but none has been accepted completely by the scientific community. Three with promise, however, are tests for mutagenesis, DNA repair synthesis, and *in vitro* cell transformation [39]. These or other similar tests depend on the fact that carcinogenesis may involve a somatic mutation. Short-term tests would be used to screen chemicals for potential carcinogenic (or mutation) ability. Those chemicals that showed such ability would be subjected to more or less conventional animal testing. Of course, before the screens are accepted, their ability to predict carcinogenicity *and its absence* will have to be convincing.

REFERENCES

[1]   R. L. CARTER AND F. J. C. ROE, "Chemical Carcinogens in Industry," *J. Soc. Occup. Med.* **25** (1975), 86–94.

[2]   G. SACCOMANNO, ET AL., "Histologic Types of Lung Cancer among Uranium Miners," *Cancer* **27** (1971), 515–23.

[3]   O. AXELSON AND M. REHN, "Lung Cancer in Miners," *Lancet* **2** (Sept. 25, 1971), 706–07.

[4]   D. J. B. ASHLEY, "Lung Cancer in Miners," *Thorax* **23** (1968), 87–91.

[5]   M. R. ZAVON, U. HOEGG, AND E. GINGHAM, "Benzidene Exposure as a Cause of Bladder Tumors," *Arch. Env. Health* **27** (1973), 1–7.

[6]   W. L. CALDWELL, "Carcinoma of the Urinary Bladder," *J. Am. Med. Assn.* **229** (1974), 1643–45.

[7]   P. E. ENTERLINE, "Respiratory Cancer among Chromate Workers," *J. Occup. Med.* **16** (1974), 523–26.

[8]   J. W. LLOYD, JR., "Long-Term Mortality Study of Steelworkers: Respiratory Cancer in Coke Plant Workers," *J. Occup. Med.* **13** (1971), 53–68.

[9]   S. F. MCCULLAGH, "The Biological Effects of Asbestos," *Med. J. Australia* **2** (1974), 45–49.

[10] S. WADA, ET AL., "Mustard Gas (Dichlorodiethyl Sulfide) as Cause of Respiratory Neoplasia in Man," *Lancet* **1** (June 1, 1968), 1161–63.

[11] I. R. TABERSHAW AND W. R. GAFFEY, "Mortality Study of Workers in the Manufacture of Vinyl Chloride and its Polymers," *J. Occup. Med.* **16** (1974), 509–18.

[12] L. MAKK, ET AL., "Liver Damage and Angiosarcoma in Vinyl Chloride Workers," *J. Am. Med. Assn.* **230** (1974), 64–68.

[13] L. B. LAVE, "Air Pollution and Human Health," *Science* **169** (1970), 723–33.

[14] E. SAWICKI, ET AL., "Polynuclear Aromatic Hydrocarbon Composition of the Atmosphere in Some Large American Cities," *Am. Ind. Hyg. Assn. J.* **23** (1962), 137–44.

[15] B. W. CARNOW AND P. MEIER, "Air Pollution and Pulmonary Cancer," *Arch. Env. Health* **27** (1973), 207–18.

[16] R. HOOVER AND J. F. FRAUMENI, JR., "Cancer Mortality in U.S. Counties with Chemical Industries," *Env. Res.* **9** (1975), 196–207.

[17] J. L. STEVENSON AND E. VON HAAM, "Carcinogenicity of Benz(*a*)anthracene and Benzo(*c*)phenanthrene Derivatives," *Am. Ind. Hyg. Assn. J.* **26** (1965), 475–78.

[18] S. S. EPSTEIN, "Carcinogenicity of Organic Extracts of Atmospheric Pollutants," *J. Air Poll. Con. Assn.* **17** (1967), 728–29.

[19] "Cancer from Mineral Oil," *Brit. Med. J.* **4** (1969), 443–44.

[20] EULA BINGHAM AND H. L. FALK, "Environmental Carcinogens: The Modifying Effect of Cocarcinogens on the Threshold Response," *Arch. Env. Health* **19** (1969), 779–83.

[21] C. A. VEYS, "Etiology of Tumors of the Urinary Bladder," *Urol. Int.* **24** (1969), 276–88.

[22] E. F. STULA, ET AL., "Experimental Neoplasia in Rats from Oral Administration of 3,3'-Dichlorobenzidene, 4,4'-Methylene-bis(2-chloroaniline), and 4,4'-Methylene-bis(2-methylaniline)," *Tox. Appl. Pharm.* **31** (1975), 159–76.

[23] H. W. GERARDE AND D. F. GERARDE, "Industrial Experience with 3,3'-Dichlorobenzidene," *J. Occup. Med.* **16** (1974), 322–44.

[24] H. F. HENNING, "Precautions in the Use of Methylene-Bis-*o*-Chloroaniline (MBOCA)," *Ann. Occup. Hyg.* **17** (1974), 137–42.

[25] C. S. WEIL, ET AL., "Experimental Carcinogenicity and Acute Toxicity of Representative Epoxides," *Am. Ind. Hyg. Assn. J.* **24** (1963), 305–25.

[26] H. SAKABE, "Lung Cancer Due to Exposure to bis(Chloromethyl) Ether," *Ind. Health* **11** (1973), 145–47.

[27] E. M. BEAVERS, W. G. FIGUEROA, AND W. WEISS, "Lung Cancer in Chloromethyl Methyl Ether Workers," *N. E. J. Med.* **290** (1974), 971–72.

[28] M. KUSCHNER, ET AL., "Inhalation Carcinogenicity of Alpha Halo Ethers—III. Lifetime and Limited Period Inhalation Studies With Bis(Chloromethyl)Ether at 0.1 ppm," *Arch. Env. Health* **30** (1975), 73–77.

[29] S. LASKIN, ET AL., "Inhalation Carcinogenicity of Alpha Halo Ethers—II. Chronic Inhalation Studies With Chloromethyl Methyl Ether," *Arch. Env. Health* **30** (1975), 70–72.

[30] J. L. CREECH AND M. N. JOHNSON, "Angiosarcoma of Liver in the Manufacture of Polyvinyl Chloride," *J. Occup. Med.* **16** (1974), 150–51.

[31] J. B. BLOCK, "Angiosarcoma of the Liver Following Vinyl Chloride Exposure," *J. Am. Med. Assn.* **229** (1974), 53–54.

[32] E. D. PALMES, L. ORRIS, AND N. NELSON, "Skin Irritation and Skin Tumor Production by Beta Propiolactone (BPL)," *Am. Ind. Hyg. Assn. J.* **23** (1962), 257–64.

[33] M. GERALD, ET AL., "Respiratory Cancer and Occupational Exposure to Arsenicals," *Arch. Env. Health* **29** (1974) 250–55.

[34] H. WITACHI, "Inhibition of Deoxyribonucleic Acid Synthesis in Regenerating Rat Liver by Beryllium," *Lab. Inv.* **19** (1968), 67–70.

[35] F. M. HASAN AND H. KAZEMI, "Chronic Beryllium Disease: A Continuing Epidemiologic Hazard," *Chest* **65** (1974), 289–93.

[36] D. F. FLICK, H. F. KRAYBILL, AND J. M. DIMITROFF, "Toxic Effects of Cadmium: A Review," *Env. Res.* **4** (1971), 71–85.

[37] H. I. LIBSHITZ, ET AL., "Asbestosis and Carcinoma of the Larynx," *J. Am. Med. Assn.* **228** (1974), 1571–72.

[38] J. L. GARGUS, O. E. PAYNTER, AND W. H. REESE, JR., "Utilization of Newborn Mice in the Bioassay of Chemical Carcinogens," *Tox. Appl. Pharm.* **15** (1969), 552–59.

[39] D. R. STOLTZ, ET AL., "Evaluation of Short-Term Tests for Carcinogenicity," *Tox. Appl. Pharm.* **29** (1974), 157–80.

11

ABNORMAL PRESSURE

Man has evolved to live and be healthy at the bottom of a sea of air. When the composition of the air breathed is varied or the pressure of that air changes, effects may occur that are dependent on the direction of the alterations. Before the effects of pressure (or composition) changes can be studied, however, some knowledge about the "normal" environment is necessary.

Normal Pressure

Basic Data

Two slightly different pressure standards are in use throughout the world: one mainly by meteorologists and the other by most of the remainder of the scientific and engineering community. Compounding confusion is a series of engineering as well as scientific units, including the most recent SI system.

The *standard atmosphere* represents an average pressure of air experienced at sea level everywhere in the world. There is nothing magical about sea level, however, and for their standard the meteorologists chose to use a multiple of a standard force per unit area, the dyne per square centimeter. The multiple chosen was 10^6, and the unit was called the *bar* (for *bar*ometric pressure). The multiple was chosen so that the bar would be close to (but is not equal to) the standard atmosphere (see Table 11-1).

Engineers quite often use gauges to indicate the pressure inside equipment. When the pressure inside is equal to that outside, pressure stress on the walls of the equipment is zero. To equate zero stress with zero pressure was a natural thing to do, resulting in *gauge* pressure, this is zero at standard (or ambient) conditions and must then be distinguished from absolute pressure, which is zero only in an absolute vacuum.

Composition of the atmosphere has not changed materially since measurement was begun in the early 1900s. Of all the gases, only carbon dioxide has increased in concentration (from about 280 to about 330 ppm by volume), and none has decreased. The concentration of water vapor is quite variable and, therefore, composition of atmospheric samples is usually given on a dry basis. In Table 11-2, that basis is contrasted with an assumption of body temperature, saturation with water vapor, and standard atmospheric pressure (abbreviated as BTPS). Concentrations are all given in ppm, parts (by volume) gas or vapor per million parts (volume) of the whole mixture. To convert any of the values to percent, divide by 10,000.

Gas Laws

Boyle's law. Sometime after 1654 Robert Boyle, an English chemist, proved with experiment that, at constant temperature,

$$Pv = \text{a constant} \tag{11-1}$$

where P is the pressure of a gas and v is the volume of the container of the gas. A generalization of that law is

$$P_1v_1 = P_2v_2 \tag{11-2}$$

where the subscripts indicate different conditions.

Dalton's law. In 1803 John Dalton published a treatise containing his law of partial pressures, which states that in a gas mixture the total pressure is equal to the sum of the partial pressures of the components of the mixture. He defined *partial pressure* as the pressure that would be found in the container if one of the components of the mixture were present alone. For the gases found in human respiration (neglecting those present in very low con-

Table 11-1 Barometric Pressure Conventions

One Standard Atmosphere	One Bar
1.000000 Ata (atmospheres absolute)	0.986923 Ata
1.01325 bar	1.000000 bar
1,013.25 mbar	1,000 mbar
1,013,250 dyne cm^{-2}	1,000,000 dyne cm^{-2}
101,325 Pa (pascal)	100,000 Pa
101,325 N m^{-2}	100,000 N m^{-2}
760 mm Hg	750.06 mm Hg
760 torr	750.06 torr
14.696 psia (lb/sq in., abs.)	14.504 psia
33.899 ft H$_2$O (feet of water)	33.456 ft H$_2$O
29.921 in. Hg (inches of mercury)	29.530 in. Hg

centration), Dalton's law is

$$P_T = P_{H_2O} + P_{CO_2} + P_{N_2} + P_{O_2} \qquad (11\text{-}3)$$

where P signifies pressure, T means total, and chemical symbols identify the components of the mixture.

Boyle's and Dalton's laws apply strictly only to ideal gases but, at pressures reasonably close to atmospheric, most gases (and all of those with physiological importance) do not deviate very far from ideal gas behavior. A

Table 11-2 Composition of Air

Component	ppm Dry	ppm BTPS
Nitrogen, N$_2$	780,900	732,610
Oxygen, O$_2$	209,400	196,450
Water, H$_2$O	0	61,842
Argon, Ar	9,300	8,725
Carbon dioxide, CO$_2$	330	310
Neon, Ne	18	16.9
Helium, He	5.2	4.9
Methane, CH$_4$	1.1	1.0
Krypton, Kr	1.0	0.9
Nitrous oxide, N$_2$O	0.5	0.47
Hydrogen, H$_2$	0.5	0.47
Carbon monoxide, CO	0.1	0.094
Xenon, Xe	0.08	0.075
Nitrogen dioxide, NO$_2$	0.02	0.0188
Ozone, O$_3$	0.01–0.04	0.0094–0.0375

consequence of behaving as ideal gases is that for mixtures

$$P_A = (\text{fraction of } A)(P_T) \qquad (11\text{-}4)$$

or knowing the fraction of a gas in the mixture enables calculation of the partial pressure of that gas when the total pressure is known and *vice versa*.

Henry's law. At about the same time that Dalton's law was published, Robert Henry discovered and published his law, which states that the concentration of a gas or vapor in a liquid is directly proportional to the partial pressure of that gas (or vapor) above the liquid surface.

$$x_A = H_A(P_A) \qquad (11\text{-}5)$$

where x_A is the mole fraction of A in the liquid and H_A is the Henry's law constant, which is not really a constant but is usually inversely proportional to temperature. H_A is specific and relates to the particular gas and to the liquid in which that gas is dissolved. Deviations from true proportionality are many, but in most physiological situations Henry's law is sufficiently precise to be useful.

Respiratory Gas Concentrations

One of the constants of human life familiar to everyone is normal body temperature. To the physiologist, that temperature is not 98.6°F but, instead, 37°C. Furthermore, any good table of the vapor pressures of water at various temperatures indicates that at 37°C the vapor pressure of water is 47.067 mm Hg, which the physiologist rounds to 47 mm Hg (6,266 Pa). As long as the person is healthy and not undergoing biothermal strain (see Chapter 13), body temperature remains constant and so, therefore, does water vapor pressure in vapor spaces within the body.

When a person inhales even completely dry air, that air becomes saturated with water vapor before it reaches the alveoli where gas exchange takes place. In this case, *saturation* implies that prior to reaching alveoli the inhaled air must always contain water vapor with a partial pressure of 47 mm Hg. Equation (11-3) further implies that all the other gases of interest must have partial pressures that sum to the total pressure minus 47 mm Hg. Under standard conditions (BTPS), then saturated inhaled air will have the following composition:

Component	Fraction	Partial Pressure (mm Hg)
Water vapor	0.0618	47
Nitrogen	0.7326	556.8
Oxygen	0.1964	149.3
Carbon dioxide	0.00031	0.24
Argon + other inerts	0.00874	6.64

Note that this table indicates composition prior to the changes in O_2 and CO_2 caused by gas exchange in the lungs.

In practice, "Argon + other inerts" is usually added to the amount of nitrogen, so that the fraction of nitrogen becomes 0.7413 and the partial pressure of nitrogen is 563.4 mm Hg. Rounding all values and adding the SI system of units, reduces the information to that contained in Table 11-3.

Table 11-3   Inhaled Gas Concentrations at One Atmosphere

Component	Fraction	Partial Pressure	
		mm Hg	*Pa*
Water vapor	0.0618	47.0	6,266
Nitrogen	0.741	563.0	75,100
Oxygen	0.196	149.0	19,900
Carbon dioxide	0.0	0.24	32

As the inhaled gases reach the alveoli, equilibration with carbon dioxide removed from (venous) blood takes place very rapidly; equilibration with oxygen in the blood is a much slower process. Therefore, even though CO_2 concentrations in alveoli do not change very much throughout the process of a breath, O_2 concentrations do change and, therefore, some kind of an average is usually used for alveolar concentrations. Furthermore, under most circumstances that do not involve heavy exercise, the volume of carbon dioxide given off is somewhat less than the amount of oxygen taken up by the blood so that the volume of gas exhaled is less than that inhaled; this results in an apparent increase in the concentration of nitrogen because the amount of that gas remains unchanged by the process of respiration. Also, even moderate amounts of exercise alter all relationships somewhat, so that the data in Table 11-4 can be regarded as only approximately correct.

Concentrations of the various gases in average expired air are similar to those in the alveoli, except for the dilution that occurs when inspired and expired air mix. Usually, however, expired air samples are obtained under conditions that allow them to approximate alveolar concentrations.

Table 11-4   Alveolar Gas Concentrations at One Atmosphere

Component	Fraction	Partial Pressure	
		mm Hg	*Pa*
Water vapor	0.0618	47.0	6,266
Nitrogen	0.749	569.0	75,860
Oxygen	0.137	104.0	13,865
Carbon dioxide	0.053	40.0	5,333

Low Pressures

Low atmospheric pressure is associated with increased altitude. Acute exposures to this condition are usually the result of flying in unpressurized aircraft or of ascending a high mountain.

Because air is a compressible fluid, ambient pressure is not a direct function of altitude. Instead, at lower elevations (below about 100 km) the logarithm of the barometric pressure is proportional to elevation. For the levels of most interest physiologically (below about 7 km), the relationship is well approximated by an equation of the form

$$\ln(P_B) = \frac{a - h}{b} \tag{11-6}$$

or, in exponential form,

$$P_B = e^{(a-h)/b} \tag{11-7}$$

where P_B is the barometric (ambient) pressure, h is the elevation, and a and b are constants.

The magnitude of the constants a and b depends, of course, on the units chosen for pressure and altitude. Table 11-5 is a compilation of values for the constants associated with various pressure and elevation units.

Acute Exposure

Many of the effects of increased altitude on the human body are caused by hypoxia, which in turn is related to the decrease in oxygen partial pressure at high elevations. Other effects such as decompression sickness are directly

Table 11-5 Values of the Constants for Equations (11-6) and (11-7)

Unit of Pressure	Unit of Elevation	a	b
mbar	km	53.72	7.75
mbar	m	53,720	7,748
mm Hg	km	51.49	7.75
mm Hg	m	51,492	7,748
mm Hg	ft	168,937	25,420
in. Hg	ft	86,709	25,420
lb in.$^{-2}$	ft	68,636	25,420
Pa	km	89.40	7.75
Pa	m	89,401	7,748

related to low pressure itself. If the exposure to low pressure is the result of mountain climbing, then the stresses of increased ultraviolet radiation and cold may be added to those associated with *hypobaria* (low pressure).

Effects associated with low pressure depend on the pressure experienced and the duration of that experience. If the exposure is to an altitude below about 1,800 m (6,000 ft), most people will experience little if any difficulty, no matter what the length of stay. Denver, Colorado, has an altitude of about 5,280 ft (1,600 m) and most visitors to that city are unaware of any particular effect except perhaps a slight breathlessness on exertion.

As the altitude is increased to 2,400 m (8,000 ft), the subject of acute exposure becomes more and more aware of heavy breathing that results from light exercise. In addition, he may experience insomnia and, perhaps, drowsiness. Although an elevation of 2,400 m appears to have little effect on mental abilities, it may impair the learning of a new task [1].

Many altitude studies have been conducted atop Pikes Peak, Colorado, at an elevation of 4,300 m (14,100 ft). At this and even lower altitudes (3,660 m, 12,000 ft), acute exposure causes headache, insomnia, drowsiness, nausea, and sometimes euphoria, all symptoms of the syndrome known as *acute mountain sickness*. The mild hypoxia causes pulmonary congestion in some people, increased cerebral blood flow, increased cerebrospinal pressure, and occasionally cerebral edema. Breathlessness upon exertion along with increased heart rate are quite noticeable at this elevation.

To accommodate for the increasing hypoxia, cardiac output is increased more or less in proportion to the elevation [2, 3]. Nevertheless, the brain receives substantially less oxygen per unit of time than at sea level, causing deterioration of several CNS functions. One of the first to be measured was vision; decrements of accommodation, convergence, and stereoacuity have been found [4] along with retinal hemorrhage especially in those experiencing headache [5]. In addition, an hour at 4,600 m (15,000 ft) appears to decrease mental proficiency by about 50%, while 18 hr at that altitude reduces ability at mental tasks about 80%. These decrements along with the physical symptoms cause essentially complete incapacity for any useful endeavor.

At about 7 km (23,000 ft) the effects of hypoxia on the CNS are so severe that a person acutely exposed may suffer convulsions and lapse into a coma. At this level, the barometric pressure is about 310 mm Hg (41,330 Pa) and the partial pressure of oxygen in inspired air has fallen to about 55 mm Hg (7,342 Pa), or about one-third of the sea level value; in alveolar gas, the P_{O_2} is about 15 mm Hg (2,000 Pa).

Sudden hypoxia at 10 km (33,000 ft) caused, for instance, by a failure of oxygen supply or decompression of a pressurized aircraft cabin, causes unconsciousness in about 2 min with little, if any, warning. Loss of consciousness is caused by hypoxia; the sudden pressure reduction may cause decompression sickness, or bends, in addition to CNS problems associated with hypoxia [6, 7]. Even if bends is not experienced, sudden loss of cabin pressure is likely to cause just as sudden expansion of intestinal gas and associated colic-like symptoms.

At a little over 15 km (50,000 ft), the ambient pressure has fallen to about 87 mm Hg (11,600 Pa), which is equal to the sum of the partial pressure of water vapor at body temperature and the partial pressure of CO_2 in the lung. In other words, the partial pressure of other gases (N_2 and O_2) in the body must be zero and, therefore, life is not possible even on pure oxygen.

Acclimatization

Animal bodies, including man's, have a remarkable ability to accommodate or acclimatize to a variety of nonstandard conditions. The period of maximum incapacity after sudden (a few hours) transition from sea level to 4,300 m (14,100 ft) begins at about 12 to 18 hr after the exposure starts and lasts for a few days at most if the person can acclimatize to that altitude. After about 3 weeks [2] most or all of the symptoms will have disappeared, especially if there has been regular exercise as soon as that is possible.

Chronic Exposure

Several million people live and work at altitudes of 3,600 m (12,000 ft) or more, and an active mining camp is located on Mount Aucanquilcha, Peru, at an elevation of 5,330 m (17,500 ft). Between 4 and 5 million Indians live on the Andean plateaus in Peru, and these are the people whose chronic exposure to low pressure has been most extensively studied. Monge began systematic studies of Peruvian Indians in 1927 and since that time several investigators have continued the work.

Most thoroughly studied are the inhabitants of Morochocha, a town of about 6,000 people that can be reached from Lima, Peru (at sea level), by automobile in a few hours [8]. Morochocha is at an elevation of 4,540 m (14,900 ft) where the average barometric pressure is 446 mm Hg (59,462 Pa) and where the average alveolar P_{O_2} is about 50 mm Hg (6,666 Pa), half that normal at sea level.

The most obvious characteristic of the race of people acclimatized by centuries of living at high elevations is that they are able to carry on strenuous activities such as pick-and-shovel mining and hours-long soccer games while lowland scientific investigators can hardly move. Studies have shown that their adaptation consists of several modifications of the way in which the body handles air and oxygen.

When the P_{O_2} in alveoli averages 50 mm Hg, blood leaving the lungs is only about 80% saturated with oxygen, in comparison to the 96 to 98% saturation experienced with a P_{O_2} of 100 to 105 mm Hg found at standard conditions. Therefore, to transfer sufficient oxygen the body must move more air through the lungs and must arrange for more oxygen to be carried in the blood. Thus the first modification is increased pulmonary ventilation. The thoracic cavity and lung volume of these Indians are larger than normal and during rest or exercise a condition of hyperventilation is found. This means that the Indian moves more air through his lungs per unit of time than does a person acclimatized to sea level.

Polycythemia (an increase in the number of red blood cells), increased hemoglobin, and increased blood volume enable more oxygen to be transported by the blood per unit of time. The average increase in blood volume of Morochocha residents appears to be about 20% when compared to normal, and the increase is almost all due to more red cells as there is no increase in plasma volume. In consequence, the viscosity of the blood of these people is increased, causing the heart to work harder even though the cardiac output is about the same as that of a sea level dweller. Nevertheless, systemic blood pressures are low [9] except that the Indians have a condition of permanent *pulmonary hypertension* [8]. This increased blood pressure in the lungs appears to be an adaptive mechanism that allows (or causes) a more effective perfusion of lung capillaries by blood, thus enhancing the gas exchange capacity of the respiratory system.

Anatomical studies have shown that the high altitude dwellers have an increased number of dialated capillaries per unit volume of perfused tissue than lowlanders, thus facilitating the transfer of oxygen from blood to tissue. In addition, the Indians appear to have an increased ability to use oxygen at the cellular level.

Several of the changes associated with chronic exposure to low pressure are similar to those experienced by anyone who lives at high altitudes for a few weeks. These include hyperventilation, polycythemia, pulmonary hypertension, and increased blood volume. On the other hand, some of the changes appear to have been the result of natural selection over an extended period of time. These include the increased vascularity of tissue, increased ability to use oxygen at the cellular level at low P_{O_2}, and increased lung volume.

Occasionally, a person who has become acclimatized to living at low pressure will lose his acclimatization. Monge first described the resulting syndrome in 1928, and chronic mountain sickness has been called *Monge's disease*. Apparently the person loses sensitivity to the buildup of CO_2 in blood and/or to a decreased P_{O_2} so that instead of hyperventilating, he hypoventilates. This, in turn, results in cyanosis from the anoxia, an accentuated polycythemia, and a more severe degree of pulmonary hypertension [8]. The symptoms experienced are usually more severe than those of one acutely exposed to similar conditions. Treatment consists of moving to higher pressure (lower altitude) where the affected person will usually recover rapidly and there be able to live a normal life.

High Pressures

People are exposed to higher than normal ambient pressure during underwater diving and in tunneling or caisson work where air pressure is used to help support an underground cavity. Caisson and tunnel workers are rarely exposed to pressures much above 3 or 4 Ata but may be exposed daily for months or years. Divers, on the other hand, are usually not exposed so often but may experience much higher pressures. Almost all experience with high pressure has been from acute or subchronic exposure; there is no group of people living at high pressure that is at all comparable to the Peruvian Indians at low pressure.

Acute Exposure

Suppose a person were to invert a bucket and place it so that the rim were just under the surface of a body of water. The volume of the bucket is v_1 and the associated pressure, P_1, is ambient barometric pressure. If the bucket were then pushed down so that it was 33.9 ft (see Table 11-1) below the surface of the (fresh) water, the pressure on the air within it would be $P_2 = 2p_1$. Then, from Eq. (11-2), the volume of air within the bucket must be $v_2 = v_1/2$. If the bucket were pushed down another 33.9 ft, the absolute pressure within it would be 3 Ata, and the volume of air would be one-third that originally present. This is one of the phenomena that limit the practice of underwater exploration with a bucket (or bell or barrel) over one's head. The other phenomenon is, of course, the very limited supply of air (oxygen) that can be carried in the container. Nevertheless, the first diving was probably done in just this manner.

After being half-drowned by the water rising in a bucket over his head, the canny inventor might then have decided that the problem lay in the fact that the water could rise without restriction into the bucket. He could prevent that by putting his whole body into a sealed container and thus the submarine was born at least in theory. In practice, sealing a container so that water under pressure of 1 or 2 atmospheres could not enter was a difficult feat prior to the advent of modern technology, as was the engineering task of building a container that would not collapse. Therefore, the inventor might have tried to pump air down to his bucket from the surface. If he succeeded, he may have been responsible for his or his diver's death through the phenomenon of *spontaneous pneumothorax* (collapse of a lung).

If one is under only a few feet of water breathing air from the surface, the pressure of that air within the lungs must be very nearly equal to atmospheric plus the pressure associated with the water depth. Lung tissue is very delicate and therefore if the person inhales and then holds his breath and rises to the surface rapidly, that extra pressure may rupture lung tissue and cause the lung to collapse. Spontaneous pneumothorax is more of a problem with emergency surfacing from deep dives with SCUBA (self-contained underwater breathing apparatus) or escaping from a submerged structure but can occur even in shallow water when using a snorkel.

Pressure increase. As diving equipment was improved, the hazard of simply drowning during a dive was reduced but that hazard was replaced by others associated with pressure changes. During descent or pressurization the main problem is that termed simply *squeeze*. Squeeze occurs because of Boyle's law; man has several cavities in his structure that tend to collapse as pressure is increased. The middle ear and nasal sinuses are two areas where squeeze can be very painful and damaging. For that reason, a person with clogged sinuses or eustacian tubes (that connect the middle ear with the throat) cannot work under pressure. Another area that has sometimes given trouble is cavities in teeth. Cavities in soft tissue such as those caused by gases in the intestine are much less troublesome.

A "hard hat" diver wears an impervious suit and a rigid helmet. Both suit and helmet are supplied with air from the surface. If the supply of air to the suit fails, the diver's body is subjected to the full pressure of the water, which squeezes him into the helmet. This is called *equipment squeeze* and cannot occur with SCUBA.

At pressure. When one breathes a gas mixture for a sufficiently long period of time, an equilibrium becomes established between gas and liquid (or semiliquid) phases in accordance with Henry's law: *When the total pressure is increased without changing the gas composition, the partial pressure of each component is increased proportionately.* In accordance with Eq. (11-5), the amount of each gas in body tissues and fluids increases, each according to its own coefficient H, specific to the gas and to the fluid (or tissue) in which the gas is dissolving. The respiratory gases are water vapor, carbon dioxide, oxygen, and nitrogen; the problems they cause range from nil to severe.

Water is always present in all tissue in the liquid phase so that the presence of higher pressure means little; the partial pressure is controlled by temperature alone.

Carbon dioxide causes few problems because the body is equipped to handle it very rapidly and efficiently, even during heavy exercise. This gas is very soluble in water but not particularly soluble in fat, so that it is present in greatest concentration in tissue that is plentifully supplied with blood vessels. Equilibration is rapid at all pressures.

Oxygen is used during metabolism in all tissue and therefore cannot accumulate to any great extent. Oxygen, in common with all other materials, is toxic, however, and that toxicity is manifested when the partial pressure is high and the exposure is long. Manifestations of oxygen toxicity range from pulmonary edema to convulsions.

Some symptoms may appear after an exposure to a P_{O_2} of 3.0 Ata for more than an hour or to a P_{O_2} of more than about 600 mm Hg for a few days. Although several studies have shown that antioxidant drugs can protect animals from some of the effects of oxygen [10, 11], the practice in long deep dives has been to reduce the fraction (and thus the partial pressure) of oxygen in the breathing air. As long as the P_{O_2} is kept well below 600 mm Hg (80,000 Pa), few if any problems are experienced.

Nitrogen is much more soluble in fat than in water and water solutions, but the fat depots in the body are not well perfused with blood vessels and equilibration is relatively slow. The equilibration problem works both ways. That is, equilibration takes several hours to become established when the pressure is increased, but it also takes a fairly long period of time to become established when the pressure is reduced. During the establishment of equilibrium when pressure is increased, nitrogen causes few problems unless the pressure is quite high (on the order of 10 Ata or more) in which case a phenomenon called *nitrogen narcosis* may appear. Also called "rapture of the depths," nitrogen narcosis can be severe enough to be incapacitating and thus fatal. Trial and error established that neon and argon probably cause similar effects but that helium does not. Subsequently, for deep dives the nitrogen of air has been replaced by helium, essentially eliminating the problem and helping to pinpoint its real cause [12].

Nitrogen is physiologically inert, and so are neon and argon; the fact that these gases could cause narcosis under high pressure has puzzled a great many people. One logical explanation for this phenomenon is that none of these gases is capable of causing narcosis at all—that the effect is much more related to the density and/or viscosity of the breathing mixture than to the identity of the inert gas employed [13]. Nitrogen narcosis is then probably a direct effect of pressure causing bronchial blockages and hypoxia and not an undiscovered biochemical property of the inert gases.

Pressure decrease. The most common and best known hazard of undergoing a major decrease of ambient pressure is the syndrome known as *bends, caisson disease,* or *decompression sickness* [14]. Symptoms of this disorder may range from itching (pruritis) anywhere and tingling sensations to severe pain in joints and along tendons. Although the exact mechanism or cause of decompression sickness is not known, it is quite obviously related to the washout of inert gases from the fluids and tissues in which they have dissolved. The easy explanation involves visualizing a bottle of carbonated water. As long as the bottle cap is in place, there is no sign that the water contains carbon dioxide in solution. As soon as the cap is removed, however, bubbles form throughout the water and escape to the surface. Bubble formation in tissue and the bloodstream following a pressure decrease must parallel that seen in carbonated water.

If the bubbles in the bloodstream become large enough because decompression is too rapid, they may lodge at any point where a blood vessel branches or becomes smaller, thus blocking the flow of blood. This is called *air embolism*. The usual place for this to occur is in the lungs, giving rise to a suffocating sensation called the *chokes*. If bubbles form in veins, the vessels become successively larger until the heart is reached, which is the largest vessel of all. After going through the right atrium and ventricle, blood is pumped to the lungs where blood vessel size decreases rapidly and at one or more of the stages of decrease the bubbles lodge, blocking flow. Bubbles formed in arteries also lodge at points of size decrease, but those points are not localized in one area such as the lungs; they are scattered throughout the body. Blockage of blood flow to the heart or brain in this manner is, of course, life-threatening.

Bubble (air embolism) formation in the bloodstream is easy to visualize and understand. On the other hand, air embolism is much less common than bends (fortunately) and finding reasons why bubbles should choose to form preferentially near the skin surface (pruritis) or in joints or along tendons and thereby cause pain has not been easy. Despite a fair amount of research [15, 16, 17], some of the facets of caisson disease are still mysteries. As mentioned earlier, the most serious complication of a pressure decrease is that of spontaneous pneumothorax caused by rupturing a lung. When one lung collapses, no respiration can take place and all must be done by the other lung. If that lung too should be ruptured, death is the usual result.

Bubble formation is not limited to soft tissue and blood during decompression. Bubbles can form anywhere. When the locus of formation is within the spine, effects may be immediate and severe, ranging from paralysis of the extremities to paralysis of the respiratory apparatus and death [15].

If any of the complications of decompression should occur, the only treatment of any value is recompression as rapidly as practical. Recompression is usually accomplished either by descending to a greater depth (in water) or by means of a recompression chamber at the surface. Many recompression chambers have been designed and built, from small portable units to ones large enough to hold several people at once. Recompression on the surface, either for divers or for caisson or tunnel workers, is preferred because the victims are then easiest to reach with medical aid.

Valuable as recompression chambers are, the best procedure is to prevent decompression sickness. Haldane is credited with first making decompression a rational procedure. He decided that decompression should take place in stages, each involving a decrease of pressure by a factor of about two, with time allowed at each stage for equilibrium of gas phases to become established (or nearly so). From his initial work and that of others has come the U.S. Navy Diving Tables and other similar regimens designed to prevent bends and other complications of decompression. That following the tables will not prevent all problems has been known for many years and several attempts have been made to devise better tables that will protect everyone and yet allow decompression to proceed as rapidly as possible. No one, from the diver to his employer to medical authorities, likes to see 20 min of work at 300 ft (91 m, corresponding to about 10 Ata) actually take 3 hr because of the extensive decompression time required, but no one wants divers or caisson workers to suffer from bends or other complications.

One of the attempted solutions to the recompression problem is to establish semipermanent work stations at depths where the divers could live for days or weeks while working and only have to go through the whole decompression process at the end of that time. This idea is an extension of the process of *saturation* diving where the diver stays at a depth long enough for complete equilibrium with the inert gas to become established. Although some problems remain to be solved [18], this approach seems to hold a great deal of promise.

Chronic Exposure

Some divers work often and extensively enough so that their exposures to high pressure may be considered to be chronic in a manner analogous to the chronic exposure of an industrial worker to an air contaminant. Using the same criteria, most tunnel and caisson workers do have chronic exposures to high pressure even though they are under pressure only about 1.5 hr twice per day in the United States.

Caisson and tunnel workers probably experience some symptoms of decompression sickness fairly often [14]. Usually the discomfort is not so severe that they want to seek help and recompression on their own time, so the minor problems are never reported to anyone in authority. One complication of chronic exposure to high pressure, then, is periodic recurrence of bends. Unfortunately, the possible cumulative effects of this kind of mistreatment of the body have not been well studied.

Another complication of chronic exposure to high pressure is called *aseptic bone necrosis,* a deterioration of bone structure mainly at the head of the humorus or femur that is unrelated to infection. The cause of this problem is chronic high pressure, but the mechanism of the necrotic process is obscure. The disease can be diagnosed radiographically [19] and can be treated successfully by the use of prosthetic implants [20] but potentially susceptible workers cannot be identified prior to developing the disease. Although the prevalence of this disorder among compressed air workers appears to be low, even that statistic is not known with certainty.

REFERENCES

[1]   G. R. KELMAN AND T. J. CROW, "Impairment of Mental Performance at a Simulated Altitude of 8,000 Feet." *Aerosp. Med.* **40** (1969), 981–82.

[2]   J. A. VOGEL, J. E. HANSEN, AND C. W. HARRIS, "Cardiovascular Responses in Man

During Exhaustive Work at Sea Level and High Altitude," *J. Appl. Physiol.* **23** (1967), 531–39.

[3] J. A. VOGEL AND C. W. HARRIS, "Cardiopulmonary Responses of Resting Man During Early Exposure to High Altitude," *J. Appl. Physiol.* **22** (1967), 1124–28.

[4] M. K. OHLBAUM, "The Effects of Hypoxia on Certain Aspects of Visual Performance," *Am. J. Optom. Arch. Am. Acad. Optom.* **46** (1969), 235–49.

[5] G. A. SCHUMACHER AND J. H. PETAJAN, "High Altitude Stress and Retinal Hemorrhage—Relation to Vascular Headache Mechanisms," *Arch. Env. Health* **30** (1975), 217–21.

[6] S. T. LEWIS, "Decompression Sickness in USAF Operational Flying," *Aerosp. Med.* **43** (1972), 1261–64.

[7] T. H. ALLEN, D. A. MAIO, AND R. W. BANCROFT, "Body Fat, Denitrogenation and Decompression Sickness in Men Exercising after Abrupt Exposure to Altitude," *Aerosp. Med.* **42** (1971), 518–24.

[8] A. HURTADO, "Man and Altitude," *Am. Ind. Hyg. Assn. J.* **27** (1966), 313–20.

[9] P. T. BAKER, "Human Adaptation to High Altitude," *Science* **163** (1969), 1149–56.

[10] A. P. SANDERS, R. G. LESTER, AND B. WOODHALL, "Hyperbaric Oxygen Toxicity Prevention with Succinate," *J. Am. Med. Assn.* **204** (1968), 241–46.

[11] A. P. SANDERS, ET AL., "Protection Against the Chronic Effects of Hyperbaric Oxygen Toxicity by Succinate and Reduced Glutathione," *Aerosp. Med.* **43** (1972), 533–36.

[12] P. B. BENNETT AND E. J. TOWSE, "Performance Efficiency of Men Breathing Oxygen-Helium at Depths Between 100 Feet and 1500 Feet," *Aerosp. Med.* **42** (1971), 1147–56.

[13] E. G. VAIL, "Hyperbaric Respiratory Mechanics," *Aerosp. Med.* **42** (1971), 536–46.

[14] R. I. McCALLUM, "Decompression Sickness: A Review," *Brit. J. Ind. Med.* **25** (1968), 4–21.

[15] S. S. SHIM, F. P. PATTERSON, AND MARIE J. KENDALL, "Hyperbaric Chamber and Decompression Sickness: An Experimental Study," *Can. Med. Assn. J.* **97** (1967), 1263–72.

[16] H. HASEGAWA, M. SATO, AND H. TSURUTA, "A Biochemical Study of Divers," *Ind. Health* **9** (1971), 113–36. Abstracted in *Ind. Hyg. Digest* **36** (Oct., 1972), 10.

[17] H. KREKELER, ET AL., "Washout of Inert Gases Following Hyperbaric Exposure," *Aerosp. Med.* **44** (1973), 505–7.

[18] P. WEBB, "Body Heat Loss in Undersea Environments," *Aerosp. Med.* **41** (1970), 1282–88.

[19] R. I. McCALLUM, "Pneumoconiosis and Caisson Disease of Bone," *Trans. Soc. Occup. Med.* **22** (1972), 63–68.

[20] H. STUTZER, "Bone Necrosis in Tunnel Workers," *Lancet* **2** (Aug. 31, 1974), 530.

12
NOISE

Noise is the only common industrial hazard that in and of itself is not life-threatening. In fact, compensation for the only injury caused by noise (hearing loss) was not paid by any state in the United States until the 1960s. Recognition of noise as an occupational hazard, then, is relatively new and the science of the evaluation and control of noise has lagged behind many others. Nevertheless, the amount of interest in noise as a hazard is increasing rapidly because the extent of noise problems is also increasing [1].

Noise in the amounts encountered today is a product of our way of life. As more and more power becomes available, that fraction wasted by being transformed into sound energy creates more problems. If the fraction of total power that becomes noise were to remain constant, sections of the planet Earth could become intolerably noisy and, therefore, efforts to evaluate the extent of problems and to learn to control them are warranted even though noise is not life-threatening as are so many of the other hazards encountered daily.

Physics of Sound

Noise is usually defined as unwanted sound. Sound has several characteristics such as pitch, timbre, and loudness, but the main parameters identified as being important to the hazard of noise are amplitude and frequency.

Frequency

The sensation we perceive as pitch is mainly a function of the frequency of the sound waves present, but pitch is much too complex to be described completely by specifying frequency. Pitch is a *psychoacoustical phenomenon* in that it represents an interaction between something having to do with sound in the physical world and how that something is perceived by the listener. Because of the complexities associated with psychoacoustics and because frequency is obviously a large part of pitch and is much easier to specify and study, most work dealing with this phenomenon has been confined to pure, complex, or random tones, leaving pitch to those more interested in perception than in physics. Furthermore, although frequency has been shown to be important in the evaluation of noise hazards, pitch has never been so identified.

A *pure tone* is sound with a single frequency and, as it happens, a single pitch. Pure tones are most easily produced with tuning forks, which also can serve to illustrate how the term *frequency* applies to sound. Imagine a tuning fork being held within hearing distance of an ear. The tuning fork is actuated by striking it gently on a firm surface. Tines of the fork are designed to oscillate back and forth at an unvarying rate. When doing so, they push air in such a way as to create periodic compactions and rarefactions that move away from the fork in all directions. The analogy usually used to explain this is that of a pebble dropped into the still surface of a pond. The waves that expand from where the pebble enters the water, however, are variations in amplitude and are present only at an interface between water and air; therefore they do not truly represent the *waves* emitted from a tuning fork.

When the periodic compactions and rarefactions of air meet the ear, they set in motion the eardrum or *tympanic membrane*, which (through a very complex process to be described later) results in the sensation of pitch. The amplitude of the sound coming from the tuning fork is related not to the spacing or size of the waves of compaction and rarefaction but to how compact and rarefied those waves are. Just a little compaction and rarefaction corresponds to a small amplitude and a small perceived loudness. As long as the tuning fork is active, its tines move back and forth exactly the same number of times each second. The distance they travel during each movement is greater, however, just after the fork is struck on a solid object than later. The quantity or amplitude of sound is related to how far the tines move and, therefore, how much they push and pull at the air while moving.

To say that amplitude of sound is a function of just how much components of the medium move together and apart is equivalent to saying that sound is a phenomenon of pressure oscillation. This is true whether the sound is traveling in air, steel, or water. For this reason, sound will not travel through a vacuum because a vacuum contains nothing to compact and rarefy.

Not all sounds are pure tones; in fact, most sounds are extremely complex. The sound emanating from an orchestra in volume is obviously complex, but the sound from a single musical instrument used in the orchestra is also far from being a pure tone. Both a flute and a violin can, for instance, be used to play the same note of the scale but anyone at all familiar with them can easily tell which instrument is being played. A violin can produce an almost pure tone when it is played with an open string (the left hand not being used) but such notes lack richness and except for special effects would only be used by a beginner. The "richness" of a well-played violin note and the difference between that note and the same one played on a flute have to do with the mixture of overtones or harmonics generated along with the fundamental frequency. Those overtones are what make the sound very complex and, incidentally, are probably responsible for much of what is called *timbre*.

Voice sounds are so complex and vary so much in both frequency and amplitude that they approximate randomness. Sound that is random in direction, frequency, and amplitude is, by some definitions, noise. If the amplitude varies randomly over the whole acoustic frequency spectrum, the noise is said to be *white* by analogy with white light. Noise that is weighted with greater amplitudes at low frequencies to approximate a random variation of energy (rather than amplitude) with frequency is called *pink* by analogy with light weighted with more energy at the low frequency end of its spectrum.

In all cases, the unit of frequency is the hertz (Hz). One hertz is equal to one cycle (of compaction and rarefaction) per second.

Velocity

The velocity with which sound travels depends on the medium it is in and on several properties of that medium including but not limited to density and physical state. Temperature of the medium can also be very important. In air at sea level, for instance, sound velocity is best expressed by means of an equation:

$$c = 1053 + 1.093t \qquad (12\text{-}1)$$

where c is velocity expressed in feet per second and t is in degrees Fahrenheit. If c is in meters per second and t in degrees Celsius, the equation becomes

$$c = 331.5 + 0.607t \qquad (12\text{-}2)$$

At 20°C (68°F), for instance, the velocity of sound in air is 343.6 m sec^{-1} (1,127 ft sec^{-1}). At the same temperature, the velocity of sound in typical specimens of water, wood, and steel is 1,430, 3,960, and 5,030 m sec^{-1} (4,700, 13,000, and 16,500 ft sec^{-1}), respectively. The velocity of sound in air is a function also of the absolute humidity so that the values calculated from Eq. (12-1) or (12-2) are only approximate.

Quantity

In acoustics, the quantity or amplitude of sound is almost always specified by use of the word *level*. A level is the logarithm of the ratio of a quantity to a reference quantity and level is never used in any other sense. The most important level for most industrial hygienists is the *sound pressure level,* which is equal to 20 times the logarithm (base 10) of the ratio of the sound pressure to the reference pressure which is usually (and always in this book) 0.00002 Pa (0.00002 N m^{-2}) or 20 μPa.

$$L_p = \text{SPL} = 20 \log P/P_0 \qquad (12\text{-}3)$$

where P = the sound pressure [usually effective (root-mean-square) pressure]
P_0 = the reference root-mean-square pressure (20 μPa)
SPL = the sound pressure level in decibels (dB)
L_p = another symbol for the SPL
Another level sometimes used is that of *sound power level:*

$$L_P = \text{PWL} = 10 \log W/W_0 \qquad (12\text{-}4)$$

where W = the sound power (watts)
W_0 = the reference power (usually 10^{-12} watt)
PWL = the sound power level in decibels (dB)
L_P = another symbol for the PWL

The only difference in form between Eq. (12-3) and (12-4) is the constant; the reason for that difference relates to the fundamental units of the parameters involved. The *bel* is the basic unit of level and the *decibel* is $\frac{1}{10}$ bel just as a centimeter is $\frac{1}{100}$ meter. If the number of bels is then found by taking the logarithm of a ratio, the number of decibels must be ten times that, as in Eq. (12-4).

Bels and decibels were first used in electrical engineering. In that field, the unit of pressure (electrical potential) is the volt and the unit of power is the watt. These units are connected through two equations:

$$E = IR \tag{12-5}$$

$$W = EI \tag{12-6}$$

where E = the electrical potential (volts)
  I = the current flow (amperes)
  R = the resistance (ohms)
  W = the power (watts)

By rearranging Eq. (12-5) (Ohm's law), $I = E/R$. Substituting in Eq. (12-6), $P = E^2/R$. That is, to determine power, pressure must be squared. If the bel, then, were to have a similar meaning for both pressure (voltage) and power, the pressure unit had to be squared. With logarithms, this is done by multiplying by 2. Hence, the constant for calculating potential levels in decibels was 20; that used for power levels in decibels was 10. For similar reasons this practice was necessary in acoustics, as in Eq. (12-3).

Using decibels for the quantity of sound is for convenience only; the decibel has no particular physiological importance. The reference sound pressure, on the other hand, does have a physiological base in that 20 μPa of sound pressure is close to an average threshold of hearing at a frequency of 1,000 Hz.

Because decibels are logarithms, adding them is equivalent to multiplication. Quite often, however, there is need to know the resultant level at the receiver if one sound is added to another. If both, one at a time, cause the sound pressure level in a certain location to be 95 dB, using the two sources together will not result in a sound pressure level of 190 dB but rather [if the sound under consideration is (random) noise] close to 98 dB. If the sounds under consideration happened to be pure tones, then out of phase the resultant would be zero (or $-\infty$ dB, no sound at all) but in phase the resultant would be 101 dB.

Most of the sounds encountered by occupational health engineers in the course of hazard evaluation at least approximate noise with a random direction of incidence. In such cases, Fig. 12-1 can be used to add one measurement to another [2]. In all cases where there is any question about the randomness of the noise, Eq. (12-3) should be solved for the sound pressure (knowing L_p) in each case and then the pressures added before resolving the equation to determine the resultant L_p.

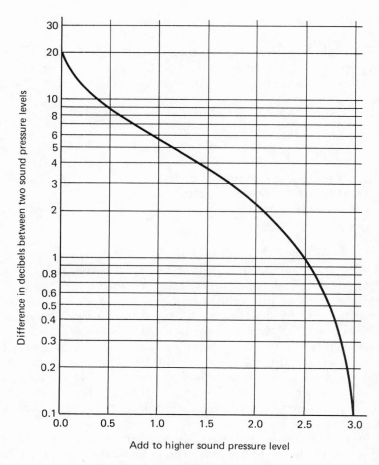

Figure 12-1   Combining decibel levels.

Noise Sources

In the total environment, the noise sources of most importance are aircraft, surface vehicles, and machinery. The environmental health engineer may become involved with evaluating and/or controlling noise from any of these source categories. Of course, each category can be further subdivided into many individual types of noise generators from the electric shaver to the punch press to the jet airliner, each with its peculiar problems. In a very rough manner, the noise produced by mechanical equipment may be proportional to the useful power or mechanical work output of that equipment. That is, even without other things being equal, one would expect much more noise from a jet plane than from a vacuum cleaner but the span of noise pressures perceived by the human ear is tremendous and, therefore, comparisons are difficult. One such comparison is that in Fig. 12-2.

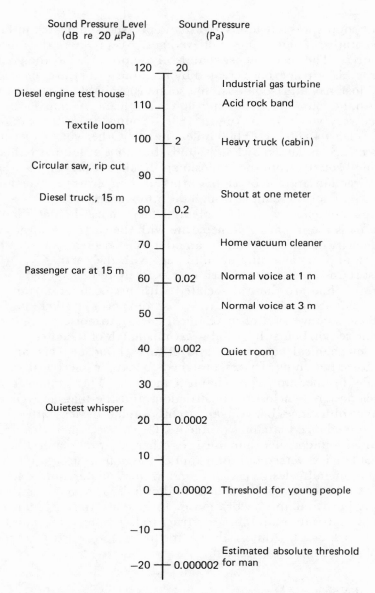

Figure 12-2 Sources of noise and the relationship between sound pressures and sound pressure levels.

The sound pressure levels and sound pressures indicated in Fig. 12-2 are representative of high values, not averages, except where the human voice is the source. The very loudest sounds, those produced at the greatest sound power levels, are not those created by man but rather are natural. Loud as the explosion of an H-bomb is, that sound is probably miniscule in comparison with the sound from the explosion of a volcano such as Krakatoa, which erupted very violently in August 1883. Sounds from the explosions were heard at least 4,800 km (3,000 miles) away. Other volcanic eruptions have also been accompanied by loud sounds, but none as loud as Krakatoa. (Perception of sounds from the explosion of Krakatoa may have been aided by wind direction and/or by *ducting* where a temperature inversion creates a favored path for the sound through the atmosphere.)

A phenomenon with the potential for a noise louder than that from Krakatoa is the impact of a meteorite with the earth. A large meteor at a high velocity carries tremendous amounts of energy, an appreciable fraction of which appears as sound upon impact with the earth. Nevertheless, the loudest sounds to which modern man is likely to be exposed over extended periods of time are those associated with his occupation, and it is these sounds that are apt to cause hearing loss and perhaps other effects as well.

Efficient conversion of mechanical energy to noise energy is very difficult, one reason being the usual great mismatch of the effective impedances of the mechanical equipment and of air. Loudspeakers are specialized impedance-matching devices designed to couple electrical energy to air through the medium of mechanical motion. The proliferation of loudspeaker designs, enclosure construction, and the like, as well as the performance differences between apparently similar devices, attest to the extent of the impedance-matching problem. Of course, when the production of unwanted sound rather than music or other program material is considered, the problem is reversed but some hint of the difficulties of quieting mechanical equipment by design can be found in this comparison. Often, discovering the true source of a particular kind of noise is a major undertaking and learning how to damp a vibration or to increase the impedance mismatch between that source and air (or structural components) may be the most difficult task of all. Almost always the first step toward a solution for this kind of problem is measurement.

Measurement of Noise

Noise measurements are made for two main purposes. First and perhaps most important is to determine the extent of a possible hazard. Second is to determine sources of the noise in such a manner that the information obtained will aid in a solution of the problem. These two purposes are often not served by the same kind of equipment or survey and evaluating one part of the problem even extremely well does not imply that any phase of the other part has even been considered. Hazard evaluation surveys are not conducted in quiet locations and are most often done to define the extent of a known problem. Were it not for the presence of people in a noise field, there would be no problem and, therefore, determining the exposure of people to noise is the first order of business. If only that is done, the information obtained will most often be completely insufficient to pinpoint actual sources and therefore will be insufficient to suggest solutions to any hazard found. All too often the tendency is to attempt to solve the problem by use of personal protective devices and/or limiting the duration of exposure to noise. One reason for that tendency is the lack of knowledge that could be gained from a survey conducted to determine as accurately as possible the main contributors to the problem and source surveys may be avoided because the proper equipment is not easily available.

Because of widespread interest in noise and because noise is measured with electronic devices that are in a very rapid stage of development and minaturization, sound-measuring equipment has undergone and is undergoing rapid evolution. One consequence has been that specifications have been tightened; measurement precision and accuracy have improved; and size, weight, and costs have dropped. Another has been the development of new kinds of devices for integrating exposure and/or dose, for determining the sound energy of impacts or impulses, and for analysis of complex spectra either in real time or from samples of noise recorded on magnetic tape.

Sound Pressure Level Meters

The sound (or noise) level (or pressure level) meter is the basic and most important instrument used in the evaluation of noise problems. This instrument may be small enough to fit into a pocket and sophisticated enough to provide data adequate for a good description of a noise hazard (see Fig. 12-3). Most versions are equipped with slow and fast meter movements and two or three weighting networks and will provide readings over much of the range of sound pressures encountered occupationally. This kind of meter, however, cannot make good measurements of impact noise and usually does not have the filter network shown in Figure 12-3 needed to separate the noise into some kind of a spectrum.

The fast meter movement is not fast enough to do a good job on impact peaks and is not slow enough to be followed by the eye easily for mental integration of readings. Therefore, nearly all sound pressure level meters now have incorporated a slow meter movement to average the momentary fluctuations and provide a more consistent reading over short time intervals.

Weighting networks labeled A, B, and C on most meters correspond roughly to human ear response at low, moderate, and high noise levels, respectively. The C network closely approximates a flat response over the noise spectrum from about 20 to 10,000 Hz and is used whenever the instru-

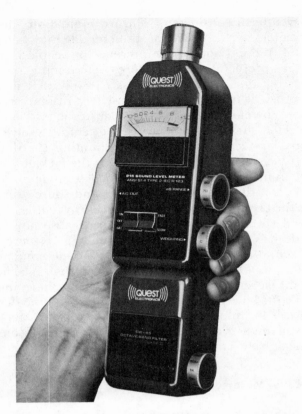

Figure 12-3 A modern sound level meter with an octave band analyzer attached. (*Courtesy* Quest Electronics, Oconomowoc, Wisconsin.)

ment is connected to a spectrum analyzer. The A network readings have been well correlated with noise-induced hearing loss in man and therefore are most widely used. The B network is used rarely and may disappear in time. Convention appends to dB the letter designation of the weighting network used during measurement so that dBA, for instance, refers to the overall sound pressure level in decibels measured with A weighting. Figure 12-4 indicates the characteristics of the three networks [3].

The greatest inaccuracies of a sound level meter are usually associated with the microphone and/or its use. Even the newer dynamic microphones have frequency and range limitations and, in addition, are vulnerable to physical abuse by being dropped, banged against solid objects, splashed with cutting oil or other fluids, and perhaps used inadvertently to test the dew point of a plant atmosphere by being brought in from the cold. Next most vulnerable is probably the meter and its movement even though many

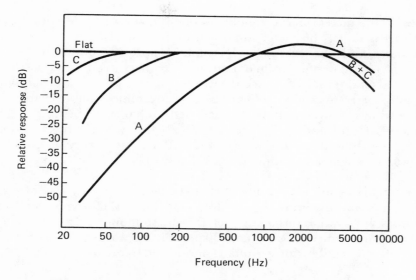

Figure 12-4 Sound pressure level weighting networks. (From the GenRad *Handbook of Noise Measurement,* 7th Ed., p. 9.)

modern meters are very rugged. Least subject to physical abuse are the electronic components that may well survive an accident disabling to a microphone or readout meter.

Sound level meters seem to be very simple devices to the casual observer and perhaps too often are used with little thought given to the fact that noise rather than a vapor or dust is being sampled and noise has properties that differ greatly from the properties of chemical agents. One property of noise is that it is directional and reflects from solid objects, sometimes in a very complex manner and, in addition, may be absorbed to a certain extent by solids. Furthermore, as far as noise is concerned, a person can be a fairly good absorber and reflector even if the person happens to be the one holding the sound level meter. Particularly where the noise has a high-frequency component, measurements at a specific location should be made with the meter as far from the body as practical and, if possible, with the observer's body in several different positions with respect to the meter.

Although noise can be reflected and absorbed, it does not drift around through a volume of air. Therefore, if hazard is being assessed, the microphone should be in the place where ears would normally be. As ears and heads tend to interfere with such measurements, noise level readings are usually best obtained in the absence of the person exposed even though this appears to be contrary to common industrial hygiene practice.

Even omnidirectional microphones are somewhat directional in their response and this is so of all microphones used with sound level meters. Directionality of response is another reason why measurements should be made with the meter operator/holder's body in several different positions. When this is done, the meter almost always must also be held in several orientations and the readings thus may approach omnidirectionality. Under the assault of noise from operating machinery in a factory the human ear is not a good sensor of source direction or intensity and a properly used noise level meter may give indications that are almost unbelievable. The meter readings, however, have a much better chance of being right than does mere listening and while that is especially true when doing a frequency analysis, it also holds for overall readings. If a noise survey gives poor results, even inexpensive equipment is less likely to be at fault than is the user of the equipment.

Even though equipment is not likely to be faulty, the only way of making sure that it is working properly is by frequent calibration. Noise level meter calibrators are available that fit over the microphone and thus test every component of the metering system. Especially because of the vulnerability of microphones, calibration should be performed frequently.

Noise Spectrum Analyzers

Noise spectrum analyzers are essentially filter networks with appropriate center frequency and passband characteristics. The simplest in common use is the octave-band analyzer, which (beginning at 62.5 or 63 Hz) separates the noise spectrum into a series of bands, each with a center frequency double the next lowest (that is, into octaves). Center frequencies are thus 62.5 (or 63), 125, 250, 500, 1,000, 4,000, and 8,000 Hz, although the highest center frequency may be 16,000 or 32,000 Hz and the lowest may be 31.25 or 31.5 Hz. Passband characteristics usually specify a cutoff at about 70 and 140% of the center frequency. Other analyzers may slice the spectrum even thinner, into $\frac{1}{3}$ or $\frac{1}{10}$ octave steps, or may even be able to achieve resolution to a single hertz.

With increasing resolution usually comes increasing price and, perhaps even more disadvantageous, increasing amounts of time necessary to analyze the noise spectrum. The narrow band frequency analyzers thus tend to be laboratory rather than field tools, often used in conjunction with recordings of noise made on a loop of tape that can be played continuously. Whether used with magnetic tape or not, the spectrum analyzer may be equipped with a strip chart recorder so that sound pressure level is plotted as a function of frequency more or less automatically. Particularly when used as a field instrument, the spectrum analyzer may be connected to a rather sophisticated averaging and/or totaling device so that averages over a period of time may be obtained for each portion of the spectrum and this device may, in turn, be connected to a strip chart recorder.

For a direct determination of the potential noise hazard an analysis of the spectrum is usually not necessary. Frequency analysis is used when loudness must be determined and particularly when the noise itself must be examined for clues concerning its source or to determine the effect of a control method that is being tested. For much of this kind of work an octave band analyzer is quite adequate and, therefore, manufacturers have made great strides in miniaturizing such devices so that, if desired, they can be physically coupled to a handheld sound pressure level meter.

A cathode-ray tube (CRT) oscilloscope may be used in conjunction with a sound level meter or with a spectrum analyzer especially when the transient characteristics of the noise are important. Work with a CRT is very likely to be done in the laboratory even though oscilloscopes can be quite portable. When used to analyze magnetic tape recorded noise, the exceptionally good response characteristics of the CRT may be limited severely by the microphone and/or the tape recorder and care must be taken to minimize that kind of potential error.

Impact Analyzers

Impact or impulse sounds pose special problems for the theorist, hygienist, and equipment manufacturer. Foremost is the lack of general agreement on the relationships of any of the impact parameters to hearing loss in man [4, 5, 6, 7]. Several theories with bearing on this problem have in common an assumption that equal amounts of energy received in the inner ear will result in equal amount of injury. At present, however, there is no way to measure the amount of energy received by the inner ear, even from steady-state noise, and if only because of the inertia of bone and membrane the impact noise is probably transmitted through parts of the ear in a manner different from steady noise. Most (but not all) theorists agree that the *threshold* concept applies to noise—that the amount of energy must exceed some value to cause injury—but agreement on the concept of a threshold does not mean or even imply agreement on a numerical value of that threshold.

Most investigators of the impact noise problem agree that the important parameters include pulse duration and rise time as well as peak sound pressure levels. Unfortunately, many impact sounds cannot be well characterized with only three parameters, and the more parameters involved, the greater the number of experiments necessary to evaluate the effect of each. The problem is formidable.

These and other difficulties are reasons why an impact analyzer is not an indispensable component of every industrial hygienist's collection of instruments [8]. Even if equipment manufacturers would decide to produce impact analyzers to one set of specifications or another (and some have), most users or potential users of that equipment would hesitate to buy because the device might be obsolete a short time thereafter. Only a good theory backed by experimental verification is likely to result in the ready availability of inexpensive impact analyzers. Until then, those that are purchased will be used mainly for research purposes in an attempt to find correlations between impact noise and hearing loss in man or experimental animals.

Noise Dosimeters

People who work with ionizing radiation have long worn film badges and dosimeters to provide records of their exposure to external sources of radiation. That concept was extended to noise in the mid 1960s with development of personal noise dosimeters to be worn by workers [9]. Several noise dosimeters are now commercially available.

Many problems must be overcome before the concept of a noise dosimeter can be put into practice [10]. The device must be physically small, light in weight, and reliable. Ideally, either it should be worn close to an ear or it should have its microphone portion near an ear. The microphone must be small, omnidirectional, and very rugged in addition to having a good predictable frequency response. The electronics involved should compensate for the microphone's characteristics and weight the noise received in a manner approximating the response of the ear (the weighting normally used now even in a high noise field is A) and should integrate the noise received over at least a full shift of work according to some preset concept that usually includes a threshold below which noise is ignored by the instrument. In accord with most recent thinking, the noise dosimeter should provide a warning if a predetermined setpoint is exceeded.

All the criteria indicated have been met with at least some success by commercial dosimeters. One problem unlikely to be solved in the near future is that posed by the close proximity of the microphone to the body (or head); this provides shielding from one direction and reflection(s) from others. Field tests have indicated, however, that the proximity problem is probably much less severe than are others related to reliability, precision, accuracy of the threshold, etc. As these problems are solved and as the relative cost of noise dosimeters goes down, more experience will determine the real extent of the proximity problem.

Effects of Noise

Any discussion of the effects of noise on man falls naturally into two catagories: effects on hearing and others or extraauditory effects. That noise affects hearing has been known since antiquity but only recently have attempts been made to quantify those effects. Most effects not related to the ear and hearing are controversial not only with regard to their relationship to health but also with regard to whether they happen at all.

Hearing

Hearing is a very complex process that begins with the external ear and ends in the brain. Some understanding of physiology and function is necessary for a discussion of noise-induced hearing loss.

The Ear. The human ear consists of three physically isolated sections, the external, middle, and inner ear [11]. The external ear consists of the *pinna* (that part outside the head normally called "the ear"), the *external auditory canal*, and the ear drum or *tympanic membrane*. Although some research indicates that the convolutions of the pinna are helpful in localizing sound, most physiologists seem to feel that in man the pinna is more decorative than functional. The canal serves to conduct sound waves from the outside world to the tympanic membrane and to act as the first stage of a very efficient transducer. Sound waves impinging on the tympanic membrane cause it to vibrate, transmitting those impulses to the middle ear.

The middle ear is an air-filled cavity between the external and inner ears. It contains three small bones called the *malleus* (hammer), *incus* (anvil), and *stapes* (stirrup). Vibrations of the tympanic membrane are transmitted to the malleus which rests against it, from there to the incus and stapes, and from the stapes to the oval window of the inner ear. The three bones move as a unit with a lever action that serves as an impedance-matching device, carrying out another step of transduction. Muscular action can affect movement of the middle ear bones and loud noises actuate a reflex that results in a stiffening of the structure, and in less-efficient impedance-matching by the middle ear. The reflex is too slow to give much protection against impact sounds, however, and prolonged exposure to high noise levels causes muscular fatigue; the middle ear was not well designed by nature for an industrial civilization. Very high-energy noises can displace the bones in such a way that they no longer aid in transmitting sounds from the tympanic membrane to the inner ear. This is called *acoustic trauma* or *middle ear deafness* and is one of the forms of *conductive hearing loss* that also may be caused by a physical obstruction such as wax in the canal or perforation or tearing of the eardrum.

In the inner ear, vibrations are actually translated into nervous impulses that, in turn, are sent to the brain along the cochlear nerve. Several theories have been advanced concerning just how this last step of transduction takes place and how prolonged exposure to noise interferes with the process. Although none of the theories offers a complete explanation, each recognizes that inner ear or sensorineural hearing loss is a consequence of overexposure to noise.

The inner ear consists of two structures: the cochlea, which is concerned with hearing, and the semicircular canals, which are concerned with equilibrium and balance. Within the cochlea, which has an appearance similar to a snail shell, are further structures immersed in fluid. The *basilar membrane* appears to function as a pitch discriminator and the organ of Corti, which lies along the basilar membrane, is where the actual conversion of mechanical energy to electrochemical impulses takes place. Sensorineural hearing loss is associated with disruption of portions of the organ of Corti: The location of disrupted sections probably fixes the portion of the spectrum where the loss takes place and the extent of disruption is associated with the extent of the threshold (of hearing) shift.

Sound waves enter the cochlea through the oval window, a membrane actuated by the footplate of the stapes. The waves travel through passages called the *scala vestibuli* and then the *scala tympani* above and below the *cochlear duct,* formed by the vestibular and basilar membranes, to exit at the round window (another membrane) to the middle ear. The organ of Corti lies within the cochlear duct on the basilar membrane. While translation of mechanical to nervous impulses takes place in the organ of Corti, hearing takes place in the brain where the signals from the ear are interpreted.

Hearing threshold. Man cannot hear sounds with frequencies much lower than about 20 Hz unless they are very loud indeed, nor can he hear sounds with frequencies much above about 20,000 Hz regardless of their intensity; his hearing ability diminishes as these areas of the spectrum are approached. In the region between about 275 and 2,500 Hz are found almost all speech frequencies, but the area of greatest hearing acuity is from about 1,000 to about 4,000 Hz.

Many animals can hear sounds that are *ultrasonic* (above hearing range in frequency) for man, and the shape of a hearing sensitivity curve may well vary from species to species; but in the range from 1,000 to 4,000 Hz no animal can hear much better than man. In this range, man's hearing acuity is not quite great enough to enable him to be aware of the noises associated with the motions of air molecules or of those caused by the turbulence of blood in vessels near the eardrum; such noises impose a limit on the ability of all species to detect faint sounds although pure tones may not be masked by broadband noises as much as 20 to 30 dB more intense.

Sounds that are *subsonic* (below hearing range in frequency) for man tend to be felt as vibrations, if their amplitude is great enough, rather than being heard. *Subliminal* (below the range of sensation) subsonics at certain frequencies are able to cause feelings of fear or impending doom at least in some people. Intense ultrasonics may contain enough energy to cause some extraauditory effects but seem to have no effect on man's hearing ability [12].

The ability to hear diminishes with increasing age, a phenomenon called *presbycusis*. This hearing loss first appears at the upper end of the frequency spectrum at about the age of 20 years. Hearing acuity decreases progressively at high frequencies and eventually involves those frequencies associated with speech. Children's high voices are usually not heard well by the aged. Even at very advanced ages, however, there is usually little loss of hearing in the range below about 2,000 Hz for men, although women's hearing may be more seriously affected. This may be partially compensated for by the fact that women tend to lose less of their high-frequency hearing ability with increasing age [13] than men do.

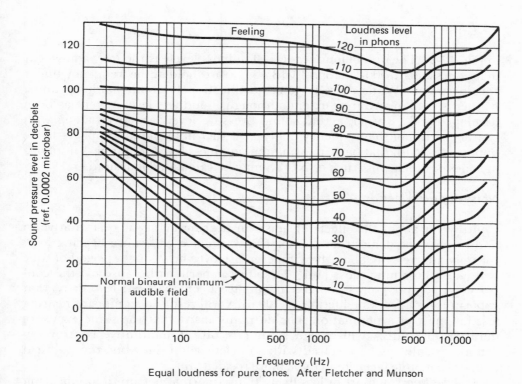

Frequency (Hz)
Equal loudness for pure tones. After Fletcher and Munson

Figure 12-5 Equal loudness curves for pure tones. (From the GenRad *Handbook of Noise Measurement,* 7th Ed., p. 21.)

Some studies appear to indicate that presbycusis may be a form of noise-induced hearing loss (NIHL) caused by chronic exposure to the noises of civilization. Whatever its cause, presbycusis is reasonably uniform in the population of the United States and must be considered when occupational NIHL is being assessed.

Some diseases (notably measles) and some drugs can cause sensorineural hearing loss and thus an increase in the threshold of hearing.

Loudness. *Loudness* is the subjective sensation associated with the quantity of sound. Loudness is not constant with either frequency or sound pressure level and can only be determined by reference to another sound. Many years ago, Fletcher and Munson assembled a group of people who rated the loudness of simple tones by comparing them to the loudness of a 1,000-Hz tone at various sound pressure levels. The result was a series of curves (called, appropriately enough, the Fletcher-Munson curves), each labeled with the sound pressure level of the 1,000-Hz note having equal loudness. The labels were given the name *phons* and called *loudness levels.* Fletcher-Munson curves based on newer data determined by Robinson and Dadson in 1956 at the National Physical Laboratory, Teddington, England appear in Fig. 12-5. The unit of loudness (not loudness level) is the *sone*, equivalent in loudness to a 1,000-Hz tone 40 dB above threshold at the frequency under consideration. While there is some justification for both loudness units, neither is used to any great extent, although a loudness level meter has been developed [14]. Rather, actual sound pressure levels are given along with characteristics of the frequency analyzer if such information appears to be important. With that basic information, loudness level or loudness in phons or sones or other units such as NOYS may be calculated if desired [3, 15].

Noise-Induced Hearing Loss

Controversy about the effects of noise on hearing acuity is quantitative in nature. No one denies that overexposure to noise will cause hearing loss; the only quarrels are about how much noise over how long a period of time will cause how much loss [16]. Despite (or perhaps because of) the controversy, several points of agreement have been reached; few deny that exposures for a working lifetime to 100 dBA will result in significant hearing loss for most of a population, nor do many insist that the same exposure duration to 80 dBA will have a similar result. Quantitative controversy can almost always be resolved with sufficient good data but the political and/or economic facets tend to be more recalcitrant.

Is a deaf person more or less likely than a normal person to lose hearing because of an exposure to noise? Folklore would probably assign noisy jobs to the deaf or, rather, to those already having some NIHL on the theory that because the ear is less sensitive to sound, it is also less sensitive to the damaging effects of sound. Furthermore, if hearing loss is complete, that idea must be correct; a person with no hearing can hardly lose any. Carrying that idea to the extreme, if noise has caused the loss of one hair cell of the organ of Corti, that cell cannot be lost again and, therefore, the loss is "protective" [17]. Some data appear to indicate that hearing loss is proportional to duration of exposure to steady noise at least in the early stages so that a loss offers no protection [18]. Later, when there has been a great deal of injury and a large hearing loss, the rate of loss appears to diminish with time so that at this stage a loss does "protect."

One of the problems involved in trying to settle a question such as the one about the protection offered by a hearing loss is that data are really necessary through an extended period of time. Such data can be obtained by classifying workers according to age and/or time on the job and making assumptions about the quality of their hearing at an earlier age when they started to work, or a group of workers can be followed through time. The two kinds of epidemiological study are called *retrospective* and *longitudinal* or *prospective*. Prospective studies are much easier to interpret and potentially much more valuable than are retrospective ones, but each has its own difficulties. If one begins a prospective study today, what percentage of the work force will be available next year? or in 5 or 10 years? Will the noise exposure for each person stay the same or change? Who can predict the future?

The retrospective study suffers mainly because of imperfect knowledge about the past. That is, usually the hearing status of a worker who had 5 years of experience 10 years ago is assumed to be similar to that of a worker who has 5 years of experience today once the effects of presbycusis are subtracted from both. That assumption may or may not be valid. Furthermore, the chances are good that the noise exposure for each worker has changed over the years, if not in intensity then in pattern. Off-the-job noise exposures have also changed over the years, lending further uncertainties to this kind of study. Nevertheless, retrospective surveys predominate in this and in all other fields because they are usually less expensive and they do not require a good "crystal ball."

Studies of NIHL show that the injury caused by noise corresponds in its effects to other causes of sensorineural hearing loss. That is, no matter what the makeup of the ambient noise to which the exposure occurs [9], loss begins at the high end of the spectrum and progresses toward the low [18]. There are two exceptions to that statement. First, industrial and/or occupational noise is almost always broad-banded so that there is probably very little exposure to loud noise having predominently low-frequency components. Therefore, "no matter what the makeup of the ambient noise" must refer mainly to noise having high-frequency components. Second, quite often NIHL begins with an attenuation in hearing ability in the range of 3,000 to 5,000 Hz, which has been called the 4,000-Hz *dip* or *notch*. Although that pattern is not absolutely characteristic of NIHL, it appears often enough to be noticed; its appearance has not been correlated with specific frequency components of the noise spectrum. A possible third exception to the general statement is that very little data have been published on the industrially related hearing losses of women and as there is a sex difference in presbycusis, there may be a sex difference in noise-induced hearing loss.

When people are exposed to a high level of noise, they almost always exhibit a transient attenuation in their ability to hear. This *temporary threshold shift* (TTS) usually vanishes a few hours after the exposure [20]. One theory of NIHL has held that a *permanent threshold shift* (PTS) is simply the result of a large number of TTS's, each superimposed on the last. According to that theory, then, avoidance of any demonstrable TTS should result in no PTS. Unfortunately, there are no good data available to support a "no TTS—no NIHL" concept. On the other hand, daily exposures resulting in TTS have been shown to result in significant PTS. Avoidance of repeated exposure to noise capable of causing a measurable or noticeable TTS is good practice no matter what exposure guidelines are used [21].

TTS can be confounded with PTS if audiometric readings are made too soon after an exposure to noise. That problem is likely to appear most prominently in severe overexposures because these are associated with in-complete recovery from the TTS. TTS is almost an all-or-none phenome-non, with the rate of recovery being quite dependent on the intensity of the exposure and the amount of TTS (measured immediately after the expo-sure) appears to be more characteristic of the individual than of the expo-sure once a particular threshold for TTS has been exceeded.

That people have different thresholds for TTS and differing amounts of characteristic TTS has been known for many years. Attempts have been made to identify persons most apt to experience PTS by means of either their TTS threshold or the extent of their TTS after a challenge exposure. Although these techniques have appeared promising, they have not led to the correlations of PTS with characteristic TTS patterns necessary for gen-eral acceptance [22]. One reason for this is that the pattern of TTS as a function of frequency may not even be similar for two individuals exposed to the same noise and, in fact, may not even repeat for a reexposure of the same individual. Another is the lack of a well-accepted theory of NIHL that fully explains TTS, PTS, and their interrelationships [20, 21, 23].

Miscellaneous Effects Related to Hearing

Excessive noise interferes with the reception of audible information by the ear. One of the ways in which people demonstrate the noisiness of an en-vironment is to speak to others with the normal and raised voice at one or two distances. Effects of noise on the audibility of speech have been the sub-ject of research that has produced indexes related to intelligibility. One such index with rather wide acceptance is the preferred speech interference level (PSIL), the average sound pressure level in the octave bands centered on 500, 1,000, and 2,000 Hz. For broad-band noise, the PSIL is about 7 dB less than the overall noise level in dBA, and for most practical purposes it can be so estimated. The relationship between the PSIL and distances at which speech can be understood has been determined [3].

Another characteristic of noise that has received rather thorough inves-tigation is that of annoyance. Annoyance is much more of a noise pollution problem than one associated with occupations, but it can be very serious nonetheless [24]. Noises that interfere with sleep [24] or cause people to write angry letters to newspaper editors range from jet plane overflights to police and fire sirens to the noise from large transformers. Public reaction to just about any sound is related much more to its unexpectedness than to its level and is also related to the kind of neighborhood in which it originates and/or is perceived, whether it is continuous or intermittant during daylight hours or at night, and especially what connotations that particular noise has for the listener. Annoyance is not related to hearing loss and, in general, cannot be evaluated just by using the methods developed to assess risk. A measurement of the sound pressure level, in dBA or even in $\frac{1}{10}$-octave bands for instance, may give absolutely no information concerning the probability or intensity of annoyance but may be a valuable benchmark upon which to gauge progress when striving for a solution [6, 27, 28, 29].

Extraauditory Effects of Noise

If noise were to be shown capable of causing a permanent change in the metabolic rate or in the frequency of respiration or to result in excitement or irritability on a chronic, predictable basis, then noise control might be more acceptable. Even if noise were shown able to add to the general stress level of a person, there might be less reluctance to spend large sums of money on control measures. For these reasons, extraauditory effects of noise have been sought assiduously [30, 31]. Occasionally a paper is written purporting to demonstrate one effect or another on animals or people, but for the most part any effects seen have been temporary, nonreproducible, or caused by noise levels considerably in excess of those tolerated by man.

A person who moves from a rural location to one in the city almost always finds that sleep is nearly impossible the first few nights because of the continual interruptions caused by unfamiliar noises. Eventually, however, the ever-alert guardian in his cerebellum learns to accept the new noises as part of the general background; he may not awaken even to the sound of a siren passing his house (but may bound out of bed at the first cough of a sick child two rooms away). Perhaps surprisingly the person who moves from the big city to the country may suffer from the opposite effect. He may not be able to sleep because of the absence of background noises; certainly a nonauditory result.

Man and animals may exhibit a *startle* reaction to a sudden loud noise, and the reaction may be accompanied by signs of sympathetic nervous system stimulation such as increased heart and respiratory rates. If the stimulus is repeated at frequent intervals, however, startle no longer occurs. If sufficient time is allowed to intervene so that startle accompanies every stimulus, then the effect is an acute one and little or no case has been made for chronic extraauditory effects. There appears to be little likelihood that effects of noise not related to hearing loss will be used as the basis for determining control levels (TLV's) [32].

Audiometry

Audiometry is the measurement of hearing. It can be done with or without mechanical/electronic aid, manually or automatically, and with or without a noise-attenuating enclosure; it can determine either air conduction or bone conduction hearing thresholds or both with or without a masking noise being applied to the ear not being tested. Audiometry is done to determine hearing levels of noise-exposed workers or potential workers, to help in the fitting of hearing aids, and to assist in the diagnosis of hearing problems by physicians.

Occupational audiometry is almost always done with a device called a *pure tone audiometer*, a tone generator equipped with a step attenuator so that several pure tones can be transmitted to the subject at several different levels. Most audiometers are equipped with switches to present the tone to either the right or the left earphone and to turn the tone off and on silently. Air conduction with conventional headphones (as opposed to bone conduction) is employed, usually without masking.

Most often, the audiometer is used in conjunction with an audiometric booth, a structure designed to attenuate ambient noises to levels below those that interfere with hearing threshold determinations. Audiometric booths can be built but are usually purchased because of the many problems of assuring good noise attenuation at all frequencies that require experience and knowledge for solutions. Whether a booth is used or not, the subject's environment during the test must be quiet. Reasonable specifications for the noise background in that environment are presented in Table 12-1.

Table 12-1   Maximum Background Levels for Hearing Threshold Determinations (From ANSI Standard S3.1–1960, R1971)

Frequency (Hz)	Sound Pressure Level dB re 20 μPa
250	40
500	40
1,000	40
2,000	47
4,000	57
6,000	62
8,000	67

Determinations should be made with any ventilation equipment in operation and an octave band (or better) spectrum analyzer must be used. Noise-attenuating headphones have been used in some situations; they are much less costly than booths but cannot always substitute [33].

No audiometer, manual or automatic, diagnostic or screening, presently being sold can be relied upon to maintain its calibration over a period of even a few weeks [34]. Although many will remain calibrated over extended periods of time, they cannot be relied upon to do so. Most often, the attenuator develops problems; the frequency synthesizer is less likely to be faulty. In addition to calibration difficulties, switches can become noisy or inoperative. The proper prophylaxis for all of these problems is for the audiometrist to determine his own threshold levels prior to using the instrument, on a daily basis if the audiometer is used that often. The only good

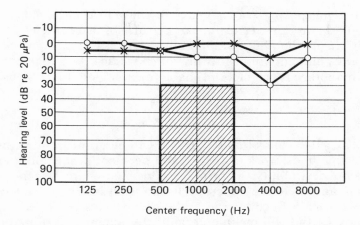

Figure 12-6 A conventional manual audiogram. Circles represent the right ear, crosses the left. The cross-hatched area represents the speech interference region.

substitute for this practice is the daily use of an electroacoustic ear and its importance cannot be overemphasized.

When received, and periodically thereafter, the audiometer should be calibrated electromechanically. Although calibration laboratories are available, this can usually be done with a good sound level meter equipped with an octave band analyzer and the proper coupler to connect its microphone to the earphones of the audiometer, one at a time. Electromechanical calibration should be done at least annually, and more often if the biological calibration indicates any possible error; every frequency, every attenuation step, and both earphones should be included. If this calibration shows any defect, repairs should be made before the audiometer is put back into use. At an interval of 2 to 5 years a more thorough testing of the audiometer should be done and should include measurements of harmonic distortion at each test frequency, accuracy of each frequency, suppression of transients, and so forth in addition to the basic measurements. An uncalibrated audiometer is an expensive but worthless piece of equipment. (In that sentence the name of practically any piece of equipment used by the industrial hygienist could be substituted for *audiometer*.)

The *audiogram,* or record of attenuator setting at threshold versus frequency, is the usual result of audiometric testing and is illustrated in Fig. 12-6. On this graph, a person with hearing equal to the norm will have both ears represented by straight lines across the top at the 0 level. Specifications of the audiometric zero have been the subject of controversy in the past because this zero is not an absolute of any kind [35, 36]. Instead, it represents the average ability to hear of a group of people having healthy ears. Deviations from that norm represent losses in the ability to hear and therefore the graph is arranged with the zero level near the top. And, because the word *loss* disturbs some people who may actually have lost nothing, *hearing level* is now used to identify the vertical axis. Because of inaccuracies of precision inherent in the method [37, 38, 39], audiograms are usually no better than ±5 dB at any point and frequently are much worse. A change of 5 dB from one time to another is therefore probably meaningless under most circumstances.

Noise Control

Techniques used to control noise hazards can be grouped into categories of engineering methods, work practices, and personal protection, more or less in the order of desirability. Control methods of any kind depend on human intervention and therefore fit very well into what is usually called a *hearing conservation program*. Such a program almost always is directly concerned with employees and will include periodic audiometric examinations, noise surveys, constant surveillance over the use of noise specifications on new equipment, proper education of all concerned regarding safe work practices, and perhaps enforcement of the use of personal protective devices.

Engineering Control

Engineering control of noise is usually achieved either by modifying the noisy equipment or process or by modifying the environment in which the equipment or process is located. Quite often both methods are applied in an effort to achieve the greatest amount of attenuation at the least cost, and there are very few absolute guidelines to follow. Examples of successful control abound in the literature of noise and constant attention to that source of information is likely to be very helpful.

Source control. Substitution of a quieter machine, process, or material should be considered first when noise control is sought. Particularly important is the increasing use of noise specifications on new equipment [40]. Although many equipment manufacturers still claim that theirs is a kind of machinery that is inherently noisy, competitors quite often have found a way to eliminate whatever was inherent, resulting in much less noise being generated. Some groups of manufacturers have gone so far as to set up specifications on how to rate the product they make. Quite often noise rating specifications will deal with maximum sound pressure levels, octave band levels, or even $\frac{1}{3}$-octave band levels at certain points and distances from a standard mounting of the product. Occasionally, these measurements are made in such a manner that the radiated acoustic power (PWL) can be determined [41, 42] and made a part of the specification. This is greatly to be desired in that noise levels may be calculated from the PWL in most situations [43]. Groups with standardized noise test codes include those making air moving equipment, transformers, cooling towers, electric motors, pneumatic tools, and others.

When the machinery cannot be changed or quieted, quite often the problem lies in the process itself. For instance, a machine might be required to make pellets from strands of hard thermoplastic material. Changing the slope of the blade doing the work from straight across to slanting will change the process from breaking to cutting, with an obvious reduction in noise. If the basic process could be changed so that pellets rather than strands were formed initially, however, there would be no need for a cutter of any sort and the noise problem would vanish.

Changing the material is another way of altering a noise source. Use of carpet on the floor of a hall, for instance, instead of wood, concrete, or vinyl tile will result in a reduction of the noise associated with foot traffic. Sometimes use of heavier metal walls on noisy equipment will result in a marked reduction of radiated noise by changing the vibrational characteristics of the surfaces.

Modification of the noise source by reducing driving forces is another way of reducing noise. If the amount of wasted power appearing as noise is a constant fraction of the power input to the device, the amount of noise will be reduced in proportion to the power reduction. Quite often the fraction of power that appears as noise is not constant but increases with increases in power, so that even a small power reduction can pay large dividends.

Other equipment modifications that can be used include reduction of the surface area available for noise radiation and the use of damping materials especially on large surfaces [44, 45]. Even if damping is not particularly successful in reducing the radiated acoustic power, its use may well change the character of the noise (from a *ping* to a *thud*, for instance) and make it more acceptable.

Control by modifying the environment. Whenever the amount of sound radiated from a piece of equipment cannot be further reduced, the next step is probably to attempt to confine the sound by means of an enclosure of some sort. This may be difficult to do, and not only because noise has a tendency to leak through even very small cracks. Equipment confined well enough to keep noise from escaping often retains waste heat as well with disasterous results. Furthermore, mere confinement is not enough; the container must be made of material able to attenuate sound adequately for the purpose [46]. Devising good noise attenuators that are also inexpensive is difficult because, for the most part, attenuation of noise is a function of the mass of the attenuator. For this reason, lead or leaded plastic sheets are used to provide what has been called a *limp mass*, one that does not easily transmit vibrations of any sort [47].

Vibration isolation is often necessary to achieve good noise control. Especially with heavy equipment, vibrations can cause a whole structure to be a source of noise and then no amount of confinement will provide much relief. Vibration isolators can now be designed for a guaranteed job of isolation [48].

Another technique with application for sources of a relatively pure-toned noise is to use a resonant cavity for its absorption. As with electrical circuits, the proper combination of circuit components can provide almost complete attenuation of a pure tone.

Last of the environment changes in overall effectiveness is the use of sound absorbers in the general area of the noise sources(s). Because this technique is often the least expensive, it is often the first used and disappointment with the results may cause other methods to be abandoned before being tried. Acoustic tile for ceilings, suspended sound absorbing units, and "acoustic" plaster all have their uses and their defenders but for the most part they should be reserved for the last touch in a program that is already almost successful. No treatment of walls or ceiling will, for instance, be of much value to the person who works a meter or two from the noise source (as usually is the case) even though such treatment may reduce exposures of his neighbors [49]. Most sound absorbers on the market today are very poor attenuators. That is, while they may provide some help in the room in which they are used by preventing reflection of noise, they will not usually reduce the amount of noise transmitted through a wall or ceiling materially. Acoustic tile treatment of the basement ceiling will not give the parents upstairs much relief if their children are having a noisy party in the basement (but complete silence might be even more disturbing in this case).

Work Practice Control

Control of noise exposure by control of work practices seeks to limit the total exposure by limiting the time allowed in areas of excessively high noise levels. This kind of exposure reduction has long been used where the hazard is ionizing radiation, but it could be applied to noise only after the data base had been expanded to the point where noise exposures could be integrated with reasonable confidence. That is, exposure-time limitations cannot be applied unless good knowledge is available concerning the additivity of exposures [21, 22]. While such knowledge is still far from perfect regarding noise [50], it may be adequate for this purpose.

The basis for noise exposure integration is the fact that the maximum no-effect doses seem to fall on a straight line when the logarithm of the duration of exposure (per day) is plotted versus the noise level in dBA. That is, if, say, one could endure an exposure to Y dBA for 8 hr/day, 5 days/week with no effect, then if the exposures were only for 4 hr/day, the safe level would be $Y + X$ dBA, and the safe exposure for 2 hr/day would be $Y + 2X$ dBA, and so forth. A perhaps unexpected result of applying this concept is that the safe level for 16 hr/day would then be $Y - X$ dBA and that for 24 hr/day would be $Y - 1.59X$ dBA. The safe continuous exposure to noise is not known for man, but the chances are good that it will not be represented by $Y - 1.59X$ for any of the values Y and X used for regulatory purposes.

Personal Protection

Devices for protection against noise are usually classed as earplugs or ear-muffs. *Earplug* is the term used for any kind of material whether in plug form or not inserted into the external auditory canal to attenuate noise. *Earmuff*, on the other hand, is used for devices applied external to the canal that cover the pinnae. Two disadvantages are shared by ear protectors of any kind. First is that the protection must be used to be effective and, fur-thermore, it must be used correctly. An ill-fitting plug or muff may offer essentially no protection against noise. Second is that these devices have an absolute upper limit on the attenuation offered. That limit is imposed by the noninfinite attenuation provided by the skull, which averages about 40 dBA but varies with frequency. The best attenuation that a plug or muff can hope to offer is to equal that of the skull [51]. If more protection than that is needed, a helmet may be used, but noise can still be transmitted through other bones, soft tissues and airways to the head so the gain from more complete enclosure of the head is not very dramatic.

Usual experience has been that muffs provide more protection than plugs. The many reasons behind that observation include the fact that muffs are far more visible than are plugs and their use is easier to police. Another is that a good fit may be easier to obtain with a muff than with a plug, and in many situations the muffs are more comfortable than well-fitting plugs. Finally, by their construction, muffs can interpose more mass between the noise and the tympanic membrane than can plugs and there-fore should be more effective [51]. That reason, however, is often not valid as tests have shown some plugs exceeding some muffs in attenuation ability. Maximum attentuation is provided by wearing both plugs and muffs.

The efficiency of any sound attenuator except a vacuum is frequency dependent, and this applies to plugs and muffs as well as to walls [53]. In all cases the protection provided is less at low frequencies and greatest at high frequencies. This is fortunate because the ear is least sensitive to injury at low frequencies. No two varieties of plugs or muffs provide exactly the same attenuation across the frequency spectrum and the opportunity exists to fit the protector to the noise as well as to the individual [54]. In practice, how-ever, the first rule (that the protector must be worn to be effective) domi-nates and employers tend to provide the protection employees will wear no matter what degree of attenuation is provided [55].

Several steps are necessary for a hearing conservation program using personal protective devices to be effective [56, 57]. They are as follows:

1. Select the people to protect. That is, use personal protective devices only where other controls fail and then identify the population of workers where this kind of protection must be used.

2. Select the type(s) of protection to use. Do this on the basis first of test results and then of employee acceptance. The first cost of all plugs is

much less than that of muffs; the overall cost of muffs may be much less, however, because they are less likely to be lost and they wear much longer than plugs.

3. Ensure that the protective devices fit each wearer and that each wearer knows the proper technique for using the product.

4. Ensure that the protective devices are used by all who need them.

5. Periodically evaluate the hearing of persons using the protective devices. If hearing is being lost, either the equipment or its use is at fault. Determine the fault and take corrective action.

6. Periodically evaluate the whole program.

REFERENCES

[1] W. ALEXANDER, "Some Harmful Effects of Noise," *Can. Med. Assn. J.* **99** (1968), 27–31.

[2] P. L. MICHAEL, "Physics of Sound," Chapter 23 in *The Industrial Environment—Its Evaluation & Control,* (Washington, D.C.: U.S. Department of Health, Education, and Welfare, Public Health Service, Center for Disease Control, National Institute for Occupational Safety and Health. For sale by the Superintendent of Documents, U.S. Government Printing Office, 1973), 299–308.

[3] A. P. G. PETERSON AND E. E. GROSS, *Handbook of Noise Measurement,* (West Concord, Mass.: General Radio Company, 1967).

[4] K. D. KRYTER, "Evaluation of Exposures to Impulse Noise," *Arch. Env. Health* **20** (1970), 624–35.

[5] G. A. LUZ AND D. C. HODGE, "Recovery from Impulse-Noise Induced TTS in Monkeys and Men: A Descriptive Model," *J. Acous. Soc. Am.* **49** Pt. 2(1971), 1770–77.

[6] G. R. C. ATHERLEY AND A. M. MARTIN, "Equivalent-Continuous Noise Level as a Measure of Injury from Impact and Impulse Noise," *Ann. Occup. Hyg.* **14** (1971), 11–28.

[7] H. McROBERT AND W. D. WARD, "Damage-Risk Criteria: The Trading Relation Between Intensity and the Number of Nonreverberant Impulses," *J. Acous. Soc. Am.* **53** (1973), 1297–1300.

[8] W. R. KINDERT, "The Impulse Sound-Level Meter—What's It All About?," *Sound Vib.* **8** (Mar., 1974), 50–53.

[9] F. W. CHURCH, "Development of a Personal Monitoring Instrument for Noise," *Am. Ind. Hyg. Assn. J.* **26** (1965), 59–63.

[10] R. G. CONFER, J. H. BLACKER, AND R. S. BRIEF, "Evaluation of Personal Noise Dosimeters," *Am. Ind. Hyg. Assn. J.* **33** (1972), 767–74.

[11] J. R. ANTICAGLIA, "Physiology of Hearing," Chapter 24 in *The Industrial Environment—Its Evaluation & Control,* Ref. 2, pp. 309–20.

[12] W. I. ACTON AND M. B. CARSON, "Auditory and Subjective Effects of Airborne Noise From Industrial Ultrasonic Sources," *Brit. J. Ind. Med.* **24** (1967), 297–304.

[13] R. GALLO AND A. GLORIG, "Permanent Threshold Shift Changes Produced by Noise Exposure and Aging," *Am. Ind. Hyg. Assn. J.* **25** (1964), 237–45.

[14] B. B. BAUER, E. L. TORICK, AND R. G. ALLEN, "The Measurement of Loudness Level," *J. Acous. Soc. Am.* **50,** Pt. 1 (1971), 405–14.

[15] R. P. OWENS, "A Computer Program for Acoustic Loudness," *Sound Vib.* **8** (Mar., 1974), 54–55.

[16] J. C. GUIGNARD AND D. L. JOHNSON, "The Relation of Noise Exposure to Noise Induced Hearing Damage," *Sound Vib.* **9** (Jan., 1975), 18–23.

[17] E. R. HERMANN, "Biomechanics of Human Hearing," *Noise Control Eng.* **6** (1976), 10–21.

[18] E. J. SCHNEIDER, ET AL., "Correlation of Industrial Noise Exposures with Audiometric Findings," *Am. Ind. Hyg. Assn. J.* **22** (1961), 245–51.

[19] P. LABENZ, A. COHEN, AND B. PEARSON, "A Noise and Hearing Survey of Earth-Moving Equipment Operators," *Am. Ind. Hyg. Assn. J.* **28** (1967), 117–28.

[20] J. H. BOTSFORD, "Theory of Temporary Threshold Shift," *J. Acous. Soc. Am.* **49,** Pt. 2 (1971), 440–46.

[21] W. D. WARD, "Temporary Threshold Shift and Damage-Risk Criteria for Intermittent Noise Exposures," *J. Acous. Soc. Am.* **48** (1970), 561–74.

[22] M. SCHMIDEK AND P. CARPENTER, "Intermittent Noise Exposure and Associated Damage Risk to Hearing of Chain Saw Operators," *Am. Ind. Hyg. Assn. J.* **35** (1974), 152–58.

[23] A. COHEN, J. R. ANTICAGLIA, AND P. L. CARPENTER, "Temporary Threshold Shift in Hearing from Exposure to Different Noise Spectra at Equal dbA Level," *J. Acous. Soc. Am.* **51,** Pt. 2 (1972), 503–07.

[24] L. S. GOODFRIEND, "Community Noise Problems—Origin and Control," *Am. Ind. Hyg. Assn. J.* **30** (1969), 607–13.

[25] J. S. LUKAS, "Effects of Aircraft Noise on Human Sleep," *Am. Ind. Hyg. Assn. J.* **33** (1972), 298–303.

[26] S. N. GOLDSTEIN, "A Prototype Index for Environmental Noise Quality," *Sound Vib.* **6** (Feb., 1972), 30–33.

[27] T. LINDVALL AND E. P. RADFORD, "Measurement of Annoyance Due to Exposure to Environmental Factors," *Env. Res.* **6** (1973), 1–36.

[28] C. CACCAVARI AND H. SCHECHTER, "Background Noise Study in Chicago," *J. Air Poll. Con. Assn.* **24** (1974), 240–44.

[29] F. J. BELGES, "Derivation, Application and Interrelationship of Ambient Noise Measurement Parameters," *J. Air Poll. Con. Assn.* **24** (1974), 1080–84.

[30] A. COHEN, "Extra-Auditory Effects of Occupational Noise. Part I: Disturbances to Physical and Mental Health," *Natl. Safety News* **108** (Aug., 1973), 93–99.

[31] A. COHEN, "Extra-Auditory Effects of Occupational Noise. Part II: Effects on Work Performance," *Natl. Safety News* **108** (Sept., 1973), 68–76.

[32] J. E. BROWN III, ET AL., "Certain Non-Auditory Physiological Responses to Noises," *Am. Ind. Hyg. Assn. J.* **36** (1975), 285–91.

[33] A. B. COPELAND AND H. J. MOWRY, "Real-Ear Attenuation Characteristics of Selected Noise-Excluding Audiometric Receiver Enclosures," *J. Acous. Soc. Am.* **49,** Pt. 2 (1971), 1757–61.

[34] W. G. THOMAS, ET AL., "Calibration and Working Condition of 100 Audiometers," *Public Health Rep.* **84** (1969), 311.

[35] H. DAVIS AND F. KRANZ, "The International Audiometric Zero," *Am. Ind. Hyg. Assn. J.* **25** (1964), 354–58.

[36] E. C. RILEY, ET AL., "Critique on the Concept of Audiometer Zero," *Am. Ind. Hyg. Assn. J.* **26** (1965), 45–51.

[37] R. E. GOSZTONYI, JR., L. A. VASSALLO, AND J. SATALOFF, "Audiometric Reliability in Industry," *Env. Health* **22** (1971), 113–18.

[38] R. W. HOWELL AND B. P. R. HARTLEY, "Variability in Audiometric Recording," *Brit. J. Ind. Med.* **29** (1972), 432–35.

[39] E. R. HARFORD, "Determining Air Conduction, Pure Tone Hearing Thresholds," *Natl. Safety News* **105** (Feb., 1972), 62–69.

[40] C. EBBING AND P. B. OSTERGAARD, "Noise Level Specifications for Machinery and Equipment," *Sound Vib.* **7** (Jan., 1973), 22–26.

[41] R. W. RAYMOND, "Automatic Measurement of Sound Power Level," *Sound Vib.* **8** (Nov., 1974), 12–16.

[42] T. J. SCHULTS, "Outlook for in-situ Measurement of Noise from Machines," *J. Acous. Soc. Am.* **54** (1973), 982–84.

[43] L. C. MARRACCINI AND D. A. GIARDINO, "Predicting Sound Levels from Sound Power Data," *Sound Vib.* **8** (Nov., 1974), 28–30.

[44] G. E. WARNAKA, H. T. MILLER, AND J. M. ZALAS, "Structural Damping as a Technique for Industrial Noise Control," *Am. Ind. Hyg. Assn. J.* **33** (1972), 1–11.

[45] M. A. K. HAMID, "Viscoelastic Materials for Panel Vibration Damping," *Sound Vib.* **7** (July, 1973), 20–26.

[46] P. D. SCHOMER, "Measurement of Sound Transmission Loss by Combining Correlation and Fourier Techniques," *J. Acous. Soc. Am.* **51,** Pt. 1 (1972), 1127–41.

[47] B. FADER, "Practical Designs for Noise Barriers Based on Lead," *Am. Ind. Hyg. Assn. J.* **27** (1966), 520–25.

[48] C. M. SALERNO AND R. M. HOCHHEISER, "How to Select Vibration Isolators for Use as Machinery Mounts," *Sound Vib.* **7** (Aug., 1973), 22–28.

[49] R. D. BRUCE, "Noise Control of Metal Stamping Operations," *Sound Vib.* **5** (Nov., 1971), 41–45.

[50] M. SCHMIDEK, B. MARGOLIS, AND T. L. HENDERSON, "Effects of the Level of Noise Interruptions on Temporary Threshold Shift," *Am. Ind. Hyg. Assn. J.* **36** (1975), 351–57.

[51] J. H. BOTSFORD, "Ear Protectors—Their Characteristics and Uses," *Sound Vib.* **6** (Nov., 1972), 24–29.

[52] J. M. FLUGRATH AND B. N. WOLFE, "The Effectiveness of Selected Earmuff-Type Hearing Protectors," *Sound Vib.* **5** (May, 1971), 25–27.

[53] T. H. TOWNSEND AND F. H. BESS, "High-Frequency Attenuation Characteristics of Ear Protectors," *J. Occup. Med.* **15** (1973), 888–91.

[54] D. L. JOHNSON AND C. W. NIXON, "Simplified Methods for Estimating Hearing Protector Performance," *Sound Vib.* **8** (June, 1974), 20–27.

[55] D. B. SUGDEN, "Some Notes on the Provision of Personal Hearing Protection for Fettlers at an Iron Foundry," *Ann. Occup. Hyg.* (London) **10** (1967), 263–67.

[56] S. PELL, "An Evaluation of a Hearing Conservation Program—A Five-Year Longitudinal Study," *Am. Ind. Hyg. Assn. J.* **34** (1973), 82–91.

[57] N. NIGHTHAND, ET AL., "A Computer-Oriented Hearing Conservation Program," *J. Occup. Med.* **16** (1974), 654–58.

13

BIOTHERMAL STRESS

Since before the industrial revolution man has sought a way to answer the question, "How hot (or cold) is it?" with a single number. The first device to be used for this purpose was the ordinary thermometer, improved by Fahrenheit who added a scale of *degrees*. Unfortunately, even Fahrenheit's degrees were unable to answer the question. Consider, for instance, the difference in one's sensations at 35°C (95°F) when the relative humidity is 90% and when it is 5%. That humidity had something to do with the effect of the thermal environment on man was recognized early, but not until the 20th century was a real attempt made to include some measure of this variable in answers to the basic question. In 1905 Haldane (cited by Hatch [1]) proposed that the wet-bulb temperature be used as an index of thermal stress instead of the dry-bulb temperature.

Although Haldane's idea had merit, it too proved not to be the whole answer. For one reason, wet-bulb temperature was shown to have little to do with sensations of cold and, for another, air velocity seemed to be at least as important as wet-bulb temperature even in the heat as the use of fans can attest. Even with the dry- and wet-bulb temperatures and air velocity the basic question could not be answered. Consider the pictures that are shown every winter of people in bathing suits apparently enjoying themselves skiing when the dry-bulb temperature may be $-10°C$ ($14°F$), the relative humidity 25%, and the breeze appreciable, but the sun is shining brightly and the people are exercising strenuously. Or, consider a person clad appropriately for an attempt on the peak of Mount Everest but sweltering in an area where the (dry-bulb) temperature is $10°C$. In sum, experience showed that the variables important to sensations of heat and cold were

> Dry-bulb (air) temperature
> Wet-bulb temperature (a measure of humidity)
> Air velocity
> Radiant heat exchange rate
> Exercise (or metabolic) rate
> Clothing

To even think of combining all of these variables into a single number proportional to feelings of heat or cold or proportional to the stress exerted by the thermal environment seems futile except that the human body is able to do just that. Why should experimenters be daunted? Research actually went two separate ways to concentrate on thermoregulation and biothermal strain or on the environment and the parameters of comfort and biothermal stress. This dichotomy still persists, reflecting the backgrounds and interests of those working on the problems.

Temperature Regulation in Man

The Hypothalamus

A part of the brain beneath the thalamus and above the roof of the mouth is called the *hypothalamus* and has the main responsibility for body temperature control in *homeotherms* (those animals with *constant* body temperature). The hypothalamus accepts information concerning the temperature status of the body, compares that information with a "set point," and initiates action appropriate for control, from sweating to shivering to pulling up the blanket while still asleep. How the hypothalamus does all these things is still somewhat of a mystery.

Information concerning the body's temperature can reach the hypothalamus in two ways. First, the hypothalamus can sense the temperature of blood reaching it and hence gain information concerning the overall status of the body. Second, it can accept information via the neural network, from sensory nerves located anywhere in the body, and gain somewhat similar knowledge in this manner. The means by which the hypothalamus weights data from the two systems and how such weighting may be altered by warming or cooling the body are under investigation and, in fact, are controversial. Nevertheless, that a central site for thermoregulation exists is well accepted and that such a site (or sites) is (are) associated with the hypothalamus is not disputed. In addition to hypothalamic thermoregulation, however, evidence also indicates that there are direct peripheral effects caused by local warming or cooling in which there may be little involvement of the hypothalamus or any other central regulator.

No control theory has been developed for any system of homeostasis from body temperature to liquid level in a tank that does not include a set point. That is, information concerning the status of the system must be compared with a set point or goal in order to determine whether or not there is any deviation from that goal. Therefore, the hypothalamus must have a set point of some kind with which to compare information about the body's temperature status. The fact that such a set point must exist has, however, been of little help in finding out just what or where it is. Somehow the hypothalamus knows that the human body should have a temperature that fluctuates in a narrow range around 37°C (98.6°F), but as yet physiologists do not know how the hypothalamus can be aware of this simple fact.

Control theory calls for a set point, *feedback* (information on the status of the controlled variable), and some kind of action to effect regulation of the controlled variable. Of the four possible kinds of control action (on-off, proportional, integral, and rate) the body appears to use all but one (integral) in the control of temperature [2].

Control Actions

The body has three basic actions with which to respond to thermal stress, namely shivering, sweating, and controlling blood circulation. Shivering, an example of on-off control, causes an increase in the metabolic rate and hence the heat production of the body and is the typical response to cold [3], although cold can cause metabolic rate increases without any shivering. No analogous response to heat exists; the metabolic rate does not decrease as a result of heat stress. Instead, the basic response to heat stress is to secrete sweat, which, by evaporation, causes heat to be lost from the body. Sweating rates increase in proportion to heat stress and also, in localized areas, in proportion to the rate of temperature rise of the skin. Sweating is therefore both proportionally and rate controlled.

Both heat and cold stress cause circulatory changes but of an opposite nature, examples of proportional control. When the skin cools, for instance, blood flow is shunted from vessels which lie near the surface to those which are deeper within tissue. This is done by reducing the diameter of surface capillaries through muscular action (*vasoconstriction*) and/or by increasing the diameter of the deeper vessels (*vasodilation*). Although the resistance of body tissues to the flow of heat is not altered materially by either heat or cold stress, the overall conductance (reciprocal of resistance) from the core to the periphery (which includes transfer of heat by blood flow) can be altered dramatically. This third control method is usually called *vasomotor* control because it is a control of blood vessels (*vaso-*) by motor nerves.

When the amount of control needed falls beyond the ability of effector systems, the inevitable result is loss of homeostasis and body cooling or heating unless voluntary actions to control the body's environment supervene. Man and animals can tolerate more cold stress than heat stress and, in addition, have a much greater range of voluntary activities in response to cold than heat. In the cold a man can voluntarily exercise, seek shelter, increase the insulation value of his garments, light a fire, and so forth. In heat, clothing removal is limited by nudity and the metabolic rate can only be reduced to whatever minimum applies to the species. Unless a body of cool water or an air conditioner, fan, dehumidifier, or other mechanical contrivance is handy, man in a hot environment may have a real problem disposing of his metabolically generated heat.

Acclimatization

Also called *acclimation* (which is probably the better term unless a true climate is involved, but which is not much used), *acclimatization* is the process by which a person or animal becomes adapted to or more tolerant of stress. Although true acclimatization of animals to cold has been shown by increases in basal metabolic rates and survival times, whether the process occurs in man is controversial. One reason for the controversy is that people are not likely to allow themselves to become very cold very often; they wear clothing and seek shelter to prevent such happenings and thus are quite unlikely to experience real cold acclimatization, however it might be measured. Nevertheless, "artificial" acclimatization in man has been demonstrated [5, 6, 7] and natural cold acclimatization probably takes place in some aboriginal peoples such as the natives of Australia and Tierra del Fuego. Cold acclimatization rarely, if ever, occurs among industrialized populations.

In contrast to cold acclimatization, which is experienced by few humans, heat acclimatization is probably experienced by most to some extent and by many to the fullest extent. Characteristic of unacclimatized work in the heat is first discomfort, which may be followed by increased heart rate, dizziness and nausea, high skin and core temperature, secretion of too little sweat containing too much salt, and possible collapse [8]. All of these problems may be associated with a task which is fairly easy to perform in cool surroundings but which becomes nearly impossible in the heat. If attempts to do the job are continued in the heat for a week or so, however, a dramatic improvement in performance results. The discomfort will probably disappear and, perhaps suddenly, the job is once again easy to do. The heart rate goes down along with the body and skin temperatures and the sweat rate goes up, but the sweat is much more dilute; its salt concentration will have dropped considerably.

Two necessities have been determined for true heat acclimatization to occur; work and heat. That is, little heat acclimatization if any occurs unless work is performed, but even the best physical condition does not of itself confer acclimatization [9, 10]. By performing work in the heat, however, acclimatization can occur very rapidly; for a person who is in good physical shape, 2 hr of hard work in the heat for 4 or 5 days will result in about 80% of the possible acclimatization to that job and heat stress. Acclimatization to one set of conditions (metabolic rate and environment) facilitates acclimatization to more severe conditions so that the best way to acclimatize to hard work under high heat stress is to begin with moderate work loads and increase them in environments that simulate the most stressful that will be encountered.

The process of heat acclimatization is very complex and involves the cardiovascular system [11], hormones associated with the pituitary and adrenal glands, and the mechanism of sweat production [12, 13]. How these systems interact to accomplish the bodily changes called *heat acclimatization* is under active investigation with few generally accepted answers [14]. Acclimatization to heat is a fact; it does occur; it is a natural defensive adaptation of man and many animals; it protects against the major consequences of excessive heat stress; it is beneficial; and it involves major changes in the body's thermoregulatory systems. Acclimatization to heat is one of the best examples of the fact that stress can and does have survival value for man.

Progress of heat acclimatization is most easily followed by observing changes in heart rates or temperatures of people undergoing the process. Measurement of electrolytes in sweat could be used as could the rate of sweat production associated with exercise but both of the latter are more difficult than heart rate or temperature to measure and/or monitor. Once acclimatized, a person will remain so for about 2 weeks even with no further work in the heat, but the extent of acclimatization begins to fall even after a few days [15]. The rate of becoming acclimatized can be reduced and the rate of losing acclimatization can be increased by dehydration and by inadequate salt intake. During work in the heat, thirst is not an adequate indicator of the amount of water needed to assure complete hydration; the best indicator is a measure of the water lost by sweating and urination. And, even though salt tablets are often available and are often used to combat electrolyte imbalance, salt is best tolerated by the GI tract when taken with food at regular mealtimes or with snacks, or with drinking water in a concentration of about 0.05% by weight.

Consequences of Control Failure

Cold stress. *Frostbite,* damage to tissue caused by overexposure to low temperatures, is the most probable consequence of excessive cooling. Usually involving the toes, nose, ears, or fingers; rarely the cheeks, feet, or hands; and even more rarely other parts of the body, frostbite can cause injury ranging in severity from quite superficial but painful to frank necrosis. A consequence of cold stress is the shunting of blood away from exposed areas in the extremities and away from the skin. Because of the decreased blood supply, temperature in those tissues can drop rapidly, especially if the overlying skin is exposed to extreme cold or wind chill. Frostbite is less likely to occur when the person is actively exercising than when he is sedentary or asleep, but even hard exercise is not certain to prevent it if the weather is cold enough. Frostbite is not particularly painful as the tissue becomes colder but may be very much so on rewarming.

Prolonged cold leading to *hypothermia* (subnormal body temperature) causes lethargy and somnolence that make any activity more difficult, leading to increased hypothermia. Hypothermia and its consequences can take place with little warning except for the cessation of violent shivering. Rewarming is the only treatment.

A combination of cold and vibration can cause *Raynaud's phenomenon* or *traumatic vasospastic disease.* This is a condition, usually of the fingers and hands, characterized by pallor caused by a greatly diminished blood supply resulting from spasm of the blood vessel walls. In addition to white fingers, the victim may also experience numbness of the affected areas. The disease is most prevalent among those who work with vibrating machinery in the cold and a typical occupation is that of chain saw operator. Both cold and vibration appear to be necessary over a prolonged period of time, but precise conditions have not been well characterized.

Heat stress. The first consequences of overexposure to heat stress are likely to be discomfort, irritation, and reduced efficiency of mental work. These psychological reactions are likely to be accompanied by psychophysiological reactions of increased error rates, reduced efficiency of physical work [16], and increased risk of accidents. All of these reactions may take place with no significant increase of body temperature and all become much less likely with good acclimatization.

A consequence of heat stress is sweating, and keeping the skin damp with sweat may cause the minor skin disorder called *prickly heat.* Another consequence of heat stress is peripheral vasodilation and shunting of large amounts of blood away from the core. This can cause *heat syncope,* characterized by dizziness and possibly fainting [17, 18].

More serious consequences of heat stress are water and salt imbalances caused by excessive sweating and complete failure of the temperature regulating system. Excessive sweating causes losses of both water and salt from the body, leading to dehydration and, because of the salt loss, electrolyte imbalances. Dehydration causes a decrease in blood volume and an increase in blood viscosity; both cause increased circulatory strain. The imbalance of electrolytes can lead to heat cramps and/or heat exhaustion. *Heat exhaustion* is extreme fatigue, with dizziness, nausea, and subnormal temperature. *Heat cramps* are painful muscle spasms accompanied by dilated pupils and a weak pulse. In neither disorder is there an abnormally high body temperature.

When the temperature regulating system becomes overloaded and gives up, the consequences are anhidrotic heat exhaustion, hyperpyrexia, or heat stroke. *Anhidrotic* means without sweat. Anhidrotic heat exhaustion was first described during World War II. Its most striking sign is the almost complete lack of sweat on the trunk and extremities but a flushed and profusely sweating face and neck. There is no dehydration and no real electrolyte imbalance, but the body temperature is elevated a degree or so (°C) [17] so that the term *hyperpyrexia*, which means excessive accumulation of heat, is also used.

Heat stroke is caused by excessive heat storage in the body and is characterized by the complete absence of sweating, a much elevated body temperature [usually 41°C (106°F) or higher], disorientation, delirium, convulsions, and/or loss of consciousness [19]. Without intervention it is usually fatal [20]. Heat stroke is most likely to occur in a small fraction of the population unusually susceptible to heat and, in that population, most often in unacclimatized persons. Acclimatization reduces the risk of the more serious effects of heat stress partially because of somatic adaptations and partially because the acclimatized person learns his limits and is less likely to exceed them than is the unacclimatized.

Heat stroke seems to be more likely in obese elderly people, particularly those who are alcoholics or who have a history of alcohol imbibition [17]. Hypertension may be a consequence of chronic exposure to heat [21].

Thermal Comfort

Although man has heated his residence and/or his surroundings (Note: He does not heat himself; the consequences of doing so are examined in the preceding paragraphs.) almost as long as he has existed, only recently has he been able to cool his immediate environment more or less at will. One consequence of being able to cool as well as heat was the realization that little data were available upon which to base a definition of thermal comfort, and the design of environmental control systems needed such a definition. Engineers design heating and cooling systems and, therefore, engineers were forced to define thermal comfort zones well before there were any thermal physiologists.

In the 1920s, research was sponsored that eventually led to the development of the *effective temperature* (ET) scale that reduced to a single number the effect of air temperature, wet-bulb temperature, and air velocity on comfort [22]. This research has continued and recent efforts have produced a scale called the *corrected effective temperature* (CET), which includes a measure of radiant heat exchange in addition to the other variables [23].

In essence, ET's are determined by having people in one room and one set of environmental conditions move to another room with a different set of conditions and then vote on whether or not the second room felt the same (thermally) as the first one, or hotter or colder. Using data from such voting, nomograms were developed that allowed an easy determination of the ET or CET for many common combinations of environmental parameters. Using the same or similar data, models of the comfort zone of air temperature and/or relative humidity were developed [24]. Such models, incidentally, are valid only for a restricted population; a model for people in England is quite different from that developed for people in the United States. The ET and CET are measures of comfort and not necessarily of the thermal stress imposed by an environment. Comfort sensations are much more a function of skin and air temperature and the presence or absence of sweat than of other parameters of biothermal stress or strain [25, 26].

Table 13-1 Windchill Chart

	Actual Thermometer Reading, °C										
	10	5	0	−5	−10	−15	−20	−25	−30	−35	−40
Wind Speed (km/hr)	Windchill Temperature, °C										
Calm	10	5	0	−5	−10	−15	−20	−25	−30	−35	−40
10	7	2	−4	−9	−15	−20	−26	−32	−37	−43	−48
15	5	−1	−7	−13	−19	−25	−32	−38	−44	−50	−56
20	3	−3	−10	−16	−23	−29	−36	−42	−48	−55	−61
25	2	−5	−12	−18	−25	−32	−39	−45	−52	−59	−66
30	1	−6	−13	−20	−27	−34	−41	−48	−55	−62	−69
35	0	−7	−14	−22	−29	−36	−43	−50	−58	−65	−72
40	−1	−8	−16	−23	−30	−38	−45	−52	−60	−67	−74
45	−2	−9	−17	−24	−32	−39	−47	−54	−62	−69	−77
50	−2	−10	−18	−25	−33	−40	−48	−56	−63	−71	−79
55	−3	−11	−18	−26	−34	−42	−49	−57	−65	−73	−80
60	−3	−11	−19	−27	−35	−43	−51	−58	−66	−74	−82
65	−4	−12	−20	−28	−36	−44	−52	−60	−68	−76	−84

Over 65 km/hr (little added effect)	Little Danger (for a properly clothed person)	Increasing Danger	Great Danger
		Danger from freezing exposed skin and flesh	

Radio and television weather programs have popularized the windchill index [23] developed by Siple and Passel in the 1940s [27]. This index is an attempt to describe the relative cooling power of air for exposed skin as a function of its rate of movement and temperature by use of an equivalent still-air temperature (see Table 13-1). One problem with this practice is that people tend to ascribe literal meaning to windchill temperatures. That is, if the windchill temperature happens to be −20°C (−4°F), the uninitiated may be led to think that a pan of water or water in an automobile radiator will freeze rapidly even at an actual air temperature of 1°C (33.8°F). Rate of cooling, although important, is not the only factor in a determination of the relative hazard of being out-of-doors on a cold day or night.

Biothermal Stress

Because of the easy availability of measures to protect against cold, cold stress is very rarely an occupational health hazard. Therefore, even though many of the relationships to be described are equally valid for cold as well as heat stress, heat stress alone will be considered.

Sitting at rest the average human has a metabolic rate of about $100W$ (100 J sec^{-1}). This amount of heat must be discharged to the environment or a buildup of heat in the body with a resultant temperature rise will occur. In still air at 20°C (68°F) a lightly clad man will transfer the major portion of that heat to his solid surroundings by thermal radiation; most of the rest will be taken up by the air through the process of convection. Interference with these mechanisms of heat transfer may be stressful to the individual in proportion to its extent. The problems are to reduce qualitative statements such as these to quantitative expressions, to reduce the expressions to a single number proportional to the biothermal stress, and to correlate that number with the body's response, the biothermal strain.

Biothermal stress is the load on the thermoregulatory system of the subject due to a displacement from thermally neutral conditions caused by a change in the metabolic rate of the subject and/or by a change in the environment. *Biothermal strain* is the displacement of a biological parameter from its value in a biothermally neutral environment produced by the action of biothermal stress. A *biothermally neutral* environment is one in which a resting subject in thermal equilibrium with his surroundings has no conscious sensations of hot or cold and in which the subject's thermoregulatory system is functioning at a minimum level [28].

Both biothermal stress and biothermal strain may be positive or negative and both are zero in a biothermally neutral environment. Positive values are used for heat stress; negative, for cold. The sign used for strain depends on whether its value is increased (positive) or decreased by action of the stress. Stress indices should require no information about the associated strains in their calculation.

The Energy Balance

A careful analysis of the heat exchanges between the body and its surroundings shows that

$$M = R + C_{re} + C + E_{re} + E + W + q \qquad (13\text{-}1)$$

where
M = rate of metabolism
R = rate of heat loss by radiation
C_{re} = rate of heat loss by convection from the respiratory tract
C = rate of heat loss by convection from the skin
E_{re} = rate of heat loss by evaporation from the respiratory tract
E = rate of heat loss by evaporation from the skin
W = rate of energy loss in external work
q = rate of heat storage in the body

At equilibrium, q is zero and, therefore, it is usually ignored unless one is trying to establish *administrative work practices* based on a limit for q. In such cases, a storage of 237,400 J is equivalent (for the average man) to a temperature elevation of about 1°C.

If C_{re}, E_{re}, and W are assumed to be insignificant (which is true in many cases) or are subtracted from estimates or determinations of M [29], giving a corrected value of the metabolic rate, M', then Eq. (13-1) reduces to

$$M' = R + C + E \qquad (13\text{-}2)$$

or

$$E = M' - R - C \qquad (13\text{-}3)$$

Note that in these equations positive values of E, R, and C are losses of heat to the environment not gains.

Indices of Heat Stress

Many attempts have been made to use equations similar to Eq. (13-3) in the formulation of an index of heat stress. Most successful were Belding and Hatch, who in 1955 developed what they called the *heat stress index* (HSI), which was the value of E (the required evaporative capacity), as calculated in Eq. (13-3), divided by the maximum evaporative capacity assuming completely wet skin, multiplied by 100 [30]. They developed means of estimating R and C as well as E_{max} (the maximum evaporative capacity) for the *standard man* as functions of the dry- and wet-bulb temperatures, the globe temperature, and the air velocity.

The *globe thermometer* was developed by Vernon in his early research. It consists of a thermometer (or more recently, a thermocouple or thermistor) with the sensing element at the center of a hollow sphere, 0.15 m (or less) in diameter [31]. Knowing the air velocity and air temperature, as well as the globe temperature, a heat balance can be written around the globe thermometer, equating the gain (loss) by convection to the loss (gain) by radiation. The only unknown variable in the equation is the average temperature of the solid surroundings, t_w, which is thereby determined:

$$t_w = [(t_g + t_K)^4 + (0.248 \times 10^9 V^{0.5})(t_g - t_a)]^{0.25} - t_K \tag{13-4}$$

where t_g = globe temperature (°C)
  t_a = air temperature (°C)
  t_K = 273.17°K
  V = air velocity at the globe (m sec^{-1})

Knowing t_w is necessary in order to determine the rate of heat exchange by radiation between a man and his environment. Most of the experimental work done on such exchanges has been with nude or nearly nude people. Clothing acts to impede the exchange but in many situations where heat stress is a problem, workers wear very little clothing so that the conservative assumption of nudity may be reasonably good. The equation from which R is determined is

$$R = ekFA(T_s^4 - T_w^4) \tag{13-5}$$

where e = the emissivity
  k = the universal radiation constant
  A = the body surface area
  F = the fraction of the surface area taking part in the radiant exchange
  T_w = absolute temperature of the solid surroundings
  T_s = absolute temperature of the skin surface

The emissivity of the skin surface and of an enclosure are both essentially unity; for a man at work, F is also close to unity. Over the temperature range of usual occupational interest, little error is added by assuming that the difference between the fourth powers of the absolute temperatures is equal to the temperature difference multiplied by a constant. Combining all the constants,

$$R = k_r A(t_s - t_w) \tag{13-6}$$

where k_r is the combined coefficient of heat exchange by radiation and t_w and t_s are the temperatures of the solid surroundings and the skin, respectively. More precise approaches are available [32].

Convective heat exchange rates are calculated in a similar manner except that no fourth powers of temperature are necessary [28].

$$C = A(a + bV^d)(t_s - t_a) \qquad (13\text{-}7)$$

where a, b, and d are constants. The constant a may be omitted from equations similar to Eq. (13-7); it is the coefficient of natural convection where the only air movement is that furnished by thermal convection currents. The constant b is the coefficient of forced convection and in most occupational environments a is much smaller than bV^d, which is the reason for neglecting a if that is done. V refers to air velocity and t_a and t_s are air and skin temperatures, respectively.

If the skin is assumed to be completely wet with sweat, a rate of heat loss may be calculated that will be the maximum possible under the given circumstances. Termed E_{max}, the equation for its calculation is similar to Eq. (13-7):

$$E_{max} = A(f + gV^h)(P_s - P_a) \qquad (13\text{-}8)$$

where f, g, and h are constants similar to a, b, and d and P refers to the vapor pressure of water at the temperature of the skin (subscript s) and at the humidity of the air (subscript a). In Eq. (13-8), the value of the constant f is probably close to zero, as without any air movement little if any water would be evaporated from the skin surface so that the equation used is actually

$$E_{max} = k_e A V^h (P_s - P_a) \qquad (13\text{-}9)$$

where k_e, the coefficient of evaporative heat exchange, is actually g from Eq. (13-8) plus any residual value of f.

In order to solve Eq. (13-6), (13-7), and (13-9), a value of the skin temperature is necessary, as is the surface area of the skin of the person exposed. Surface areas of men and women are in the neighborhood of 1.8 to 2.0 m²; in their development of the HSI, Belding and Hatch used 1.86 m² for the *standard man*. The temperature of the skin is not constant in heat stress, with time or with location on the body. Some kind of average is necessary and the average used affects the calculated rates of exchange greatly. Belding and Hatch used 35°C (95°F) for the HSI but later determined that this assumption was one reason why the HSI failed to evaluate heat stress truly [33]. Actually, under most heat stress circumstances, the mean skin temperature will vary between about 30°C (86°F) and 37°C (98.6°F).

Hatch, in 1963, published a new index of heat stress, the predicted sweat rate (SRI) [34]. In the development of that index he formulated equations for prediction of the skin temperature in two zones of free and restricted evaporation. *Free evaporation* refers to circumstances where all sweat formed evaporates freely from the skin surface; *restricted evaporation* implies that some sweat is lost by dripping, thereby removing little heat from the body. By means of a series of equations, he then was able to calculate a value of the predicted sweat rate for the zone that applied to the exposure if values of two subject-specific parameters—t_b and K_b—were known. The first, t_b, is the skin temperature at which the subject's sweat and metabolic rates are equal. K_b is the rate of increase of the sweat rate minus the metabolic rate with skin temperature.

Sibbons, in 1965, used Hatch's idea of different zones of sweat evaporation, but expanded it from two to four zones. Then, using concepts long applied in meteorology and attempting to be as precise as possible (Hatch attempted to generalize instead), he formulated two new indices of heat stress: the equivalent operative skin temperature (A_{os}) and the equivalent operative ambient temperature (A_o) [35].

Many attempts have been made, both before and since the ones outlined, to develop a single-number index that will truly answer the question, "How hot is it?" but as yet none is completely acceptable. In 1968, Peterson correlated seven different indices of heat stress including Hatch's predicted sweat rate, the two Sibbons indices, and the required evaporative capacity, with 29 measures of heat strain [36]. He found that in order to predict the "best" strains well, three of the seven stress indices were necessary, the SRI, the HSI based on actual (or predicted) skin temperature (HSI2), and A_o. These three indices were those that required the most information about the environment and about the people exposed: actual surface area, t_b, and K_b. Even using all three stress indices and the best set of heat transfer coefficients [37], significant differences between subjects

Table 13-2 Correlation of Heat Stress Indices with Best Strains [28]

Heat Strain	Stress Index	Correlation Coefficient
Actual sweat rate	SRI	0.97
Incremental ear temperature	A_o	0.94
Skin temperature	(SRI)*	(0.92)
Rectal-to-skin conductance	SRI	0.88
Incremental rectal temperature	HSI2	0.86
Ear-to-skin conductance	SRI	0.87
Incremental heart rate	SRI	0.82
Duration of systole (Q-T interval on an electrocardiogram)	HSI2	−0.79

*Predicted not by the SRI but using the methods embodied in calculation of the SRI.

were found for most strains. Those heat strains best predicted by the three heat stress indices and the best predictor in each case are summarized in Table 13-2. In that table the simple correlation coefficient varies from 0.97 for a prediction of the actual sweat rate from the SRI to -0.79 for a prediction of the duration of systole from the HSI2. None of the other stress indices did so well, and no other strains were predicted as well as those in Table 13-2. (An r value of 0.79 indicates that about 62% of the variation was explained by the correlation.)

In the early 1970s a new stress index appeared, the WBGT; it had been used previously to determine the extent of hard outdoor activity allowed for armed services training camps [38]. This index has the virtue of extreme simplicity, requiring only determinations of the wet-bulb and globe temperatures for most environments and the air temperature for some others. It is calculated as

$$\text{WBGT} = 0.7t_{wb} + 0.3t_g \quad \text{(indoors)} \tag{13-10}$$

$$\text{WBGT} = 0.7t_{wb} + 0.2t_g + 0.1t_a \quad \text{(outdoors)} \tag{13-11}$$

where t_{wb} is the natural (unaspirated) wet-bulb temperature, t_g is the globe temperature, and t_a is the dry-bulb or air temperature. Unfortunately, attempts to correlate the WBGT with heat strains have been almost uniformly unsuccessful so that more than simplicity may be required for a useful index [39, 40, 41]. Those heat stress indices best correlated with strain are those that require the most information, and although that is not surprising, it is disappointing.

Another observation is that the ability of a stress index to correlate well with strain and minimize between-subject differences is a function of the heat transfer coefficients used in the equations for determining rates of heat exchange by radiation, convection, and evaporation. Even though there is, again, no real agreement on "best" values to use, where exchange rates are in watts, areas in square meters, temperatures in degrees Celsius, air velocities in meters per second, and vapor pressure in millimeters of mercury, the following equations should give reasonable values for standing nude or very lightly clothed men:

$$R = 5.08A(t_s - t_w) \tag{13-12}$$

$$C = A(2.27 + 7.44V^{0.67})(t_s - t_a) \tag{13-13}$$

$$E_{\max} = 13.8AV^{0.63}(P_s - P_a) \tag{13-14}$$

Refinements in the coefficients appear in the literature with regularity and the values in Eq. (13-12) to (13-14) should not be regarded as final or absolute [42, 43]. Ordinary work clothing may reduce all heat exchange rates by as much as 40%.

Another approach to the problem of assessing heat stress on workmen has been to develop devices designed to readout a number proportional to the stress. The first device so-used was the Kata thermometer, a conventional thermometer but with a large silvered bulb. The black globe thermometer was first developed as a stress index meter [44], and recently a wetted black globe has been used [45]. In addition, electronic circuitry has been developed to record values of the pertinent variables [46] and to calculate and display automatically a simple index such as the WBGT [47]. None of these devices has been well correlated with biothermal strain except over very small ranges of the stress index.

Evaluating the Thermal Environment

The first step in evaluating the thermal environment is to decide what must be controlled. Because heat strain is the true subject of control, heat stresses that do not produce measurable strains are seldom controlled. There are probably dozens of variables that could qualify as heat strains by being altered in accord with heat stress. These range from various measures of mental performance to physiological parameters such as body temperatures or measures of circulatory strain to the appearance of signs or symptoms of impending collapse. When the environment so lacks stress that heat strain is minimal, then measures of comfort such as the effective temperature are probably more rational than any heat stress index.

Table 13-3 Effects of Heat Stress on the Actual Sweat Rate [28]

Physiological Reaction	*Hatch's Zone*	*Sibbons' Regime*
Increasing Heat Stress		
Skin completely wet with sweat; loss by dripping increases in proportion to the increase in heat stress.	Restricted evaporation	Wet skin
Skin partially wet with sweat; most sweat evaporates, but some is lost by dripping.	Restricted evaporation	Restricted evaporation
Insensible sweating; skin is not wet.	Free evaporation	Free evaporation
No sweating; all heat transfer is by radiation and convection.	Free evaporation	Operative

No one biothermal strain appears able to give complete information concerning any other biothermal strain. That is, the correlation between various measures of strain is far from perfect. The best measure of overall heat strain, however, is probably the sweat rate and the best predictor of the sweat rate is the SRI [36]. Table 13-3 shows the effect of heat stress on the sweat rate. The SRI is a good stress index; it deserves further use and exploitation.

Most necessary are benchmarks of the SRI associated with measures of psychological and psychophysiological performance. Physical performance limits associated with the SRI should be relatively easier to obtain because this index is well correlated not only with the sweat rate but with the conductance of heat from the body core as represented by either the rectal or the ear (canal) temperature to the skin and with the heart rate. Sweat rate, skin temperature [48], and heart rate indices have been proposed as measures of strain. Most popular of the heart rate indices is the *recovery pulse sum* first used by Brouha [49]. This parameter is associated with heart rates in the period immediately following a task in the heat. It, as well as the time required for the heart rate to return to base line, are reasonably well correlated with the SRI (values of the correlation coefficient of 0.70 and 0.66) [36].

If the ear temperature were chosen to represent the core temperature of the body (and it can be measured with much less fuss than can the rectal temperature), it is well predicted by A_o and responds to heat stress more rapidly than the rectal temperature. The rectal temperature is best predicted by the HSI2 and several authorities have proposed rectal temperature limits, usually on the order of a 2°C elevation from normal. Once a psychophysiological or physiological limit has been chosen, then by means of the correlation between that limit and a stress index a maximum value of that index can be set, the actual statistics being used to determine the amount of safety factor needed.

Control of Biothermal Stress

Cold Stress

Cold stress is controlled by the use of proper clothing and shelter [50]. Experiment has shown that several layers of clothing that allow the formation of dead air spaces are better than one heavy layer and that the clothing must allow sweat to evaporate and leave the skin of all areas of the body, including the feet. Providing for these factors while still keeping wind and water from the skin has been a difficult objective that has been attained by civilized man only recently. People who live in very cold weather areas of the world seem to have solved these problems centuries ago.

Heat Stress

The goals of heat stress control are (1) no disturbance of physiological functions that impair the health or longevity of workers and (2) as much thermal comfort as possible. Most important of all control methods are *engineering* controls and their use and installation depends on a good knowledge of the parameters of heat stress. That is, if the main problem is radiant heat, reducing the humidity of the air or increasing the velocity of the air are poor means of control. On the other hand, shielding of infrared radiation will do little good when the average temperature of the surrounding solids is close to the air temperature. Therefore, a good heat survey is the first requirement of control and that survey should be conducted in such a manner that the main sources of heat (stress) are determined. Once that has been done, controls should be instituted in accord with the survey results, with good engineering practice, and with the goal of eliminating adverse reactions to heat.

Work practice controls have been used when engineering controls were impractical. This kind of control can involve limiting both time [51] and frequency of exposure in areas of high heat stress and providing cool-off rooms where an accumulation of heat in the body can be shed rapidly and in relative comfort [49]. In order to practice good time-of-exposure limitations, either physiological measures (temperature, heart rate) [52] must be used or sufficient information must be available from a heat survey to calculate heat storage rates by use of Eq. (13-1) [53].

Personal protective devices ranging from head coolers [54] to water-cooled suits [55] and aluminized insulated garments resembling space suits have been developed [56, 57, 58], and each has its own area of utility [59, 60]. As with any kind of personal protection, its use should be much more of a response to emergencies than standard operating procedure. Nevertheless, when people must work in areas of extreme heat stress, the correct personal protective equipment can change the problem from incipient exhaustion to annoyance.

REFERENCES

[1] T. F. Hatch, "Heat Control in Hot Industries," in *Industrial Hygiene and Toxicology* by F. A. Patty, D. D. Irish, and D. W. Fassett, ed., 2nd rev. ed. (New York: Interscience Publishers, Inc., 1958), pp. 789–807.

[2] J. D. Hardy and H. T. Hammel, "Control System in Physiological Temperature Regulation," Chapter 54 in *Temperature, Its Measurement and Control in Science and Industry*, C. M. Herzfeld, ed.-in-chief, vol. 3, J. D. Hardy, ed. (New York: Reinhold Publishing Corporation, 1963), pp. 613–25.

[3] N. Glickman, et al., "Shivering and Heat Production in Men Exposed to Intense Cold," *J. Appl. Physiol.* **22** (1967), 1–8.

[4]   M. MARMOR, "Heat Wave Mortality in New York City, 1949 to 1970," *Arch. Env. Health* **30** (1975), 130–36.

[5]   T. R. A. DAVIS, "Acclimatization to Cold in Man," Chapter 38 in *Temperature, Its Measurement and Control in Science and Industry.* See Ref. 2, pp. 443–52.

[6]   M. C. LAPP AND G. K. GEE, "Human Acclimatization to Cold Water Immersion," *Arch. Env. Health* **15** (1967), 568–79.

[7]   C. S. NAIR, M. S. MALHOTRA, AND P. M. GOPINATH, "Effect of Altitude and Cold Acclimatization on the Basal Metabolism in Man," *Aerosp. Med.* **42** (1971), 1056–59.

[8]   D. E. BASS, "Thermoregulatory and Circulatory Adjustments During Acclimatization to Heat in Man," Chapter 28, pp. 299–305. See Ref. 2.

[9]   C. GISOLFI AND S. ROBINSON, "Relations Between Physical Training, Acclimatization and Heat Tolerance," *J. Appl. Physiol.* **26** (1969), 530–34.

[10] R. W. PIWONKA AND S. ROBINSON, "Acclimatization of Highly Trained Men to Work in Severe Heat," *J. Appl. Physiol.* **22** (1967), 9–12.

[11] L. B. ROWELL, ET AL., "Central Circulation Responses to Work in Dry Heat Before and After Acclimatization," *J. Appl. Physiol.* **22** (1967), 509–18.

[12] C. H. WYNDHAM, "Effect of Acclimatization on the Sweat Rate/Rectal Temperature Relationship," *J. Appl. Physiol.* **22** (1967), 27–30.

[13] W. HOFLER, "Changes in Regional Distribution of Sweating During Acclimatization to Heat," *J. Appl. Physiol.* **25** (1968), 503–06.

[14] E. SHVARTZ AND D. BENOR, "Heat Acclimatization by the Prevention of Evaporative Cooling," *Aerosp. Med.* **42** (1971) 879–81.

[15] C. G. WILLIAMS, C. H. WYNDHAM, AND J. F. MORRISON, "Rate of Loss of Acclimatization in Summer and Winter," *J. Appl. Physiol.* **22** (1967), 21–26.

[16] J. A. VAUGHAN, E. A. HIGGINS, AND G. E. FUNKHOUSER, "Effects of Body Thermal State on Manual Performance," *Aerosp. Med.* **39** (1968), 1310–15.

[17] D. MINARD AND L. COPMAN, "Elevation of Body Temperature in Disease," Chapter 25. See Ref. 2, pp. 253–73.

[18] J. P. KNOCHEL, "Environmental Heat Illness: An Eclectic Review," *Arch. Int. Med.* **133** (1974), 841–64.

[19] Y. SHAPIRO, T. ROSENTHAL, AND E. SOHAR, "Experimental Heat Stroke," *Arch. Int. Med.* **131** (1973), 688–92.

[20] G. H. A. CLOWES, ET AL., "Heat Stroke," *N. E. J. Med.* **291** (1974), 564–67.

[21] K. KLOETZEL, ET AL., "Relationship Between Hypertension and Prolonged Exposure to Heat," *J. Occup. Med.* **15** (1973) 878–80.

[22] F. C. HOUGHTON AND C. P. YAGLOU, "Determining Lines of Equal Comfort," *J. Am. Soc. Heat. Vent. Engs.* **29** (1923), 165–76.

[23] B. A. HERTIG, "Thermal Standards and Measurement Techniques," Chapter 31 in *The Industrial Environment—Its Evaluation & Control,* (Washington, D.C.: U.S. Department of Health, Education, and Welfare, Public Health Service, Center for Disease Control, National Institute for Occupational Safety and Health. For sale by the Superintendent of Documents, U.S. Government Printing Office, 1973), 413–29.

[24] P. O. FANGER,."Assessment of Man's Thermal Comfort in Practice," *Brit. J. Ind. Med.* **30** (1973), 313–24.

[25] A. P. GAGGE, J. A. J. STOLWIJK, AND B. SALTIN, "Comfort and Thermal Sensations and Associated Physiological Responses During Exercise at Various Ambient Temperatures," *Env. Res.* **2** (1969), 209–29.

[26] A. P. Gagge and R. R. Gonzalez, "Physiological and Physical Factors Associated with Warm Discomfort in Sedentary Man," *Env. Res.* **7** (1974), 230–42.

[27] P. A. Siple and C. F. Passel, "Dry Atmospheric Cooling in Subfreezing Temperatures," *Proc. Am. Phil. Soc.* **89** (1945), 177–99.

[28] J. E. Peterson, *Human Biothermal Strain in Relation to Environmental Stress Parameters*, Thesis submitted in partial fulfillment for Ph.D. degree, University of Michigan, Ann Arbor, Michigan. Available from University Microfilms, Ann Arbor, Michigan, 1968.

[29] S. M. Horvath and M. O. Colwell, "Heat Stress and the New Standards," *J. Occup. Med.* **15** (1973), 524–28.

[30] H. S. Belding and T. F. Hatch, "Index for Evaluating Heat Stress in Terms of Resulting Physiological Strains," *Heat. Piping Air Cond.* **27** (1955), 129–35.

[31] K. W. Graves, "Globe Thermometer Evaluation," *Am. Ind. Hyg. Assn. J.* **35** (1974), 30–40.

[32] R. L. Harris, "Computer Simulation of Radiant Heat Load and Control Alternatives," *Am. Ind. Hyg. Assn. J.* **35** (1974), 75–83.

[33] H. S. Belding, B. A. Hertig, and M. L. Riedesel, "Laboratory Simulation of a Hot Industrial Job to Find Effective Heat Stress and Resulting Physiologic Strain," *Am. Ind. Hyg. Assn. J.* **21** (1960), 25–31.

[34] T. F. Hatch, "Assessment of Heat Stress," Chapter 29. See Ref. 2, pp. 307–18.

[35] J. L. H. Sibbons, "Assessment of Thermal Stress from Energy Balance Considerations," *J. Appl. Physiol.* **21** (1966), 1207–17.

[36] J. E. Peterson, "Experimental Evaluation of Heat Stress Indices," *Am. Ind. Hyg. Assn. J.* **31** (1970), 305–17.

[37] Jean Colin and Yvon Houdas, "Experimental Determination of Coefficient of Heat Exchanges by Convection of Human Body," *J. Appl. Physiol.* **22** (1967), 31–38.

[38] C. P. Yaglou and D. Minard, "Control of Heat Casualties at Military Training Centers," *AMA Arch. Ind. Health* **16** (1957), 302–16.

[39] B. D. Dinman, et al., "Work in Hot Environments: I. Field Studies of Work Load, Thermal Stress and Physiologic Response," *J. Occup. Med.* **16** (1974), 785–91.

[40] N. L. Ramanathan and H. S. Belding, "Physiological Evaluation of the WBGT Index for Occupational Heat Stress," *Am. Ind. Hyg. Assn. J.* **34** (1973), 375–83.

[41] J. D. Ramsey, et al., "Heat Stress Limits for the Sedentary Worker," *Am. Ind Hyg. Assn. J.* **36** (1975), 259–65.

[42] A. P. Gagge and J. D. Hardy, "Thermal Radiation Exchange of the Human by Partitional Calorimetry," *J. Appl. Physiol.* **23** (1967), 248–58.

[43] D. Mitchell, et al., "Radiant and Convective Heat Transfer of Nude Men in Dry Air," *J. Appl. Physiol.* **26** (1969), 111–18.

[44] A. D. Hosey, "A Recording Thermistor Globe Thermometer," *Am. Ind. Hyg. Assn. J.* **20** (1959), 1–4.

[45] J. H. Botsford, "A Wet Globe Thermometer for Environmental Heat Measurement," *Am. Ind. Hyg. Assn. J.* **32** (1971), 1–10.

[46] A. D. Hosey and A. C. Mendenhall, Jr., "The Envirec—An Instrument for Continuous Recordings of Environmental Air, Wet-Bulb and Globe Temperatures, and Air Velocity," *Am. Ind. Hyg. Assn. J.* **20** (1959), 121–30.

[47] L. A. Kuehn and L. E. MacHattie, "A Fast Responding and Direct Reading WBGT Index Meter," *Am. Ind. Hyg. Assn. J.* **36** (1975), 325–31.

[48] P. F. IAMPIETRO, "Use of Skin Temperature to Predict Tolerance to Thermal Environments," *Aerosp. Med.* **42** (1971), 396–99.

[49] L. BROUHA, *Physiology in Industry* 2nd ed. New York: Pergamon Press, (1967), 97–114 and 127–38.

[50] J. V. MORRIS, "Developments in Cold Weather Clothing," *Ann. Occup. Hyg.* **17** (1975), 279–94.

[51] J. D. WALTERS, "Hot Environments—Measurement and Tolerance Estimation," *Ann. Occup. Hyg.* **17** (1975), 255–64.

[52] J. J. VOGT, ET AL., "Evaluation of Workload and Heat Stress Under Actual Working Conditions: Principles and Application of a New Method," *Travail Humain* **33** (1970), 125–39. Abstracted in *Ind. Hyg. Digest* **36** (May, 1972), 21.

[53] R. R. STEPHENSON, ET AL., "Work in Hot Environments: II. Design of Work Patterns Using Net Heat Exchange Calculations," *J. Occup. Med.* **16** (1974), 792–95.

[54] S. A. NUNNELEY, S. J. TROUTMAN, JR., AND P. WEBB, "Head Cooling in Work and Heat Stress," *Aerosp. Med.* **42** (1971), 64–68.

[55] J. W. HILL, "Water Cooled Suit for Protection Against Hot Environments," *Proc. Roy. Soc. Med.* **63** (1970), 1014–15.

[56] A. J. VAN RENSBURG, ET AL., "Physiological Reactions of Men Using Microclimate Cooling in Hot Humid Environments," *Brit. J. Ind. Med.* **29** (1972), 387–93.

[57] S. KONZ, ET AL., "Personal Cooling with Dry Ice," *Am. Ind. Hyg. Assn. J.* **35** (1974), 137–47.

[58] G. W. CROCKFORD AND M. A. AWARD EL KARIM, "An Assessment of a Dynamically-Insulated Heat-Protective Clothing Assembly," *Ann. Occup. Hyg.* **17** (1974), 111–21.

[59] P. WEBB, "Measuring the Physiological Effects of Cooling," *Human Factors* **13** (1971), 65–78.

[60] T. MIURA, ET AL., "Effect of a Local Cooling Vest with CO_2 Ice as Cooling Material on the Physiological Functions of Man in a Hot Environment," *J. Sci. Labour* **47** (1971), 309–38. Abstracted in *Ind. Hyg. Digest* **35** (Nov., 1971), 26.

14

NONIONIZING RADIATION

In a vacuum all electromagnetic radiation travels at the same velocity, 2.99781×10^8 m sec^{-1}, but the amount of energy associated with that radiation varies tremendously. *Nonionizing* radiation is that electromagnetic radiation with insufficient energy to ionize water. The relationships are

$$c = \lambda\nu = 3 \times 10^8 \text{ m sec}^{-1} \qquad (14\text{-}1)$$

$$E = h\nu = \frac{hc}{\lambda} \qquad (14\text{-}2)$$

where c = velocity of electromagnetic radiation in the medium through which it travels (3×10^8 m sec^{-1} is used for air or a vacuum)
    λ = wavelength of the radiation (m)
    ν = frequency of the radiation [sec^{-1} (Hz)]
    h = Planck's constant (6.626×10^{-34} J sec)
    E = energy [joules (J)]

Another important item of information is that 1.0 electron volt (eV) is equal to 1.6×10^{-19} J.

Experiment has shown that the least energetic photon that will cause ionization in water and hence in tissue has an energy of about 12 eV (or about 19×10^{-19} J). Rearranging Eq. (14-2),

$$\lambda = \frac{hc}{E} = (6.63 \times 10^{-34}) \cdot \frac{(3 \times 10^{8})}{19 \times 10^{-19}}$$

$$\lambda = 1.047 \times 10^{-7} \, m = 104.7 \times 10^{-9} \, m$$

$$\lambda = 105 \, nm$$

Therefore, *non*ionizing radiation is electromagnetic radiation having a wavelength *greater* than about 100 nm.

Table 14-1 contains a listing of the names of various portions of the electromagnetic spectrum and their wavelengths. By substituting in Eq. (14-1), the associated frequencies were found. Several of the ranges overlap and, especially in those regions, terminology may be confusing; the name used is dictated by the method of generation rather than by wavelength or frequency alone.

Table 14-1 The Electromagnetic Spectrum

Type of Radiation	Wavelength Range (nm)	Frequency Range (Hz)
Gamma rays	$< (0.14)$	$> (2.1 \times 10^{18})$
X rays	0.006 to 100	(5×10^{19}) to (3×10^{15})
Ultraviolet	180 to 400	(1.7×10^{15}) to (7.5×10^{14})
Near UV	315 to 400	(9.5×10^{14}) to (7.5×10^{14})
Mid-range UV	280 to 315	(1.1×10^{15}) to (9.5×10^{14})
Far UV	200 to 280	(1.5×10^{15}) to (1.1×10^{15})
Visible	400 to 750	(7.5×10^{14}) to (4.0×10^{14})
Infrared	700 to (3.0×10^5)	(4.4×10^{14}) to (1.0×10^{12})
Near IR	700 to 1400	(4.4×10^{14}) to (2.1×10^{14})
Far IR	1400 to (3.0×10^5)	(2.1×10^{14}) to (1.0×10^{12})
Microwaves	(1.0×10^5) to (1.0×10^{10})	(3.0×10^{12}) to (3.0×10^7)
Radio	(1.0×10^{10}) to (3.0×10^{13})	(3.0×10^7) to (1.0×10^4)
Very-low frequency	$> (3.0 \times 10^{13})$	$< (1.0 \times 10^4)$

Throughout most of the spectrum, wavelength is used as the basic designation but, beginning with the microwave portion, frequency is usually used instead. There is no esoteric reasoning behind the use of either wavelength or frequency, merely the availability of easy-to-use prefixes for powers of ten. The largest prefix designation in common use is *giga-*, meaning 10^9 (although *tera-*, 10^{12}, is available); on the small-prefix side, *nano-*, *pico-,* and *femto-*(10^{-9}, 10^{-12}, and 10^{-15}, respectively) are all commonly used. The angstrom unit, Å (10^{-10} m), may be encountered in older literature. Frequency and wavelength are interchangeable, in any medium, by using Eq. (14-1) and the correct speed of light.

The effects of electromagnetic radiation on man are functions of several variables; most important of those related to the radiation itself are the energy content of each photon (quantum), the power in the beam, and the ability of the radiation to penetrate tissue. In addition, throughout most of the spectrum, radiation may consist of a single wavelength (be monochromatic) or be broad-banded, it may be coherent or incoherent, and the beam or field may be diffuse or concentrated. Despite rather extensive experimentation, there is little agreement about the effect of each of these variables on the hazard of the radiation, especially when variables related to the exposure and to the species must also be taken into consideration. Progress has been made, however, and some aspects of the overall picture can be examined and discussed [1, 2].

Units of Measurement

The literature of nonionizing radiation (NIR) can be very confusing because there has been essentially no agreement on terminology. Three main concepts are most important. They are energy, flux density (energy per unit area), and irradiance (flux density per unit of time). *Energy* is the capability of doing work; it is expressed in joules (abbreviated as J and equal to 1.0 newton-meter) or ergs (1.0 erg is equal to 10^{-7} J). *Flux density* is expressed in joules per square meter ($J\ m^{-2}$), and *irradiance* is expressed in joules per square meter-second ($J\ m^{-2}\ sec^{-1}$). Unfortunately, energy, flux density, and irradiance are given several other names even by the same author, some of which may be regarded as synonyms and some of which must be regarded as mistakes. In addition to the *basic* units of measurement, others are used, further confusing an already confused situation.

Energy may be called *energy content* or *radiant energy,* and it may be expressed in joules, watt-seconds, or ergs or multiples or submultiples of these units. If the energy in a beam (or some fraction thereof) is absorbed, the term *absorbed dose* is appropriate, but it is rarely used.

Flux density is a more useful concept in the evaluation of the hazards of NIR than is energy and, therefore, it receives more synonyms. Flux density may be called *illumination,* a proper synonym that is rarely used, or *energy density.* Because the amount of energy incident upon a unit area of tissue is proportional to the dose received by that unit area, flux density has also been called *exposure dose, radiant exposure,* and *exposure,* terms that are somewhat ambiguous.

Irradiance is flux density per unit time and is also called *radiation intensity.* Because irradiance is often expressed in units of watts per unit area and the watt is a unit of power, it is often called *power density*. One watt is equal to one joule per second and therefore, instead of using the joule per square meter per second as the unit of irradiance, the watt per square meter is usually substituted, but both may be found in the literature. Irradiance is given many incorrect names; often several are used in the same publication. Table 14-2 contains a collection of synonyms, units, and conversion factors useful in descriptions of NIR.

Table 14-2   Units of Measurement and Conversion Factors

Energy	*Flux Density*	*Irradiance*
Synonyms		
Absorbed dose	Energy density	Effective irradiance
Energy content	Exposure	Exposure dose rate
Radiant energy	Exposure dose	Intensity
	Radiant exposure	Power density
		Radiant intensity
Equivalents		
1.0×10^7 erg	1.0×10^5 J m^{-2}	1.0×10^5 J sec^{-1} m^{-2}
1.0 J	1.0×10^7 erg cm^{-2}	1.0×10^7 erg sec^{-1} cm^{-2}
1.0 W sec	1.0 J cm^{-2}	1.0 J sec^{-1} cm^{-2}
0.2389 cal	1.0 W sec cm^{-2}	1.0 W cm^{-2}
0.2778 mW hr	0.2389 cal cm^{-2}	0.2389 cal sec^{-1} cm^{-2}
	0.2778 mW hr cm^{-2}	1,000 mW cm^{-2}

Ultraviolet

Effects on Man

Ultraviolet means *beyond violet*, referring to the frequency of electromagnetic radiation. Although *hard* ultraviolet (or *soft* X-ray) radiation exists in the wavelengths between 100 and 200 nm, it is of very little significance to human health. Radiation in this wavelength range is completely absorbed by passing through very small thicknesses of air, mainly in the production of ozone, and very little can reach the surfaces of a person. As wavelength increases from about 315 or 320 nm to the lower end of the visible spectrum, the amounts required to cause any kind of injury increase rapidly. Therefore, the region of most concern to industrial hygienists lies between 200 and 320 nm and encompasses the portions of the ultraviolet (UV) spectrum called *far* and *mid-range.*

UV radiation in the region of most concern is very rapidly absorbed by water and tissue through a series of reactions that are not at all well understood. Because of the rapid absorption, UV radiation does not penetrate very deeply and, therefore, effects are confined to the skin and eye.

Skin effects. Most noticeable of all effects on the skin, and also the first one seen, is the production of redness or erythema. When erythema is caused by overexposure to the rays from the sun, it is called *sunburn,* and may vary from a barely perceptible redness with little or no pain on contact to frank blistering and severe pain.

Ultraviolet radiation from the sun decreases in intensity with shortening of wavelengths and therefore the most intense radiation in this part of the spectrum is in the longer wavelengths. Sunburn is caused mainly by radiation in the region from about 280 to 320 nm. In this region of the spectrum, light with a wavelength of about 297 nm has long been considered to be one of the most effective in the production of erythema. Radiation in the region from about 300 to 400 nm is that mainly responsible for suntans and also for the production of photochemical smog from nitric oxide, reactive hydrocarbons, and their reaction products.

Above about 320 nm and below about 240 nm, UV radiation is capable of producing little or no erythema. Although certainly not all the answer, the shortest wavelengths are absorbed too near the skin surface and the longer wavelengths contain too little energy so that the effectiveness of both is small. Between those extremes, UV radiation is most effective in the production of injury signaled by erythema in the region near 270 nm. Particularly when the radiation wavelength is shorter than about 260 nm, the erythema appears after a delay of a few hours, peaks in intensity quite rapidly, and remains at the peak intensity for a day or so. When the wavelength is longer than about 260 nm, the erythema may fade rapidly after a rapid peaking but a tan may well result.

The minimal erythema dose (MED) varies considerably throughout the effective region of the spectrum and is subject to considerable controversy as there is no real agreement on how to define minimal erythema. Nevertheless, at about 270 nm, a flux density much greater than a few tens of joules per square meter can be expected to produce some erythema on white, untanned human skin following an 8-hr exposure; the exposure dose at 315 nm may have to be several hundred times larger to produce the same effect. Particularly if the source of UV is the sun, wind may augment the erythemal effect [3].

Photoallergy is skin sensitization to light (particularly UV radiation), usually caused by the action of some chemicals. Among others, coal tars and 4-chloro-2-phenylphenol [4] have been found to cause the skin to respond to orders of magnitude less UV than normal unsensitized skin. Typically, the response in this case resembles acne rather than simple erythema.

Epidemiological evidence has shown that chronic ultraviolet radiation may cause skin cancer in man and other work has demonstrated a similar effect in experimental animals. Although there is probably a threshold for this effect, its demonstration is extremely difficult as is a real definition of the *action spectrum* for the effect [5]. Most experts in the field of NIR feel that

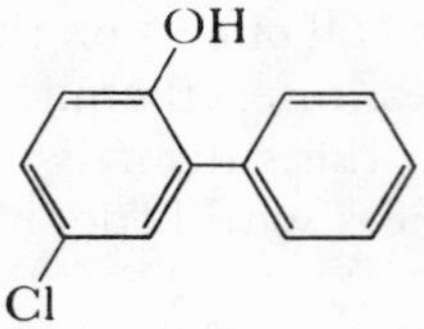

OH

Cl

4-Chloro-2-phenylphenol

effectiveness of UV for the production of skin cancer probably parallels effectiveness for production of erythema if only because there appear to be no good reasons why it should not. Largely because of major differences between the skin of any laboratory animal and that of man, this question may continue unresolved for some time.

Eye effects. Snowblindness, one form of *photokeratitis* (inflammation of the cornea by light) has long been known as an effect of sunlight. Most of the UV radiation reaching the eye is absorbed in the outer layers of tissue, the cornea, and the conjunctiva, with little reaching the lens or the interior of the eye. Because there is no sensation of light attending the exposure (or of immediate pain), an overexposure to the UV portion of sunlight or other light can occur unknowingly. After a latent period of a few minutes to many hours, depending on the exposure severity, the conjunctiva become inflamed, accompanied by a painful "sand in the eyes" feeling. Photophobia, blepharospasm, and an inability of the iris to open adequately in subdued light may occur with rather sudden onset. The victim may thus be incapacitated for several hours, but in most cases the signs and symptoms last only about 2 days. Also known as *flash burn, welder's flash,* or *welder's eye,* photokeratitis is a rather common occurrence in some arc-welding shops.

Greatest effects on the eyes are seen from UV radiation with a wavelength of about 288 nm (a range of 280 to 305 nm), but photokeratitis will occur with the least radiation at a wavelength of about 270 nm [6]. The action spectrum for this effect is thus quite similar to that for the erythemal effect on skin. Flux densities associated with photokeratitis are similar to those associated with minimal erythema but the eye has no protective ability similar to tanning of the skin.

Measurement

Meters of several kinds with and without filters designed to simulate an action spectrum of one sort or another are commercially available but as yet are not used to any great extent to evaluate UV hazards [7, 8]. One of the reasons is that the good meters are very expensive and, in essence, require large amounts of the time of well-trained people for their use. Another is that no good means of calibrating field-type meters has been developed and marketed so that even if inexpensive, simple meters were available, their utility would be extremely limited. Finally, the development of an ideal hazard meter would include a filter designed to make the meter respond just as a person's skin and/or eyes respond to UV radiation. While such a filter might be developed on the basis of an erythemal or keratitis effect, there is no assurance as yet that the meter would accurately reflect the cancer hazard. To protect against sunburn while possibly allowing cancer to develop is recognized as an exercise in futility. Furthermore, despite seeming agreement implied by the promulgation of standards, even the MED and photokeratitis action spectra are controversial; a cancer action spectrum is even more so.

Sources

The most intense source of UV radiation to which the most people are exposed is the sun. Man-made sources include the obvious ones such as sun lamps, arc welding [9], and UV lamps designed to kill germs or to elicit fluorescence with black light.

Perhaps not so obvious are quartz-halogen lamps used as intense sources of visible light for special photographic and other purposes such as catalyzing chemical reactions. Plasma torches may create temperatures equivalent to those on the surface of the sun and thus cause emission of radiation similar to that from the sun unmodified by a layer of ozone [10].

The spectrum of light emitted from these sources varies between and among the sources, and the intensity at any particular wavelength varies from nil to high. Because good inexpensive hazard meters are not available, each source must be rated on the basis of some kind of average for that class of device and precautions devised accordingly.

Control

Most solids are opaque to ultraviolet radiation, including many that are transparent to visible light. Moreover, a solid that is opaque to visible light is almost certain to be opaque to UV. Shielding, then, becomes rather easy in most cases and extremely effective.

Spectacles have been devised that are opaque to all electromagnetic radiation with wavelengths much shorter than about 420 nm and they are commercially available [11]. Window glass transmits UV quite well from about 320 nm up but is almost completely opaque to those wavelengths known to be best able to cause erythema and keratitis. Ordinary safety glasses thus offer reasonably good protection for the eyes from UV even if not from visible light.

Skin creams and ointments have been developed that offer varying degrees of protection ranging from essentially complete to almost nil [12].

Those that offer complete protection include a material such as the oxide or carbonate of magnesium, titanium, or zinc. These inorganic materials absorb, reflect, or scatter essentially all light including UV; they transmit very little, if any. Those lotions based on *p*-aminobenzoic acid or benzophenone appear to protect the skin well against the erythemal effects of UV and are quite adequate if used properly to protect even very fair skin against sunburn. A person with already burned skin or one suffering from photoallergic dermatitis, however, will probably not be protected adequately by this type of cream.

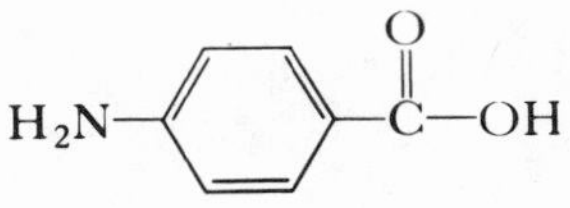

p-Aminobenzoic Acid

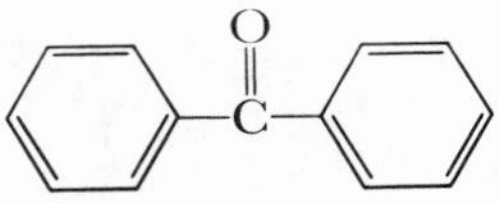

Benzophenone

Visible

Visible light energy has a wavelength range from just below 400 nm to about 750 nm. In this range the colors, in reverse order, are red, orange, yellow, green, blue, and violet; that is, red is the color with the longest wavelength. Reverse order was used because that corresponds to the color coding of resistors and capacitors from 2 (red) to 7 (violet). (Black and brown mean 0 and 1 and gray and white mean 8 and 9, respectively).

Effects on Man

In daylight, the action peak for seeing is about 555 nm (yellow-green) with 1% of the peak value at 430 and 680 nm. Color vision with plenty of light available is associated with cone-shaped receptors in the retina. The *retina* is the light-sensitive layer that lines the posterior or vitreous chamber of the eye [1]. Within the retina, the cone-shaped cells are concentrated in the *fovea,* the area of the retina lying at the focal point of the cornea-lens system, which provides maximum seeing acuity. Surrounding the fovea is an area called the *macula* where both cone- and rod-shaped receptors are located, and farther from the fovea yet are found only the rods.

Rod-shaped photoreceptor cells are primarily responsible for night vision or seeing under conditions where the light level is low and also seem to be more sensitive in detecting motion than the cone cells. Under difficult seeing conditions where only the rod-shaped cells are active, the eye is most sensitive to a wavelength of 510 nm (green), with 1% of the peak value at 390 and 620 nm.

The cornea, lens, and the aqueous (between the cornea and lens) and vitreous (behind the lens) chambers are essentially transparent to light in the range from 400 nm to about 1,400 nm but red light with wavelengths much above 700 nm does not contain enough energy to excite the photo-receptors. If the intensity is sufficiently great, light having a wavelength in the 700 to 750 nm region can be sensed, although efficiency drops rapidly as wavelength increases.

When the eye is focused on an object, the light from that object is con-centrated on the fovea by the cornea-lens system. Irradiance of the fovea may be 10^4 to 10^6 times irradiance of the cornea. This is the mechanism by which visible light (or light in the near IR) can injure the eye. If the light is intense enough and focused well enough, portions of the retina especially in or near the fovea can be damaged even though the cornea and lens com-pletely escape injury [13]. Injury to the retina may be either thermal or photochemical in nature, depending on circumstances, and can be caused by the absorption of a few millijoules of energy [1].

Production of injury not only is a function of the amount of energy il-luminating the eye but also depends on the pupil diameter, the time, and the size of the irradiated spot on the retina. Much of the research on this subject has been conducted with animals using the "worst case" of the maximum pupil diameter, which in man is about 7 mm. Expressing the irradiation as power instead of energy, the threshold for injury from visible light is on the order of a few milliwatts when that power is delivered in a short enough period of time that heat conduction between portions of the retina is essentially nil.

Seeing

Seeing as a task has been studied quite thoroughly and the ability to see has been found to be a function of four parameters: size, contrast, time, and brightness [14]. As the size of the object diminishes, seeing becomes more difficult. Decreasing contrast between the object and its background, de-creasing the time allowed, and decreasing the brightness all have similar effects. Traditionally, the most difficult seeing task has been to inspect black thread against a background of black velvet as rapidly and with as little light as possible. Despite the tales of old wives and the strictures of mothers, reading in the dark will not and cannot injure the eyes. Attempt-ing to perform difficult seeing tasks can cause headache, mistakes, and other symptoms of frustration and eye strain but will not cause any injury.

Over the years as seeing was studied more and more thoroughly, rec-ommended illumination levels were increased for almost all areas and see-ing tasks. The most recent recommendations will be found in the *IES Lighting Handbook* published by the Illuminating Engineering Society, 345 East 47th St., New York, N.Y. 10017.

Measurement

Light levels for seeing are usually determined with instruments ranging in size and complexity from the light meter used in photography (which may be incorporated into the camera) to elaborate laboratory devices. In general, as the measurement becomes more difficult, the instrumentation becomes more complex and expensive. One of the problems is that light incident on the task is not all that must be measured; glare is often quite important and the brightness of luminaires, their positions, and the shadowing caused all may be important factors.

Infrared

Infrared (IR) means *below red,* the *below* referring to frequency rather than wavelength in a manner analogous to the naming of ultraviolet. Wavelengths associated with infrared begin in the range of 700 to 750 nm (0.7 to 0.75 μm) and extend to the point where they become confused with the shortest wavelengths of radio (microwaves). The area of transition from IR to microwaves is from 1.0×10^5 to 3.0×10^5 nm (0.1 to 0.3 mm). For purposes of hazard delineation, the boundary between near and far IR is about 1,400 nm, that is, at about the wavelength where the radiation no longer penetrates much beyond the surface of the skin or eye [1].

Effects on Man

Near-IR radiation reaches the retina and penetrates the skin also to a certain extent. Far-IR radiation is absorbed by surface layers of the skin and by the cornea. In both cases, the effects are thermal in nature, to the lens and retina for near IR, to the cornea for far IR, and to the skin throughout the whole spectrum.

On the skin, IR radiation may cause increased vasodilation and increased pigmentation with long persistence [2]. If the intensity is great enough, a burn may result, but there appears to be no chronic effects from long-term exposure that differ significantly from the acute effects.

Near IR (wavelength of less than about 1.4 nm) has long been known able to cause *cataract,* or clouding of the lens of the eye. Cataract has been an occupation-related disease of glassblowers and others who handle metal or refractories at red heat. The lens absorbs some small amount of the incident radiation and becomes warmer. The iris surrounding and overlying the lens is pigmented and therefore absorbs much more incident radiation than does the lens itself. Because of this, not only is the iris not available as a heat sink for the lens, it is the main source. Heat from the iris enters the lens, producing a temperature rise [1]. Because the lens is supplied with no nerve endings at all, injury occurs with essentially no warning. After chronic overexposure to near IR, the lens begins to opacify, and if the exposure is continued, a "glass-worker's cataract" results.

Because near IR can penetrate all the way to the retina, retinal damage may well occur whether or not sufficient energy has been absorbed by the lens to raise its temperature enough to cause clouding. Injury to the retina is thermal in nature and usually occurs in discrete areas. Because the radiation is not sensed, however, the eye may not be focused in such a way as to maximally concentrate the radiation. Nevertheless, absorption of a few millijoules in a small (100-μm diameter) area can cause a retinal lesion.

Far IR (wavelengths greater than 1.4 nm) affects the cornea of the eye in a manner analogous to the thermal effects of UV radiation. A temperature rise of 25 to 35°C for a few seconds is likely to cause denaturation, although the injury is accompanied by sufficient pain to actuate the blink reflex [15]. An irradiance on the order of 0.1 W cm^{-2} will probably cause little, if any, detectable injury even if continued for several seconds.

Effects of IR radiation on the skin are accompanied by pain so that if there is the time and/or opportunity to avoid the radiation, any effect will be minimal or nonexistent. The threshold for minimal erythema appears to be nonlinear with exposure duration and, in fact, with the area illuminated so that the hazard may be difficult to ascertain. There is no reason to believe that the skin is more sensitive than the eye, however, so that control to eliminate eye damage will also assure no skin injury.

Sources

Sources of IR are almost all thermal in nature, although plasma torches, quartz-halogen lamps, and other sources of full incandescent spectra emit IR as well. With most such sources, the more energetic radiation is the main hazard, however, so IR tends to be ignored.

Measurement

Infrared radiation is difficult to measure because few materials are transparent over the wide spectral range associated with this type. Devices that measure heating effects are available but are, of course, unable to resolve the spectrum into various wavelengths [8]. Filters have been developed that give excellent resolution in the range of 2.5 μm (2,500 nm) to about 14.5 μm (14,500 nm). These filters are used mainly in infrared spectrophotometric instruments for the analysis of organic compounds but could be adapted to a hazard meter. Their success implies that filters could probably be constructed that would be capable of adequate resolution in other portions of the IR band, above and below the area of analytical interest.

Control

Glass doped with neodynium has a purple cast and is almost completely opaque to near IR. Safety glasses made from this glass are available for people chronically exposed to solids at red heat.

Most materials opaque to visible radiation are also opaque to near and most far IR and can be used for shielding. The main sources of infrared radiation are thermal (except for lasers) and also emit in the visible range so that a rough estimate of the hazard potential is usually simple. Because the skin is well able to sense small temperature rises [1], adequacy of shielding of large sources is easy to test.

Microwaves

Beginning with the microwave portion of the spectrum at about 10^5 nm (0.0001 m), frequency is used much more than wavelength to characterize the area of interest. Microwaves and radar have frequencies in the 3,000 GHz (3×10^6 MHz) to 30 MHz range; followed by shortwave radio, from 30 to 1.8 MHz; the radio broadcast band from 1.6 to 0.55 MHz (550 kHz); and below that the low and very low frequencies. Television is broadcast in two bands called very high frequency (VHF, 54 to 88 MHz and 174 to 216 MHz) and ultrahigh frequency (UHF, 470 to 890 MHz), both well into the microwave portion of the spectrum as is FM radio (88 to 108 MHz).

Effects on Man

Frequencies higher than about 10 GHz are completely absorbed by the outer layers of the skin and behave much like far IR. That is, all effects are confined to the skin, and those effects are thermal in nature.

In the range of 1 to 10 GHz, microwaves penetrate through the skin in significant amounts, reaching the fatty layers beneath. Below 1 GHz, penetration into deep muscle occurs, with depth of penetration increasing rapidly with small decreases in frequency. By the time the radio spectrum is reached somewhere between 100 and 30 MHz, penetration is deep enough so that even with complete absorption, heat is dissipated throughout the whole structure rather than in localized tissue. The main area of interest, then, is in the frequency range from a little below 100 MHz to a little above 1000 MHz (1 GHz). In that area especially, claims have been made for both thermal and nonthermal effects.

Thermal effects. Microwaves have been used therapeutically since the early 1930s to heat muscle tissue deep within the body [16]. The power densities used for this purpose have been on the order of 100 mW cm^{-2} (1,000 W m^{-2}) and up, causing deep tissue (especially that with a high water content such as muscle) to become warm. Because of complex interactions with tissue of varying density and water content, standing waves may be produced that cause local hot spots and subsequent discomfort. For whole-body exposure, the eyes and testes appear to be the critical organs [17]. There is nothing in the literature, however, that indicates any deleterious effects from prolonged use of diathermy treatment indicating that if injury from acute effects is avoided, there should be no chronic effects [18]. General agreement over several years has indicated that irradiances at or below about 100 W m^{-2} for a few minutes will not cause harmful thermal effects in the frequency range from 10 MHz to 100 GHz [2, 16, 17]. For longer exposures, the guide number may be changed from an irradiance to a flux density on the order of 1 mW hr cm^{-2} or 3.6 × 10^5 J m^{-2}. Such values may be quite conservative at the lower (radio) frequencies where few local effects are expected but they are reasonably well accepted in the United States and many other countries.

Nonthermal effects. Claims of many different nonthermal effects of microwave radiation have appeared in the literature and elsewhere almost from the development of diathermy equipment. That such effects may occur at power densities well above 100 W m^{-2} is not really disputed; lens opacities have been produced in rabbits, for example, at irradiances of 1,500 W m^{-2} for an hour or longer and have been found in some studies of microwave workers [19] but other studies have shown no such effects [20]. The real question has been whether or not nonthermal effects occur at irradiances of less than 100 W m^{-2} because if they do, then the guide number or set point should be based on them rather than on thermal effects.

Schwan and others have reviewed the literature quite thoroughly and have concluded that there is little or no conclusive evidence for any non-thermal effect of microwaves when the power density was kept at or below the 100-W m^{-2} value [2, 16]. Theoretical considerations based on calculation and an examination of several possible effects of the interaction of electromagnetic radiation with matter indicate that such effects cannot be expected at low power densities [16]. Standards regulating exposure to microwaves are likely to remain based on thermal effects in the United States.

Sources

The main sources of occupational microwave exposure are radar and broadcast (TV and FM) transmitting facilities. Nonoccupational exposure occurs from medical treatment (diathermy) and, at least potentially, from microwave ovens, which may pose hazards for repairmen [21].

Measurement

Special devices (barreters or thermistors connected to calibrated horn antennas) to determine power densities are available commercially. Calibration is a real problem, and this is especially true for high power densities.

In addition, if the field is not uniform, the measurement problems become very complex [8]. All the available measurement devices are expensive and all require well-trained knowledgeable people for their use.

Control

Without exception, all potentially hazardous sources of microwave radiation are man-made so that at least rough approximations of power densities are available from calculations based on known or assumed amounts of radiation. Exposure limits are usually based on such calculations with appropriate safety factors used to compensate for imprecision.

Lasers

One of the breakthroughs in the development of microwave technology was the discovery of a way to amplify signals by the technique of stimulated emission of radiation. Called *microwave amplification by the stimulated emission of radiation,* the acronym MASER was soon used almost exclusively. Later, methods were developed to achieve amplification of light by somewhat analogous methods first with solids such as artificial rubies and later with liquid and gaseous systems. By further analogy, this new technique was called *light amplification by stimulated emission of radiation,* or LASER. Because of widespread use, capital letters were omitted: *maser* and *laser* became accepted nouns and *lase,* an accepted verb ("mase" is not used).

Microwaves are not produced with intensities sufficient to cause concern about health effects unless some form of amplification such as the maser is used. Light, on the other hand, has many sources, most of which involve no amplification except that provided by concentration as with mirrors and/or lenses. While the maser was just another radio-frequency amplifier, the laser was the first means of directly amplifying light. In addition, the laser brought new meanings to the terms *monochromatic, directionality, coherence* (all wavefronts in phase), and especially *intensity* when applied to light.

Lasers inherently operate on a single wavelength (or frequency), a rarity in nature, and produce beams with very little dispersion (typically, divergence ranges from a fraction of a milliradian up) [22] unless specifically designed to do otherwise. The small divergence means little attenuation with distance except in highly absorbent media and, therefore, in clean air the hazard a few miles away from the device may not differ significantly from the hazard at the device. Monochromaticity and coherence of the beam appear to have little effect on laser hazards, except that with lasers the intense beams of monochromatic radiation have been produced from the visible spectrum up through the ultraviolet and down through the infrared.

Small divergence of the beam is coupled with intensity to produce hazard. The intensity of the light from a laser is usually expressed as power (in watts) for continuous wave (CW) operation or as energy content (joules) for pulsed beams. Power density or irradiance is used to express dose rate from CW devices and flux density is used for doses from pulsed beams [2]. In general, CW lasers are operated at relatively low power levels (typically, milliwatts) and while the average power from a pulsed laser may not be much greater, the power per pulse may be very high because the energy is concentrated into very small time intervals. Pulsed lasers are operated in a normal multiple-spike mode with pulse durations ranging from 100 μsec to a few milliseconds or in a Q-switched mode with pulse durations from about 100 nsec down [22].

Effects on Man

Light from a laser differs from ordinary light mainly in size and power of the beam. Effects therefore depend on pulse duration (if pulsed), time interval between pulses, wavelength, and energy density of the beam. For the same wavelength and absorbed energy, there appears to be no distinction of effect on tissue between the coherent light of a laser and the incoherent light from other sources. In fact, much of our present knowledge about the effects of light from the far UV to the far IR has been gained from research with lasers. The reason for this is that use of a laser enables the experimenter to deliver specific amounts of energy to very small areas of tissue in a time frame of his own choosing.

As with most other portions of the electromagnetic spectrum, the main hazards of light from a laser are to the eye, although skin burns can also be severe. If the precautions taken are adequate to prevent eye injury, the skin will almost always be adequately protected.

Almost all lasers are operated between wavelengths of 400 to 1,400 nm where the eye is sufficiently transparent to ensure that effects will be on the retina rather than the lens or cornea. Ruby lasers operating in a frequency-doubled mode produce ultraviolet radiation, however, that may preferentially affect the cornea. Similarly, a CO_2 laser produces light at a wavelength of 10,600 nm and erbium lasers emit in the range of 1,500 to 1,650 nm, all in the far infrared where again the cornea is much more likely to be affected than the retina.

Retinal effects are likely to be the most serious of those associated with exposure to lasers and these effects are of either a photochemical (photocoagulation, for instance) or a thermal nature. In either case, injury is repaired very slowly or not at all by the body and is usually regarded as permanent. Small (a few tens of micrometers in diameter) burns in the retina cause little, if any, decrease in visual acuity unless they are many or are in the macula or fovea. A burned area in the fovea can interfere seriously with vision in proportion to the size of the area and, unfortunately, the fovea appears more likely to be burned than any other area of comparable size in the retina. Injury to the retina is dependent mainly on energy distribution as a function of image diameter and exposure duration [23].

Sources

Those people most likely to be exposed to laser beams capable of causing injury are some military personnel, researchers in various establishments, and a few others [22].

Measurement

Although the output of some lasers could be measured by more or less conventional means (some CW lasers, for instance) [24], this is rarely done. Instead, the known power (or energy per pulse) output of the device [25] is used to calculate probable absorbed doses after making several assumptions about the geometry of the whole device-absorbing tissue system [26]. Based on those calculations, controls of varying severity may be instituted [2].

Control

Although "Don't ever look down the beam" is good advice, it rarely prevents accidents. Most accidental exposures of eyes and skin occur from reflections of the beam. If the reflecting surface is shiny, attenuation may be negligible. Controls [27] therefore include shielding the beam from direct vision, using light-absorbing drapes to cover possibly reflective surfaces, avoidance of the possibility of exposure (warning signs, indirect viewing by TV, etc.) wearing light-absorbing clothing but no jewelry, and using safety glasses designed to be opaque to the wavelength of light [28, 29]. Periodic eye examinations are the rule but have the disadvantage of providing only hindsight.

Large power outputs require large power supplies; associated electrical hazards may be more severe than those of the laser itself but neglected in favor of the more exotic problems. Other hazards associated with laser research range from those of the flash tube(s) [30] to inhalation of aerosols formed by disruption of solids and semisolids by the beam [31].

REFERENCES

[1]   W. F. VAN PELT, W. R. PAYNE, AND R. W. PETERSON, *A Review of Selected Bioeffects Thresholds for Various Spectral Ranges of Light* [Rockville, Maryland 20852: Electro-Optics Branch, Division of Electronic Products, U.S. Department of Health, Education, and Welfare, Public Health Service, Food and Drug Administration, Bureau of Radiological Health, Publication (FDA) 74–8010, 1973].

[2]   S. M. MICHAELSON, "Standards for Protection of Personnel against Nonionizing Radiation," *Am. Ind. Hyg. Assn. J.* **35** (1974), 766–84.

[3]   D. W. OWENS, ET AL., "Influence of Wind on Ultraviolet Injury," *Arch. Derm.* **109** (1974), 200–201.

[4]   R. M. ADAMS, "Photoallergic Contact Dermatitis to Chloro-2-phenylphenol," *Arch. Derm.* **106** (1972), 711–14.

[5]   H. F. BLUM, "On Hazards of Cancer from Ultraviolet Light," *Am. Ind. Hyg. Assn. J.* **27** (1966), 299–302.

[6]   D. G. PITTS AND T. J. TREDICI, "The Effects of Ultraviolet on the Eye," *Am. Ind. Hyg. Assn. J.* **32** (1971), 235–46.

[7]   R. NAGY, "Application and Measurement of Ultraviolet Radiation," *Am. Ind. Hyg. Assn. J.* **25** (1964), 274–81.

[8]   J. H. FANNEY, JR. AND C. H. POWELL, "Field Measurement of Ultraviolet, Infrared, and Microwave Energies," *Am. Ind. Hyg. Assn. J.* **28** (1967), 335–42.

[9]   J. A. DAHLBERG, "The Intensity and Spectral Distribution of Ultraviolet Emission from Welding Arcs in Relation to the Photodecomposition of Gases," *Ann. Occup. Hyg.* **14** (1971), 259–67.

[10]   C. H. POWELL, L. GOLDMAN, AND M. M. KEY, "Investigative Studies of Plasma Torch Hazards," *Am. Ind. Hyg. Assn. J.* **29** (1968), 381–85.

[11]   A. E. SHERR, "Plastics with Selective Wavelength Absorption for Safety and Comfort," *Am. Ind. Hyg. Assn. J.* **33** (1972), 583–95.

[12]   G. KAHN AND M. C. CURRY, "Ultraviolet Light Protection by Several New Compounds," *Arch. Derm.* **109** (1974), 510–17.

[13]   C. J. BARTLESON, "Retinal Burns from Intense Light Sources," *Am. Ind. Hyg. Assn. J.* **29** (1968), 415–24.

[14]   S. K. GUTH, "Lighting Research," *Am. Ind. Hyg. Assn. J.* **23** (1962), 359–71.

[15]   F. GULLBERG, ET AL., "Carbon Dioxide Laser Hazards to the Eye," *Nature* **215** (1967), 857–58.

[16]   H. P. SCHWAN, "Microwave Radiation: Biophysical Considerations and Standards Criteria," *IEEE Trans. Biomed. Eng.* **BME-19** (1972), 304–12.

[17]   D. I. McREE, "Biological Effects of Microwave Radiation," *J. Air Poll. Con. Assn.* **24** (1974), 122–27.

[18]   S. M. MICHAELSON, J. W. HOWLAND, AND W. B. DEICHMANN, "Response of the Dog to 24,000 and 1285 MHz Microwave Exposure," *Ind. Med. Surg.* **40** (Aug., 1971), 18–23.

[19]   S. F. CLEARY AND B. S. PASTERNACK, "Lenticular Changes in Microwave Workers," *Arch. Env. Health* **12** (1966), 23–29.

[20]   B. APPLETON, "Microwave Cataracts," *J. Am. Med. Assn.* **229** (1974), 407–08.

[21]   V. E. ROSE, ET AL., "Evaluation and Control of Exposures in Repairing Microwave Ovens," *Am. Ind. Hyg. Assn. J.* **30** (1969), 137–42.

[22]   W. T. HAM, JR., ET AL., "The Eye Problem in Laser Safety," *Arch. Env. Health* **20** (1970), 156–60.

[23]   E. S. BEATRICE AND G. D. FRISCH, "Retinal Laser Damage Thresholds as a Function of Image Diameter," *Arch. Env. Health* **27** (1973), 322–26.

[24]   J. C. ROCK AND J. L. UNMACK, "Self-Calibrating Technique for Measurement of Continuous-Wave Laser Beam Power Density Distributions," *Aerosp. Med.* **41** (1970), 1187–89.

[25]   R. H. JAMES, ET AL., "Calibration Systems for Laser Power or Energy Measuring Apparatus," *Am. Ind. Hyg. Assn. J.* **35** (1974), 327–32.

[26]   W. D. BURNETT, "Evaluation of Laser Hazards to the Eye and the Skin," *Am. Ind. Hyg. Assn. J.* **30** (1969), 582–87.

[27]   L. GOLDMAN AND P. HORNBY, "Personnel Protection from High-Energy Lasers," *Am. Ind. Hyg. Assn. J.* **26** (1965), 553–57.

[28]   W. J. SCHREIBEIS, "Laser Eye Protection Goggles Based on Manufacturers' Information," *Am. Ind. Hyg. Assn. J.* **29** (1968), 504.

[29]   R. L. ELDER, "Lasers and Eye Protection," *Science* **182** (1973), 1080.

[30]   M. S. LITWIN, ET AL., "Hazards of Laser Radiation: Mechanisms, Control and Management," *Am. Ind. Hyg. Assn. J.* **28** (1967), 68–75.

[31]   T. K. WILKINSON, "Health Aspects of Laser Use. Air Contamination and Lasers," *Arch. Env. Health* **18** (1969), 443–47.

15

IONIZING RADIATION

Early in the atomic age, those people who took upon or were assigned responsibility for the safe use of ionizing radiation and radioactive isotopes were called *health physicists*. This was a fairly obvious development because most of these people were physicists by training and they were doing their work in a health-related field. Today, most health physicists have received their training in a field that may be called *radiation biology, biophysics,* or something similar rather than simply in physics, but the original name is still used.

Fundamentals of Health Physics

Although the number of identified subatomic *particles* has proliferated to the point where a table of their names and properties is larger than the conventional chart of the elements, most are of no interest in health physics. With a few exceptions, the health physicist can look upon an atom as if it were composed of a nucleus containing only neutrons and protons and an external cloud of electrons in one or more discrete shells, orbits, or energy levels. In each atom of an element the number of electrons exactly equals the number of protons, but the number of neutrons in the nucleus can vary. Variation in the neutron number produces atoms of the same atomic number (the number of protons or electrons) but having different atomic weights. Atoms of the same element that differ in atomic weight are called *isotopes* of that element. Each element in the periodic table (with the exception of those most recently discovered) has at least 3 known isotopes, and most of the elements have 17 or more.

Isotopes are identified unambiguously with atomic number (Z) and either the number of neutrons (N) or the mass of the nucleus ($A = Z + N$). That is, two numbers suffice but in order to aid man's weak memory, the chemical symbol is also always used even though it and Z provide the same information. Atomic number (Z) is put in subscript location before the symbol and mass of the isotope (A) is put as a superscript, also before the symbol. Examples of this practice are $^{10}_{6}C$, $^{11}_{6}C$, $^{12}_{6}C$, $^{13}_{6}C$, $^{14}_{6}C$, $^{15}_{6}C$ for the seven isotopes of carbon; $^{131}_{53}I$ for a commonly encountered isotope of iodine; and $^{137}_{55}Cs$ for a very useful isotope of cesium. To shorten the process, the atomic number (6, 53, and 55 in these examples) is sometimes omitted and occasionally the superscript notation is abandoned entirely, resulting in C-10, C-11, etc.; I-131; and Cs-137.

One or more isotopes of every element is radioactive. The word *radioactive* means that processes taking place within the atom can be sensed at a distance. Forces within the nucleus become unbalanced enough so that something must be ejected in order to attain a new equilibrium level. There is no way of telling in advance which of many atoms of the same isotope will be the first to reach such a state of instability that a breakdown or disintegration is inevitable. Instead, the process is a statistical one, the probability of *radioactive decay* being spread evenly among all atoms of the same isotope.

Not all isotopes have the same probability of the nuclear disruption called *disintegration* or *decay;* instead, that probability varies over a tremendous range. One way of expressing the probability of decay is with the *half-life,* which is the time for a quantity of radioactive material to decay to one-half of its starting activity. If N_o is the number of atoms present originally, and N is the number remaining after time t,

$$\frac{N}{N_o} = e^{-\lambda t} \qquad (15\text{-}1)$$

or, taking the logarithm of both sides;

$$\ln \frac{N}{N_o} = -\lambda t \qquad (15\text{-}2)$$

where λ is the *decay constant* or *disintegration constant*. Setting N equal to $\frac{1}{2} N_o$ satisfies the definition of half-life; that is, half of the original atoms remain. Substituting in Eq. (15-2),

$$\ln \frac{N_o/2}{N_o} = \ln \frac{1}{2} = -\lambda t$$

Inverting the $\frac{1}{2}$ changes the sign on the other side of the equation. Doing this and solving for t gives an expression for the half-life, T:

$$\ln 2 = \lambda t$$

or

$$T = \frac{0.693147}{\lambda} \tag{15-3}$$

The half-life T is characteristic of the isotope and is not altered by temperature, pressure, or chemical combination. It ranges from a few microseconds or less to more than 10^{10} years. In most cases, the half-life rather than the decay constant is measured and by rearranging Eq. (15-3) the decay constant may be found.

If the radioisotope is incorporated into a biological system of some sort, the amount remaining in the system at any time may be a function of the elapsed time in a manner similar to that expressed by Eq. (15-1). Whether the material is radioactive or not, it may exhibit something like a biological half-life. If so, the value calculated is not likely to be as invariant as the physical (radioactive) half-life and may, in fact, change with time. Nevertheless, the concept of a biological half-life may find utility; its symbol is usually T_b. But if the isotope having a biological half-life is also radioactive, the two kinds of half-life must be added to determine the *effective half-life* (T_{eff}) or the time required for half of the original atoms to no longer be in the body.

If T_b actually can be said to be a half-life, it can be added to T by means of an equation similar to that used to find the effective resistance of two parallel resistors:

$$T_{eff} = \frac{1}{(1/T) + (1/T_b)} \tag{15-4}$$

Caution is advised when using Eq. (15-4), however, as biological systems are rarely completely described by a single number such as a half-life.

If the physical half-life is long compared to T_b, then T_b will control and T_{eff} will be close to T_b. On the other hand, if T is small compared to T_b, most of the radioactivity may have disappeared by physical decay before excretory activity has had a chance to eliminate much of the material biologically. In that case, T_{eff} is about equal to T.

There are many nuclear processes that atoms may undergo in the act of decaying from one state to another but, for the purpose of hazard evaluation, consideration can be restricted to the external manifestations of those processes. That is, no matter how complicated the transitions are in the inner workings of an atom, the visible results are restricted to the emission of gamma rays along with α and β particles as the ionizing *radiations* and neutrons, which are not ionizing.

Gamma (γ) rays are electromagnetic radiations having wavelengths less than about 0.14 nm (Table 14-1). Beta (β) rays are high-speed electrons and (α) rays are actually the nuclei of helium atoms and consist of two protons and two neutrons. Most of the β emissions encountered are (negative) electrons or *negatrons* but some atoms decay by the emission of a positive electron or *positron*. A positron, however, in the process of speeding away from the atom, loses energy by the production of ions in matter and finally encounters an electron; the electron-positron pair is annihilated, producing a pair of oppositely directed gamma rays, each with an energy of 0.51 MeV (million electron volts—recall from Chapter 14 that the energy required to ionize water is about 12 eV). Therefore, even though positron emission does occur, its manifestation is simply a trail of ions and a pair of gamma rays with specific energy levels.

When an α particle is ejected from a nucleus, that nucleus has lost a mass of 4 and two positive charges. The mass loss alone would allow the element to remain the same, but losing two protons causes a transmutation from the original element to one of two atomic numbers less. This kind of process can be illustrated by means of a plot of Z versus N as in Fig. 15-1 [1]. Also illustrated in this chart is the consequence of ejecting either a β^- or a β^+ particle from the nucleus—a process that also results in transmutation from one element to another but does not change the mass of the isotope. Atoms having the same mass (but different elements) are called *isotones*.

Alpha particles ejected from a nucleus all have the same energy characteristic of the particular decay. The same is true of gamma rays, but in a slightly different manner. That is, gamma rays are ejected along with or following many processes taking place in or near the nucleus, so that the radioactive decay of an isotope may result in gamma rays having many discrete energy levels (*many* can be several tens), but the number of energy levels associated with α-particle ejection is far fewer. Beta particles are ejected by processes associated with discrete energy levels as are gramma rays, but β particles are observed to have a spectrum of energies, from zero up to a maximum level (β_{max}) characteristic of that particular reaction. As with all processes taking place in the nucleus, the process that results in β emission in each case has a discrete energy level that is, in fact, β_{max}, but few of the β particles ejected have that energy. Instead, much of the energy is carried

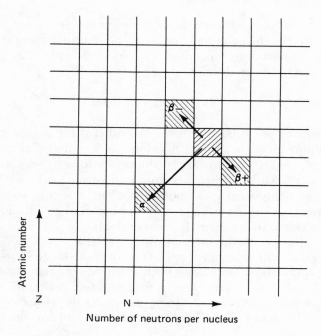

Figure 15-1 Transmutation by radioactive decay.

away by neutrinos or antineutrinos, particles having negligible mass and no charge so that they are of no interest in health physics except that their presence assures that β particles will have a spectrum of energies rather than discrete levels.

Properties of Radiation

Penetrating Ability

If the source of radiation is outside the body, the radiation must penetrate into the body to cause injury. The α, β, and γ radiations differ considerably in their penetrating abilities and a knowledge about this property is essential to an evaluation of external source hazards.

Even highly energetic α particles have little ability to penetrate matter. Their range is very short even in air. Skin is a complete barrier to α particles as is a sheet of paper.

Each time an energetic electron (or positron) passes close to an atom, it interacts with the electron cloud, causing ionization. At some point that electron has lost enough energy so that it can be captured, and the distance it travels before that occurs is proportional to its initial energy and to the mass of matter through which it has passed. *Range* of an electron is the linear travel distance from point of origin to point of capture, and this is usually correlated as the product of density and distance so that the units of range are grams per square centimeter ($= g \ cm^{-3} \times cm$). An empirical expression of range has been developed [1]:

$$R = 0.542E - 0.133 \tag{15-5}$$

where R is in grams per square centimeter and E is the actual energy of the electron in million electron volts. Equation (15-5) is useful at energies above about 0.6 MeV; other relations are available for lower energies. In tissue, the range of a 1.0-MeV electron is equal to $0.54 - 0.13 = 0.41 \ g \ cm^{-2}$, which, divided by the density of about $1.0 \ g \ cm^{-3}$, yields about 0.4 cm or 4 mm. This range is sufficiently great so that external β radiation can be expected to burn skin if its intensity is great enough.

Gamma radiation causes ionization by several processes as it passes through matter, usually losing energy in one or more large steps rather than the countless small steps associated with the loss of energy by α or β particles. Gamma radiation is very penetrating and will reach all organs in the body from an external source. At high-energy levels, an external γ source will deliver to all parts of the body about equal amounts of energy.

Absorption in Matter

If the source of radiation is within the body, then the problem is turned around. Alpha radiation penetrates only a few micrometers of tissue and therefore expends all its energy in a very localized area, within a few cells, typically. Beta radiation, on the other hand, is sufficiently penetrating so that tissue even a few millimeters away from the source may be irradiated. Gamma radiation is so penetrating that much of it leaves the body, causing some ionization in the process.

Radiation Energies

The common unit of energy used in health physics is the million electron volt, although 1,000 electron volts (keV) is often used when X rays are discussed: $1 \ MeV = 1.6 \times 10^{-13}$ J. Because of the relationships discovered by Einstein, the energy content of electromagnetic radiation can be thought of in terms of equivalent mass; in this case,

$$1.0 \ MeV = 1.074 \times 10^{-3} \ amu \tag{15-6}$$

where amu is the abbreviation for *atomic mass unit*.

All α particles resulting from the same transition in the nucleus have the same energy level and, therefore, α radiation is said to be *monoenergetic*.

Typically, the energy associated with an α particle emitted from a nucleus will be on the order of a few million electron volts.

Beta particles resulting from the decay of a nucleus have energies from zero up to β_{max} and the average level varies considerably; but typically it is about one-third of β_{max}, at least for purposes of health physics. Maximum β energies commonly encountered range from 0.0186 MeV for tritium, the radioactive isotope of hydrogen, to a few million electron volts, but many are less than 1.0 MeV.

Energies associated with gamma rays vary from a few thousand electron volts up to about 3 MeV. Gamma rays are emitted as a consequence of so many nuclear reactions that a radioisotope is likely to emit γs of many different but discrete energies. The spectrum of energy can thus be used to identify radioisotopes on devices called gamma-ray spectrometers. Even though gamma rays are monoenergetic when emitted, peculiarities of their interactions with matter degrade that energy in very complicated ways. That, coupled with the many different intensity gamma rays usually emitted from a radioisotope, tends to make this radiation behave as if it were associated with a continuous spectrum rather than with discrete energy levels.

Ionizing Ability

Carrying a pair of positive charges and a mass of 4, an α particle is a relatively slow-moving, heavy, and heavily charged object. As such, it creates many ions in a very short path before it finally comes to rest, acquires a pair of electrons, and becomes an atom of helium. Alpha particles are said to have high *specific ionization* (determined as the energy loss per unit of path length) ability, creating a high density of ions along their paths. In the creation of ions, free electrons are formed that roam through the local matter, in turn causing ionization. For the most part, those electrons formed by α ionization are not particularly energetic and therefore do not penetrate very far from their points of formation, a fact that tends to increase the local ionization associated with an α particle.

Beta particles, high-speed electrons, have less ability to ionize than α particles. Specific ionization is a function of the charge on a particle and is also an inverse function of velocity. Beta particles have half the charge of an α particle and much higher velocities, at least initially, so that their ability to ionize is at least an order of magnitude less than that of an α particle. When a β particle interacts with matter, it causes the ejection of electrons from atoms it encounters, usually from the electron cloud rather than from the nucleus. These electrons, in turn, have the ability to ionize atoms in their paths. Despite the rather complex manner in which the original β particle causes ionization, the overall effect is that about the same number of ions are created per unit distance traveled in all portions of its path, from the start to nearly the end, although that path is usually a very erratic one. Near the end when the velocity of the β particle is approaching a minimum, specific ionization increases sharply and then falls to zero [1].

Electromagnetic radiation in and of itself does not create ions no matter what its energy level. When electromagnetic radiation with a sufficient energy interacts with matter, however, ions are produced. Rather than losing its energy in a series of many small increments as with an electron, a gamma ray tends to lose energy in large packets, transferring all of it to an electron, perhaps, which then proceeds to ionize as if it were a β particle. Such *photoelectric absorption* is only one of three main mechanisms by which a gamma ray loses energy; the others are called *Compton scattering* and *pair production*.

Compton scattering, discovered by Arthur Compton, is similar to photoelectric absorption, except that in this case the photon does not give up all its energy to the electron it encounters. Instead, it gives up some and retains some so that the net result is a photon with reduced energy and an electron with increased energy. In the process, the photon may change direction or scatter.

When the energy of the photon is greater than 1.02 MeV, it can disappear and in the process create an electron-positron pair by a process which takes place in or near an atomic nucleus and which is called *pair production*. Energy above the minimum of 1.02 MeV is transferred to the pair, with the positron receiving a little more than the electron. The extra energy is kinetic so that both halves of the pair tend to have velocities in the direction of the incident photon. The electron causes ionization in the adjacent matter in a normal manner; the positron does the same until it loses most of its energy and then, meeting an electron, annihilates both with the production of two oppositely directed γ photons, each having an energy of 0.51 MeV.

Absorption of γ radiation in matter thus taking place by a series of discrete events, each involving the transfer of rather large amounts of energy, does not cause ionization as efficiently as does either β or α radiation. That is, the specific ionizing ability of γ radiation is lowest of the three types considered.

Bremsstrahlung

Whenever an electron is accelerated (or decelerated), electromagnetic energy is radiated. As an electron or β particle (of either sign) interacts with matter, it bounces from atom to atom, changing direction often. Each direction change is an acceleration that causes the emission of radiation and a loss of energy from the electron. Most of the accelerations take place by interactions with the electron cloud of atoms rather than with or near the nucleus, and each emission of energy causes a reduction in the velocity of the electron. For this reason, the radiation emitted is called *bremsstrahlung,* German for *breaking radiation,* and as the amount of curvature or acceleration per bounce must vary from zero up to a maximum, so must the energy of the photons produced. That is, bremsstrahlung exhibits a continuous spectrum from zero to a peak of intensity at the energy associated with the most probable acceleration and then declines in intensity, becoming close to zero at a maximum value equal to the kinetic energy of the electrons [1].

An electron might be expected to undergo the most violent maneuvers (and thus produce the most bremsstrahlung) in traversing materials with the largest electron clouds packed in the smallest volume, that is, with those materials with the highest atomic densities. This in fact is the case and is the reason why the targets of X-ray machines are made of materials with high atomic numbers. X rays are nothing other than bremsstrahlung produced by the interaction of accelerated electrons with a high-Z target anode. The phenomenon is just as real, of course, when the electrons are β particles from nuclear reactions instead of electrons boiled from a hot filament.

Tissue Damage

The damage done to living (or, in fact, any) tissue is a function of many variables, including several having to do with the tissue itself. Those variables related to the particle or energy that appear to be most important in assessing probable injury include penetrating ability (and thus the location of the source), average and maximum energies, and specific ionization.

Ionizing Radiation Units

Historically, one of the first radioactive materials available for laboratory work was radium, isolated by the Curies in 1898. As more and more people began working with radioisotopes, the need for a unit of radioactivity became apparent. That need was met by referring everything possible to the activity associated with one gram of radium and later, in particular, with ^{226}Ra, which was readily available and had a half-life of 1,602 years. One problem with ^{226}Ra and most other naturally occurring radioisotopes is that they are associated with their *daughter products,* which are also radioactive but with different radiations and differing half-lives, forming whole families of many members. The family of which ^{226}Ra is a member begins with ^{238}U and ends with ^{206}Pb (which is stable) and contains 15 radioactive isotopes. Even beginning with pure ^{226}Ra, nine more radioactive daughters may be present. Most of the radioactivity is associated with the short-lived components following ^{226}Ra in the chain so that equilibrium occurs fairly rapidly. Nevertheless, variability of preparation and of measurement are great enough so that the International Commission on Radiation Units (ICRU) finally defined the *curie* (Ci) as equal to exactly 3.7×10^{10} disintegrations per second (dps or simply, sec^{-1}), which is quite close to the average value associated with one gram of ^{226}Ra.

Note that the unit itself contains no information on how it is determined, or, in fact, how any particular device or instrument responds to the disintegrations (of nuclei). The unit has nothing to do with the kind of radiation emitted.

The curie is a rather large unit of radioactivity, and most often the health physicist deals with millicuries (mCi), microcuries (μCi), and even picocuries (pCi), 10^{-3}, 10^{-6}, and 10^{-9} Ci, respectively.

One of the pioneers with ionizing radiation was Roentgen, and his name is used for the unit of exposure produced by the absorption of photons in air. Numerically, 1.0 R = 2.58×10^{-4} coulomb kg^{-1} and the definition now particularly states that the unit deals with the production of ions in air by photons and the (eventual) absorption of the energy of ionization in the air. As with the curie, the roentgen is rather large, and milliroentgen (mR) is the unit more often used in health physics.

The roentgen applies only to photons. Beta particles are not photons and yet there are exposures to β particles as there are exposures to gamma rays. Usually, when an animal such as man is exposed to a source of ionizing radiation, the problem is not so much to describe his exposure but to estimate the effect of the dose of radiation he has received (or will receive). For this reason, the unit of absorbed dose was derived. Called the *rad* (for *radiation absorbed dose*), it is almost equal to the dose delivered to tissue by 1 R of reasonably *hard* photons.

$$1.0 \text{ rad} = 100 \text{ ergs g}^{-1} = 1.0 \times 10^{-2} \text{ J kg}^{-1} \qquad (15\text{-}7)$$

The rad does not depend on the kind of radiation at all and is just as applicable to the absorption of β particles as it is to gamma rays. This somewhat alleviates the problem of defining exposure to β and α particles but does not solve it because dose and exposure are not the same thing and exposure is easier to measure.

One rad of α radiation is not equal in effect on biological organisms to 1 rad of γ radiation. For many reasons, including that of specific ionization, α particles cause much more injury per rad than do β particles or gamma rays. Dose and effect are related through a *dose equivalent* (DE), calculated as

$$DE = (D)\,(QF)\,(DF)\ldots \qquad (15\text{-}8)$$

where
D = the absorbed dose (rad)
QF = the quality factor
DF = the distribution factor, which depends on spatial orientation of source and receptor
DE = the dose equivalent (rem)

The dose equivalent is actually equal to the dose of 200 keV (peak) X rays, which causes the same effect, and the unit (the *rem*) is an acronym for *roentgen equivalent man.* Quality factors vary with the kind of radiation and, more precisely, with values of the linear energy transfer (LET), the energy of a particle divided by its range, an expression of the amount of ionization associated with a unit distance of travel. *QF* values are about 1.0 for β particles and gamma rays, about 20 for α particles, 10 for protons, and from 5 to 10 for neutrons.

Effects of Ionizing Radiation

Research with high dose levels (perhaps hundreds or thousands of rads) administered to experimental subjects ranging from cell cultures to monkeys has answered many questions about the effects of radiation on living things. First, and perhaps most important, no injury or disease from ionizing radiation has been detected that is unique to radiation. Ionizing radiation may cause a disease as specific as sarcoma or as general as life-span shortening (or lengthening, for that matter), but all such effects are seen in unexposed individuals. Almost equally as important is the experimentally observed fact that a dose lethal when administered acutely may have no observable effect when fractionated or administered gradually over a long period of time. That is, many of the injuries caused by ionizing radiation are repairable by normal life processes.

Mechanisms of Injury

Much of the research that has led to an understanding of injury mechanisms in living organisms has been concerned with the effects of ionizing radiation on water. This is so because water is the major component of all living cells and any changes in water induced by the radiation may be responsible for changes in other components of the cell. In addition, the direct interactions of radiation with basic components of life such as chromosomes and DNA molecules have been investigated.

When a free electron speeds among a group of water molecules, it may cause several different events. One event is the formation of a positive ion by displacing an electron from an atom's electron cloud and then the displaced electron (or even the original one) can combine with another atom, producing a negative ion. That is, a possible event is the production of an ion pair, in which case the ionizing radiation has, indeed, produced ionization.

Another possibility is that the electron can, probably through a series of fast reactions, interact with a neutral molecule in such a way that the molecule (and others also) is left in an excited state. This occurs when energy is transferred to the atom or molecule, but the energy is not sufficient either to displace an electron or to add one so that ionization does not occur. The excited structure can decay to the ground state by emitting a low-energy photon or, more importantly, by dissociating to produce a pair of free radicals:

$$M_1^* \rightarrow M_2^0 + M_3^0 \qquad (15\text{-}9)$$

where M is any neutral molecule; the asterisk indicates excitation, and the superscript 0 indicates a free radical—an atom or molecule which is not ionized (it is neutral) but which has an unpaired electron. In the case of water, the free radicals are likely to be H^0 and OH^0. Free radicals are extremely reactive and are the means by which many polymerization reactions occur, as discussed in Chapter 7. When a free radical attaches to a neutral molecule (such as DNA, for instance), that molecule, in turn, becomes a free radical and very reactive with any other molecules present. Furthermore, about the only way in which a free radical can disappear is by reaction with another free radical, resulting in a neutral molecule as

$$OH^0 + OH^0 \rightarrow H_2O_2 \qquad (15\text{-}10)$$

The hydrogen peroxide formed may be the mediator of many effects of ionizing radiation on cells. Hydrogen peroxide is a very good oxidizer and is capable of reacting with many cellular components, particularly when it is formed in the water phase within the cellular membrane.

Ions and free radicals formed in water appear to be the main sources of chemical reactions that injure and kill cells as a result of ionizing radiation. And, because ions are always present in any water solution and free radicals are also but to a much smaller extent, the injury done must mimic that caused by many other agents that work through the same mediators. Most damage caused by ionizing radiation to living tissue, then, is through mechanisms already present but accentuated by the radiation.

In addition to the indirect effects caused by reacting with water or other nonliving substances dissolved in the water, an electron (or α particle or proton, etc.) can affect components of the cell directly. A direct effect can be the result of either ionizing a molecule such as DNA or RNA and making it more reactive or transferring so much energy to it that bonds between atoms are disrupted, tearing the molecule apart. Breaks in chromosome chains have been observed following exposure to ionizing radiation, more or less in proportion to the (large) dose. When the genetic structure of a cell is altered, the probable effect is death of the cell; less likely are several kinds of nonfatal injury, including mutation. If the mutation happens to be one that turns off the system inhibiting cell growth and mitosis, the result may be cancer and may not be evident in the individual for many years, depending on the normal growth rate of the cell concerned, how many cells were mutated, and so forth. These changes in the cell can be the result of either indirect or direct radiation-induced injury.

The injuries discussed so far affect only the person (or animal or plant or bacterium, etc.) exposed to the radiation. Such effects are called *somatic* after *soma*, which refers to the body. If the effects of the radiation (or indeed, other sources of similar action) are manifested in the reproductive organs and cause one or more mutations in either sperm or ova, the effect of the injury can be passed on to the offspring. In this case, the damage is genetic, having to do with heredity. Even genetic injury is not peculiar to ionizing radiation; it can be caused by a great many chemical agents as well. Any mutations so formed add to the already present mutation rate of the species concerned. Most genetic (as well as somatic) mutations are lethal and thus are not transmitted, and almost all the nonlethal mutations are deleterious. Occasionally a mutation will occur that enhances survival of the individual; transmittal of such mutations is the basis of evolution.

Radiosensitivity

Not all cells are equally sensitive to the effects of ionizing radiation. Sensitivity is a function of a great many variables, not all of which are known and/or understood; however, metabolic rate, degree of development, function, and rate of mitosis are all extremely important. In a very general way, those cells that are the least active are the least sensitive to the effects of radiation and those cells that are most active are most sensitive. This statement must be modified, however, because an active cell having one function (say, a nerve cell) may be more resistant to radiation than an inactive cell with a different function (production of blood cells, for instance). Another general statement is that any cell is probably most sensitive to radiation during one or more stages of mitosis (cell division).

When acute injury only is considered, a ranking of organs and organ systems of vertebrates with respect to their radiosensitivity is possible:

MOST SENSITIVE: Blood and blood-forming system
Skin and GI tract
Lungs and kidneys
LEAST SENSITIVE: Bone, muscle, nerve cells

This ranking assumes that a mature adult is exposed. Both the very young and the very old are more sensitive to ionizing radiation than those of middle age, and the radiosensitivity of their organ systems may not remain unaltered throughout life. If the exposure is chronic rather than acute, the same order may not apply because of the variable effects of age, somatic mutation, and repair.

Effects of ionizing radiation differ quantitatively from species to species with no discernible reason; there appears to be no correlation among mammals with metabolic rate or body size. The LD_{50} (30-day observation period) for whole-body photon irradiation of a dog is about 350 rads and that for a burro is about 700 rads, but the obvious size effect becomes more obscure when the information is added that the LD_{50} for a hamster is about 700 rads and that for a mouse is about 500 [1]. The LD_{50} for a man is known less well; it may range from about 300 to about 500 rads.

Acute Somatic Injury

Radiation sickness has been known since the early days of experimentation and therapy with radiation. The symptoms are very similar to those of stomach flu and consist mainly of fatigue, nausea, vomiting, and anorexia with or without a fever. This syndrome occurs following a dose of about 100 rem to the whole body administered over a period of a few hours or less. Symptoms appear a few hours after the exposure and may last for a day or two and be followed by essentially complete recovery. Both the severity of symptoms and the latent period between exposure and first symptoms are functions of the exposure severity and have been used to estimate the absorbed dose.

One of the first and main actions of radiation is on the blood and blood-forming system. A consequence of this is that in the period following development of radiation sickness the victim may become seriously ill from a bacterial or viral infection because his defenses, from leucocytes to the immune system, are not capable of dealing with the disease agents. For the most part, the diseases are caused by organisms already present in the intestinal tract or elsewhere in the body although exogenous sources can be important if present. Not even modern antibiotics can truly substitute for man's own protective mechanisms.

An acute dose of radiation that may be fatal is very likely to cause anemia because of effects on the blood-forming system and to cause alopecia because of effects on the skin and skin structures. The loss of hair can be dramatic, especially following whole-body irradiation of a furred animal. If the animal (or man) survives the exposure, hair will usually grow back but both color and texture may be altered.

A consequence of effects on the blood and/or skin is the appearance of *purpura,* splotchy red-to-purple "bruises" just beneath the skin surface. This sign is caused by hemorrhages from capillaries that have ruptured or whose permeability has been increased by the radiation.

If the dose of radiation has been greater than that normally lethal for 100% of the species, the usual cause of death is likely to change from the consequences of infection to what has been termed *gastrointestinal death* because the cause is damage to the lining of the GI tract. Bloody diarrhea is the most obvious sign and death usually occurs in less than a week, whereas death from infection is most likely to occur about 2 weeks after the exposure.

Even doses orders of magnitude larger than required for 100% lethality do not kill immediately. Such gross overexposures, however, affect the CNS in addition to other, more radiosensitive areas of the body with signs and symptoms ranging from excitement to deep coma depending on dose and time after exposure. Death follows in a day or so.

Chronic Somatic Injury

In addition to acute incidents of overexposure to ionizing radiation, a person may be subject to smaller doses absorbed over a fairly long period of time. This was especially true in the early days of experimentation with Roentgen's new discovery and with the isotopes of uranium and radium and their daughters. Most characteristic of the experimenter with roentgen (X) rays was extensive damage to the left hand (of right-handed persons) caused by the usual practice of determining operation of the equipment by placing that hand between the source and a fluorescent screen held in the right hand. Slow-to-heal ulcers and dry, cracked skin preceded cancer.

Most of the effects experienced by radiographers were confined to the skin. X rays and other bremsstrahlung characteristically have a continuous spectrum of energies ranging from zero up. Early X-ray equipment was very inefficient and a large part of the output was in soft radiation, much closer to ultraviolet than to gamma rays in character. Because of its deficiency in penetrating ability, much of this radiation was absorbed by the skin. Today X rays are produced by higher voltages and then are filtered through metal to remove the low-energy radiations and reduce skin exposure.

Extensive irradiation of internal organs of man awaited the use of radium-actuated phosphors for watch, clock, and instrument dials. The workers (usually women) who applied the paint inhaled excessive amounts of radon (^{222}Rn) during their work and also ingested some of the radium because of pointing the brush with pursed lips. Radium and several of the daughters of radon are handled by the body almost as if they were calcium, as is true of several heavy metals. The main depot for calcium in the body is in the bone, and that is where a large portion of the radium and daughters were deposited. Most blood cells are formed in bone marrow, and the radium was deposited where its α and β emissions would be absorbed by the most radiosensitive tissue in the body. Anemia, leukemia, bone necrosis, and bone cancer resulted if the amounts of radioisotopes absorbed were great enough. Data relating to the exposures of these people, some gathered long after the fact, were the first and best available upon which to base a TLV for radiation. (Because of its long physical half-life and because the rate of turnover of bone and material deposited there is low, the amount of radium in bone can be determined quite accurately even long after the exposure has ceased. The usual means is to determine the amount of ^{222}Rn exhaled and back-calculate to amounts of ^{226}Ra.)

Most chronic exposures to ionizing radiation are much less severe than those to the first radiographers and dial painters. Basic data on the probable consequences of chronic overexposure have been gathered from the many animal experiments conducted since the beginning of the atomic age. These data indicate that the most serious consequences to a large population from radiation exposure are likely to be genetic although individuals may experience effects that to them are much worse than a simple mutation that may appear in their great-great-great grandchildren.

The two most important somatic injuries are life-span shortening and cancer. Although some animal experiments have shown increased longevity in animals receiving doses of 0.1 rad day^{-1}, most experimenters have found some life-span shortening (with higher doses, however). How these animal experiments relate to humans is not clear because of the difficulty in extrapolating from the 2-year life-span of a mouse to the 70-year life-span of a man.

That ionizing radiation can cause cancer in man and in experimental animals is not disputed. What is disputed is the cancer dose-effect relationship for man. That is, how much radiation delivered over how long a period will increase the incidence of cancer by, say, a factor of two? Despite much experimentation and some data from early exposures to man and guesses about exposures caused by the nuclear bombs dropped on Hiroshima and Nagasaki at the close of World War II, more accurate information is needed [2]. One of the problems is that cancer rates in man change fairly rapidly with time for no known cause (see Chapter 10). A speculation regarding the increase seen in lung cancer among cigarette smokers is that the cancer is caused by the inhalation of ^{210}Po along with the smoke, thus irradiating the lungs.

Genetic Effects

The most conservative estimates of doses required to double the incidence of cancer in man result in figures much lower than those associated with life-span shortening in animals. Similarly, the best estimates about the doses required to double the natural mutation rate result in figures much less than those associated with even moderate increases in cancer incidence. Therefore, the genetic hazard has been the one upon which the most recent exposure control values have been based [3].

Population genetics is concerned with the gene pool of a species, in this case, man, and therefore is concerned about the total dose of radiation received by the total population of people of reproductive age or lower. No available data give any indication of the presence (or absence) of a threshold for genetic effects of radiation, and the conservative approach is to assume none. On that assumption, *any* increase in the background level of radiation to which the human species is exposed will result in some increase in the mutation rate, whether one person of reproductive age or the whole population is exposed. Geneticists therefore caution that all radiation exposures, including those for medical (and dental) [4] purposes be kept to a minimum consistent with good practice although so far no agency has seen fit to attempt regulation of medically related exposures [5].

Permissible exposure levels are based upon (1) a method of extrapolating from the known effects of moderate dose levels to the unknown effects of very low dose levels and (2) a presumably acceptable effect such as a doubling of the mutation rate. There is little agreement and essentially no knowledge about the effects of slight increases in the total radiation absorbed by the total population, and even little agreement on what would be an acceptable increase in the mutation rate. In the face of these deficiencies, committees of experts must make decisions that may have profound economic as well as health effects, and then political groups must implement those decisions and suffer the consequences of mistakes [6]. In this area especially, conservatism is more than idealistic; it is almost necessary for survival.

Measurement

Measurement of radioactivity has been refined and developed at a remarkable rate, particularly since the early 1940s. Precision and accuracy extreme in relation to the usual industrial hygiene methods of detection and determination are not only possible but practical. In some cases the techniques have been extended from health physics to industrial hygiene (the film badge is a case in point) but most are unique to the field.

Photographic Emulsions

Radioactivity was discovered by A. H. Becquerel in 1896 because of its ability to affect a photographic emulsion through paper, wood, and metal and photographic emulsions have been used to detect and determine radioactivity ever since. The first technique is called *radioautography* where a film is put in proximity to an emitting source so that an image of the source is formed on the film. That method still has utility for biological specimens in tracer experiments and also for air samples when identification of active particles is necessary [7]. Most measurements of ionizing radiation by this method, however, are made with film badges.

Well before health physics was a profession, dentists used X rays to detect hidden tooth defects. They placed a piece of photographic film enclosed in a suitable holder inside a patient's mouth and directed the beam of radiation through the teeth to the film, just as is done today. Early workers with ionizing radiation chose dental film as the most convenient and inexpensive way of detecting cumulative exposure to external sources. Eventually film holders were developed able to screen β and low-energy γ radiation from portions of the film so it could be developed and read for several radiation energies. Later, film was used to detect (nonionizing) neutron exposure.

Film badges are issued to people who work with ionizing radiation and are collected and evaluated at intervals that may vary from almost nil if an accidental exposure is strongly suspected to 1 or 3 months [8]. For most radiation, the criterion for exposure evaluation is the extent of fogging in the various areas of the film representative of radiation energies and penetrating ability. With neutrons, after development the film is searched microscopically for evidence of tracks caused by radiation emitted from nuclei with which the neutrons collided.

The main advantage of film badges is that they provide a permanent record of exposure if they are saved. They are small, rugged, light in weight, and cheap; they can record several types of radiation and they have a wide range of response to that radiation. These advantages are sufficient to make the film badge a useful adjunct to almost any radiation protection program.

Film badges have several disadvantages. They are not direct-reading so that after a suspected exposure some time (a few days, usually) must elapse before the badge is or can be developed and read. Under normal circumstances, the badge may not be examined until more than a month has passed since the exposure occurred. For this reason, film badges alone are almost never regarded as sufficient instrumentation for radiation detection.

Film badges are subject to improper use more often than are most other methods of detecting radiation exposure. They do little good if left in the storage location, for instance, instead of being worn by people who may be exposed. Usually they are pinned or clipped to a jacket or shirt, eventually tearing the fabric and providing a good excuse not to wear the badge.

Other disadvantages are that the inherent accuracy of film badges is lower than that of most other ionizing radiation measurement equipment [9], that processing and even storage variables are critical [10], and that exposure to one type of radiation may interfere with evaluation of the badge for other types [11]. These problems are recognized by most people using film badge services so that the disadvantages are not critical.

Gas Ionization Instruments

When most people think of ionizing radiation and radioactivity, their next thought is likely to be of the Geiger counter clicking loudly under unusual circumstances depicted, perhaps, on the late, late television show. Produced by the tens of thousands for civil defense work, these instruments can be very inexpensive, very rugged, and very reliable but at the same time not very precise nor very accurate. Some are even used by health physicists and other people who detect and determine β and γ radiation. They [more properly called Geiger-Müller (G-M) counters] are the best known of the gas ionization instruments.

If a dc voltage is present between two electrodes, a current will flow carried by ions of the gas in which they are immersed. When the voltage is low, the current is small and proportional to voltage as indicated by Ohm's law, Eq. (12-5). Because the current is carried by gas ions, its flow is proportional not only to the impressed voltage but also (for a constant voltage) to the relative number of ions present. Ionizing radiation increases the number of current carriers and hence the current flow rate. Several instruments used in the field take advantage of these basic facts.

The first instrument known to respond to ionizing radiation was the electroscope. In its simplest form, it consists of a pair of gold foil strips suspended from a conductor inside a metal vessel. The conducting electrode is touched with a carrier of static electricity, giving the strips charges of the same kind. They immediately spring apart, to come together again slowly as the charges drain off, largely by air conduction. Ionizing radiation will cause the leaves to collapse at faster rates in proportion to the amount of ionization. If, instead of a gold leaf, one uses a small quartz fiber and arranges to view movement of the fiber across a scale through a microscope, the result is called a *dosimeter*. Dosimeters are used in much the same way as film badges but have the advantage of providing an immediate readout (if a microscope is built in) of the integrated radiation exposure (not dose), usually in milliroentgens or roentgens over the time period they are worn and thus are indicators of dose to identically-shielded tissue. Some dosimeters must be read on a combination charger-reader, usually a minor inconvenience. There is, of course, no permanent record nor is much attempt made to distinguish between radiation energies.

If the two electrodes have opposite signs and are arranged so that the positive (anode) is a wire surrounded by the other, electrometric arrangements to monitor the very small current flow between them are not difficult. A handheld battery-powered instrument of this kind, usually employing an air chamber 5 to 8 cm in diameter is called by several names, including *ionization meter, ionization chamber,* and *rate meter;* but it is best known as a *cutie pie*. This instrument has a meter calibrated to read directly in fractions of milliroentgens per hour to several roentgens per minute. It responds very rapidly to changing exposure rates; the surrounding cathode walls can be made thin enough so that β radiation can be detected and a window thin enough to allow α penetration is possible. Voltage impressed between the electrodes is kept low so that with no radiation present the current flow is very small and leakage through the necessary insulator between the electrodes is kept to a minimum.

Many ionization chambers are constructed to allow the source of radiation to be inserted directly into the sensitive volume so that there is no shielding between the source and detector. Such instruments are usually laboratory-based rather than handheld and usually do not use air in the chamber but rather a good counting gas. These instruments are then arranged so that each primary ionization event is counted individually rather than being summed as a current flow. This is done by increasing the applied voltage between the electrodes to a few kilovolts, perhaps, into the region where the total number of ions formed subsequent to the primary ionization is proportional to the specific ionization of the primary event. That is, an ionizing β particle, say, emitted from the source, might cause the formation of a pulse of current an order of magnitude or so smaller than the pulse produced by an α particle; devices arranged to discriminate between the two pulse sizes are called *proportional counters* and are found in many laboratories.

G-M counters are operated at even higher voltages than proportional counters. In the Geiger region, a single ion pair formed within the sensitive volume causes an avalanche of ions and thus a large pulse of current. The number of pulses (or counts) can be displayed on a meter as a count rate, or each pulse can be used to actuate headphones or a small loudspeaker with the well-known click. Geiger-Müller counters can be used as well-calibrated laboratory instruments or as handheld survey meters. The handheld meters are much more sensitive to hard β and γ radiation than are cutie pies and thus have great utility in detecting the presence of these radiations.

Scintillation Counters

Some materials have the ability to absorb an energetic photon or particle and then emit a flash of light, the intensity of which is proportional to the energy of the photon or particle. Many of the materials with this property are crystalline salts such as NaI or organic compounds (anthracene is one). The *scintillators* (energy-absorbing, light-flashing materials) may be used as solids or may be dissolved in an appropriate liquid; in either case, most or all of their volume will be viewed by a photomultiplier tube or other light-sensing device. Each flash of light seen by the photomultiplier tube ultimately causes a pulse proportional to the energy of the initial ionizing event.

Scintillation counters are more sensitive than gas ionization meters for two basic reasons. First, the scintillator is always more dense and is usually composed of higher atomic number materials than the counting gases and therefore it is a more efficient absorber of energy than the gases. Second, photomultiplier tubes have been developed to such a degree that they provide greater and more reliable amplification than is found in the gas amplification characteristic of the proportional and Geiger regions of the gas ionization instruments.

Scintillation counters are found as handheld survey meters and as laboratory-based instruments perhaps connected to very sophisticated counting/recording devices. The survey meters usually employ solid scintillators, but liquids are often used in the laboratory so that the sample can be mixed with or dissolved in the fluid medium, thus providing very intimate contact between the emitter and absorber. The most used liquid scintillators are PPO (2,5-diphenyloxazole) and POPOP [1,4-bis(5-phenyl-oxazole) benzene]; the usual solvent is toluene.

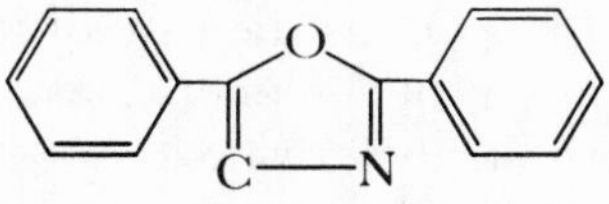

The sophisticated laboratory counting/recording equipment can be and is used to transform a simple scintillation counter into a full-fledged gamma-ray spectrometer. Because pulse size is proportional to the energy of the initial event, a spectrogram of count rate versus energy can be recorded. Spectra obtained in this manner can be used to identify many of the radio-isotopes formed, for instance, by irradiation of a sample in a nuclear reactor. Neutron activation analytical methods have, in turn, led to similar techniques that use various charged particles (protons, for instance) to effect the necessary transmutation of stable elements into radioisotopes identifiable by gamma-ray spectrometry.

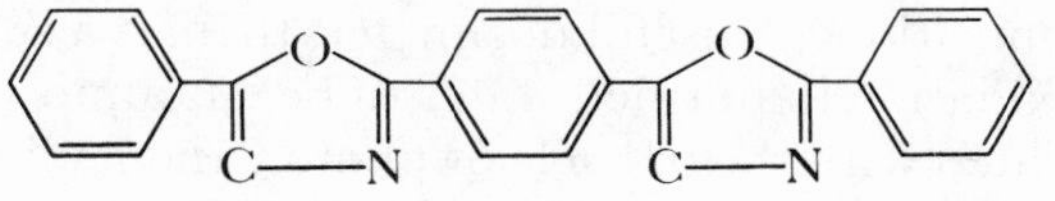

Thermoluminescent Devices

Some solid materials have the peculiar ability to "remember" their radiation history. After being exposed to a flux of ionizing radiation, these substances are outwardly unchanged until they are heated, typically to about 300°C. They then emit light in proportion to the total dose of radiation absorbed. Called *thermoluminescence* (TL), this property has been found amenable to a number of applications. For instance, the TL material ^{7}LiF (^{7}Li is the stable, most prevalent isotope of lithium) has been incorporated into a polytetrafluoroethylene matrix that allows about any shape from flat discs to finger rings to be formed [12]. TL can thus be used to monitor radiation exposure to the hand, or it can be used as a badge. The lithium fluoride used is not at all sensitive to energy of the radiation to which it is exposed over a rather large range from about 20 keV to 2 MeV, but to a certain extent it is sensitive to UV and visible light.

Several compounds and mixtures have been found to exhibit TL subsequent to ionizing radiation exposure and this technique is being applied in many areas including that of personnel monitoring [13]. By the use of suitable shielding, TL detectors can be used to discriminate between radiations of several energies in a manner analogous to that of the conventional film badge. Reading and reactivation can be automated to provide precision and accuracy at least equal to that of a photographic emulsion at a lower price. TL devices appear to have a very promising future even though they share several of the disadvantages of photographic emulsions.

Miscellaneous Methods and Devices

Ferrous ions are oxidized to the ferric state by any process that produces ionization. A useful dosimeter, then, is a solution of $FeSO_4$, as the rate of its oxidation in purified water to Fe^{3+} is almost independent of the energy associated with the particle or photon and its linear energy transfer rate.

Samples of air for radioisotopes are complicated by the constant presence of radon and its daughters so that the samples obtained must be counted over a period of time to correct for natural background or, in fact, to determine radon concentrations. Several systems have been developed that allow for good determinations of radon and its solid daughter products, particularly when encountered in mines [14, 15, 16].

Special monitors for tritium (which omits a very low-energy β particle) and α-emitting contamination [17] have been devised as have probes and techniques for use with medicinal radioisotopes. Transistor and integrated circuit electronics have simplified the amplification-readout problem and digital computers have aided tremendously in the analysis of data.

Control

More than any other area of industrial hygiene and industrial hygiene-like activities, the area of ionizing radiation has been subject to legal restrictions and control. This is just as true in almost every other country as it is in the United States. As sealed and loose sources of radioactivity began to proliferate and the possible consequences of uncontrolled release of such materials penetrated to the scientific, legal, and technically oriented population, ever-tightening controls were instituted. Therefore, anyone who seeks to use sources of ionizing radiation is subject to some sort of licensing procedure designed to, among other things, assure some knowledge of control methods. Because radiation hazards are unique in several respects, some of the consequences of that uniqueness bear examination.

Uniquenesses

Common mode of action. Although the materials involved vary across the whole spectrum of inorganic and organic chemistry, the sources are machines as well as materials, the energies vary by orders of magnitude, and a hazardous material may be within or without the body, the toxic effect is manifested through only one mechanism, that of direct energy transfer, or *ionization*. This has facilitated study of both hazard and effect so that probably more is known about the ionizing radiations and their effects than about any other hazard in the occupational (or general) environment even though, as mentioned previously, knowledge is still far from complete.

The metals, for instance, vary dramatically in their chemical toxicity, and yet in almost every case (thorium and uranium may be exceptions) the toxicity of the ionizing radiation of a radioactive metal is so great that the chemical toxicity can be disregarded in evaluations of chronic hazard. Certainly, the metals do vary in their radiation hazards but in the final case that hazard reduces to one of energy transfer.

Almost the same statement can be made about organic compounds except that the radioisotopes usually associated with them (mainly ^3H and ^{14}C) are among the least hazardous of all so that many labeled organic compounds have higher chemical than radioactive toxicities, especially for acute exposure.

Hazardous amounts. In almost all cases, the limiting factor in any application of a radioisotope that involves human exposure will be the radioactivity of the material. Some isotopes are incredibly toxic, even compared to that often-cited yardstick, botulin toxin where "one drop could wipe out a city" (or some such nonsensical expression).

The saving grace for many organic compounds labeled with ^{14}C or ^3H is that the *specific activity* (Ci g^{-1}) is purposely kept low by only labeling one or two atoms per molecule rather than all possible atoms. (Specific activity can be kept low by simple dilution with nonradioactive isotopes, too.) That the main reason for keeping specific activity low is economic does not alter the conclusion.

Action at a distance. For a chemical "over there" to affect someone "over here" with its toxicity, it must usually diffuse through the air or be carried by a wind current. In either case or in others that can be imagined, some relatively slow and easily interruptible process must take place before the material can act at a distance. In the case of radioisotopes, the same limitations hold unless the hazard is from β or γ radiation. In those cases, all the ventilation possible does exactly nothing and even a physical barrier such as a pane of glass or even a concrete wall may not provide adequate protection.

Intense radiation by β particles can produce intense bremsstrahlung, particularly when the radiation strikes high atomic number material. A piece of polyethylene or polystyrene (highest $Z = 6$) is a much better β shield than is a piece of glass (highest $Z = 14$), steel (highest $Z = 26$), or lead ($Z = 82$) for that reason. Moreover, the high Z, dense metals such as lead or tungsten (wolfram) are the best shielding materials for γ or X radiation—but those radiations produce their effects through secondary electrons that are best shielded with low Z materials. Shielding calculations can become very complicated and in many cases the optimum is found to be a combination of materials. Concrete is often used and not only because it is relatively inexpensive.

Another aspect of action at a distance is the induction of radioactivity. This is the aspect that seems to be most frightening to the general public but in most cases it is by far the least likely of all problems. Radiation is not contagious. If a person or object is contaminated with a radioisotope, the only way for another person or object to become radioactive is by direct transfer of the contaminant involved. Furthermore, a person suffering from radiation sickness caused by external radiation cannot possibly pass it on to another person as if it were a cold or the flu.

Some radiations can induce radioactivity in other materials. Most common is by the reaction of a neutron with the nucleus of an atom to cause transmutation. Neutron beams are not that easy to produce, however, unless a nuclear reactor happens to be handy and operating, so this hazard is usually a small one, at best. Neutrons can be produced in other ways such as by the action of α particles on beryllium but neither good sources of α particles nor beryllium targets are found that easily nor are proton accelerators that can also be used to produce neutrons. Under very limited circumstances, radioactivity can be induced by γ radiation but in all such cases the γ radiation is the hazard, not the transmutation.

Failure of the senses. Man has no sense that responds directly to any kind of ionizing radiation (or to neutrons or neutrinos, etc.) although some animals shy away from intense beams or become nauseated when exposed. Ionizing radiation is not unique in this respect, but the possible consequences of unknowing overexposure are serious enough to cause many people to worry. Worry in itself is rarely productive; actions designed to evaluate and then to control the hazard if control is necessary are much more appropriate responses to this aspect of ionizing radiation.

Sensitivity of analytical methods. Because hazard is associated with such small amounts of radioactivity, the ability to detect the presence of small numbers of atoms or radiated waves or particles is necessary. At least partly for that reason, man has developed methods to quantitate radioactivity in very small amounts indeed. The use of radioactive *tracers* is a reflection of this superlative ability to detect and determine emitters of ionizing radiation in amounts small by any standard of comparison. Actual sensitivity is limited more by the time available for counting and the presence of background radiation than by the ability of instruments to detect events on the subatomic scale.

Biological Sampling

Sampling of biological fluids and other material to diagnose disease has been practiced at least since a physician first tasted a drop of his patient's urine to detect the sweetness characteristic of diabetes mellitus. Analogous methods to determine the presence of foreign materials in the body, however, have not been used to any great extent except when those materials are radioactive. Then, urine [18, 19], breath [20], or other samples, evaluated by appropriate methods including the use of computers to fit decay curves enabling retrospective calculation, have been used. This technique is quite well developed for several isotopes so that a few breath or urine samples suffice to recapitulate an exposure quite well.

An extension of biological sampling is the identification of various compartments in the body and the characteristic rates at which isotopes are cleared from each. Compartmental analysis, however, is exceedingly complex and usually requires a large computer and a very sophisticated program with which to evaluate data. Much more of this kind of work has been done with and for animals than for man, but many of the methods have possible direct applications. An interesting, relatively new method of indirectly assessing X-ray exposure is by determining changes in the kinetics of the uptake of ^{86}Rb chloride by erythrocytes [21].

Leak Testing

Another aspect of ionizing radiation that has little parallel among the usual variety of occupational hazards is the use of sealed sources of radioactive isotopes. Many of these sources of radiation are used for purposes such as irradiation experimentation, thickness gauging, and liquid level indication. Man is fallible and so are the seals he puts on sealed sources. For that reason, wipe testing to detect traces of minute leaks is practiced routinely, the swabs being checked for evidence of above-background activity.

Wipe (or swipe) testing may be practiced as routine in any laboratory or other area where loose (not really loose, just not sealed) radioisotopes are handled. Again, the method is used to detect traces that represent leaks of material from where it should be to where it should not be. Wipe testing can also be used to determine the efficacy of cleaning contaminated surfaces. In cases where the contamination has been gross, both the cleaning and the testing must be extraordinarily extensive [22] with attention also paid to possible inhalation hazards [23].

Miscellaneous Control Methods

Most of the methods for controlling exposures to ionizing radiation are similar to or extensions of methods used elsewhere. Reasonably obvious are injunctions such as "use the properly determined maximum permissible dose" (MPD—comparable to a TLV or a biological TLV), "monitor exposures carefully" [24], "control the duration of exposure," or "dispose of wastes properly" [25]. Precautions peculiar to this field might be "maintain sufficient distance between the source and the subject," "provide adequate shielding" [26], or "limit source emissions," although these could apply to any source of radiation, ionizing or not.

The main lessons in control of the ionizing radiations actually can be reduced to two: Obey all applicable laws and practice good safety and industrial hygiene [27, 28, 29].

REFERENCES

[1]   H. L. ANDREWS, *Radiation Biophysics,* 2nd ed. (Englewood Cliffs, New Jersey: Prentice-Hall, Inc., 1974), pp. 36, 105, and 238.

[2]   R. W. MILLER, "Late Radiation Effects: Status and Needs of Epidemiologic Research," *Env. Res.* **8** (1974), 221–33.

[3]   L. A. SAGAN, "Human Radiation Effect: An Overview," *Health Phys.* **21** (1971), 827–33.

[4]   W. E. NOLAN AND S. BLOCK, "Radiation Hazards Associated with Dental Roentgenography," *Am. Ind. Hyg. Assn. J.* **20** (1959), 118–20.

[5]   K. Z. MORGAN, "Medical X-Ray Exposure," *Am. Ind. Hyg. Assn. J.* **24** (1963), 588–99.

[6]   A. N. B. STOTT, "Here Be Dragons," *J. Soc. Occup. Med.* **23** (April, 1973), 49–52.

[7]   J. T. QUAN, "A Technique for the Radioautography of Alpha-Active Aerosol Particles on Millipore Filters," *Am. Ind. Hyg. Assn. J.* **20** (1959), 61–65.

[8]   R. O. CAMPBELL AND P. K. LOVELL, "A Study to Determine the Effects of Increasing the Exchange Interval for Personnel Dosimetry Films," *Am. Ind. Hyg. Assn. J.* **28** (1967), 354–56.

[9]   D. E. BARBER, B. W. BROWN, AND K. E. JAMES, "A Statistical Analysis of Data from Film Badge Performance Tests," *Am. Ind. Hyg. Assn. J.* **29** (1968), 482–89.

[10] R. L. KATHREN, P. R. ZURAKOWSKI, AND MARGARET COVELL, "Effect of Humidity and Dose on Latent Image Stability," *Am. Ind. Hyg. Assn. J.* **27** (1966), 388–95.

[11] D. E. BARBER, "Beta-, Gamma-, and X-Radiation Sensitivity of a Nuclear Track Film," *Am. Ind. Hyg. Assn. J.* **29** (1968), 358–63.

[12] R. L. KATHREN, L. F. KOCHER, AND G. W. R. ENDRES, "Thermoluminescence Personnel Dosimetry at Hanford," *Am. Ind. Hyg. Assn. J.* **32** (1971), 230–34.

[13] *Solid State Dosimetry. Bibliographic Series No. 23.* (Vienna, International Atomic Energy Agency, 1967). Available from National Agency for International Publications, Inc., 317 East 34th Street, New York, N.Y. 10016.

[14] J. C. STRONG AND M. J. DUGGAN, "A Monitor for the Measurement of Radon in Mine Atmosphere," *Ann. Occup. Hyg.* **16** (April, 1973), 27–31.

[15] K. J. SCHIAGER, "Integrating Radon Progeny Air Sampler," *Am. Ind. Hyg. Assn. J.* **35** (1974), 165–74.

[16] T. L. OGDEN, "A Method for Measuring the Working-Level Values of Mixed Radon and Thoron Daughters in Coalmine Air," *Ann. Occup. Hyg.* **17** (Aug., 1974), 23–34.

[17] W. F. SPLICHAL, JR., "A Floor Monitor for Alpha Contamination," *Health Phys.* **13** (1967), 411–12.

[18] K. R. DOREMUS, M. G. O'BRIEN, AND H. T. MERIWETHER, "Rapid Monitoring of Tritium and Carbon-14 in Urine," *Am. Ind. Hyg. Assn. J.* **28** (1967), 488–90.

[19] O. H. HOWARD, "Simultaneous Determination of Uranium, Its Isotopes, and Alpha Activity in Urine by Mass Spectrometry," *Am. Ind. Hyg. Assn. J.* **29** (1968), 355–57.

[20] K. R. DOREMUS, ET AL., "A Rapid Monitoring Procedure for Carbon-14 in Breath," *Am. Ind. Hyg. Assn. J.* **30** (1969), 161–64.

[21] K. G. SCOTT, ET AL., "Occupational X-Ray Exposure," *Arch. Env. Health* **26** (1973), 64–66.

[22] R. L. HOOVER, "Industrial Hygiene Techniques in the Decontamination of a Building Contaminated with Radium," *Am. Ind. Hyg. Assn. J.* **22** (1961), 83–85.

[23] M. J. DUGGAN, ET AL., "The Inhalation Hazard to Workers Engaged in the Demolition of Buildings Contaminated with Radioactivity," *Ann. Occup. Hyg.* **17** (Aug., 1974), 15–22.

[24] R. F. COWING, "Radiation Dosage to Medical Personnel," *Am. Ind. Hyg. Assn. J.* **21** (1960), 169–72.

[25] B. M. BOWEN, J. J. SELBY, AND J. H. EDGERTON, "Radioactive Waste Disposal at the Georgia Nuclear Laboratory," *Am. Ind. Hyg. Assn. J.* **22** (1961), 119–23.

[26] Y. TAKAKU AND T. KIDA, "Radiation Dose to the Skin and Bone of the Fingers from Handling Radioisotopes in a Syringe," *Health Phys. J.* **22** (1972), 295–97.

[27] W. C. REINIG AND L. L. ALBENESIUS, "Control of Tritium Health Hazards at the Savannah River Plant," *Am. Ind. Hyg. Assn. J.* **24** (1963), 276–83.

[28] A. BRODSKY, "Determining Industrial Hygiene Requirements for Installations Using Radioactive Materials," *Am. Ind. Hyg. Assn. J.* **26** (1965), 294–310.

[29] G. X. KORTSHA, "Industrial Hygiene and Safety Controls at a 1.5 MeV Electron Beam Accelerator Facility," *Am. Ind. Hyg. Assn. J.* **36** (1975), 154–58.

16

HAZARD
EVALUATION
AND CONTROL

In very general terms, the two items of information needed for an evaluation of hazard (likelihood of injury) are the dose and the probable response to the dose. *Dose,* of course, can refer to a material inhaled over a period of many years or to the sound energy impinging on the ears in a 5-min period or to other durations and energies and routes of exposure. *Probable response* must consider not only the dose itself but also the time period that may or may not allow some recovery, as well as characteristics of the individual or population exposed.

Hazard Evaluation

An industrial hygiene survey is usually conducted for purposes of hazard evaluation and the survey proper consists of estimating dose in as much detail as seems to be required by the situation [1]. Once that has been done, probable response can be determined and from that the necessity and perhaps urgency of control measures. If control measures are required and

287

must be suggested, the methods used for dose estimation may well give some clues about where the controls are most necessary [2], but this need not be the case; then good observation is the best data source, especially when coupled with experience.

Results of the survey must be communicated to people with line responsibility and authority who are in a position to take any necessary action. This kind of communication is almost always both oral and written and is probably the most important step of the whole process. The best survey ever conducted will do absolutely no good if its results and suggested controls are not communicated effectively to someone in a position to do something about the problem(s). Even a walk-through survey can accomplish much, however, if its results are imparted effectively. The industrial hygienist must learn to speak to groups as well as to individuals and must learn to write rapidly and well if he is to be successful in his profession.

Dose Determination

Determination of dose consists of integrating the movements or fluctuations of the incident flux (using that term for airborne materials as well as pressures and energies) with those of the absorber. Concentration of a material in air, for instance, at any particular spot in a plant, may vary considerably depending on the action of its source(s) as well as upon any incidental or deliberate ventilation [3]. The pattern of a worker's movements may also vary considerably even in those cases where a work station is assigned. The industrial hygienist can rarely confine his efforts to defining the exposure of one or even a few people; most often he must estimate the exposure of many so that even if a person is fairly well confined to a specific area, that area may not be representative of areas occupied by others with the same job classification [4].

Environmental sampling, whether for noise, heat stress, or particulate concentration, can be done in either of two ways: centered on the person or on the area. Personal sampling attempts to determine dose for one or a few individuals by sampling in their immediate vicinity for periods of time that can be used to estimate whole-shift values. If the workers whose exposures are determined in this manner are well chosen and truly representative of their job classifications, this kind of sampling will yield the best possible exposure (and dose) estimates. General area sampling may be done in the vicinity of no specific individual; instead of being person-oriented it is machine-, equipment-, or area-oriented. Properly done, general area sampling can give adequate estimates of exposures and at the same time can furnish clues about exposure sources. A combination of the two methods is often desirable.

Biological sampling is a general method applicable to any stressor that produces a quantifiable sign in man which relates to the dose [4, 5]. To be useful, the sign should be caused by doses that are not harmful to the person. Analysis of body fluids and/or tissues for chemical substances, their metabolites, or evidence of effect on enzymes and other systems has long been used for dose quantitation. The same principle can be applied to almost any imaginable stressor from noise to lasers.

General area sampling. Long before personal sampling devices were available, general area sampling was used by industrial hygienists to evaluate exposure. Ideally, the sampling would be thorough enough so that in any occupied space the pattern of incident exposure would be defined so well that a knowledge of a worker's activity would be then sufficient to determine his exposure. If, for instance, vapor concentrations around equipment were known well enough to enable the drawing of *isopleths* (lines of equal concentration) on a floor plan, then a person's exposure could be determined from observing his movements and the frequency and duration he was exposed to each concentration. His daily exposure could then be found by adding exposure increments to find the time-weighted average, Eq. (3-2).

This method has several advantages, some of which are not at all obvious. In any location where a person may be exposed, the incident flux often varies with time, perhaps over astonishing ranges. The fluctuations may have a short period or time base (seconds, minutes, or fractions thereof) or one that is long in relation to a shift (weeks or months), or both [6]. Only general area sampling can hope to provide information about such fluctuations as they pertain to specific locations. These data can, perhaps, be correlated with other variables to provide a method for interpolating or extrapolating exposure data to other times or places [7]. Such data can also provide valuable clues about the main sources of exposure and hence can hint at the applicability of controls.

Another advantage of general area sampling is that the equipment used has few size or shape limitations and can depend on "mains" electrical power. The sampling equipment can be made rugged, long-lasting, and even almost tamper-proof; with adequate recording devices, it can often be allowed to run unattended for days or even weeks, providing data pertinent to all shifts over extended periods of time. Centralized analytical devices can be attached to remote probes so that data can be acquired from many areas simultaneously and, in addition, can alarm if some preset limit is exceeded. General area sampling equipment can be connected to a strip chart recorder and thus can be used for historical records of environmental data or can even be connected to a computer, perhaps over telephone lines, and thus may be able to record and retain exposure estimates for each individual in a relatively large work force [6, 7].

Disadvantages of general area sampling are also numerous. Even though equipment can be made rugged and reliable, it often is not and as equipment sophistication increases, so does its demand for attention by well trained, highly paid technicians. This disadvantage is not peculiar to general area sampling equipment but as the complexity of such equipment can grow much more rapidly than can that of personal samplers, the problems of complexity appear there more often.

Even though general area sampling may, at least theoretically, provide a good estimate of *incident flux* at many locations, the number of sampling stations often is not sufficient to describe completely the exposure of any one person. Even if the equipment is adequate, the observation most often is not, as patterns of activity may change rather rapidly and activity under the scrutiny of an industrial hygienist may not be identical to activity with no one watching. Most often, observation of work patterns is conducted on the day shift (if, in fact it is conducted at all) and patterns on other shifts may be different not only because of less watchful supervision but also because of differing job demands. In general, even though determinations of incident flux may be poor with general area sampling, the weakest point in the process is the determination of work patterns. Few people, even industrial hygienists, enjoy watching people work especially when other important jobs must be neglected to do so.

Perhaps the epitome of poor general area sampling is approached with detector tubes. These devices, which indicate vapor concentrations by means of changes in color or length of stain of a gel through which air is drawn, are probably misused more often than any other type of equipment in this field. A proper use of detector tubes is to determine the magnitude of peak concentrations when the area or time of such peaks can be well estimated from other information. That is, even though detector tubes can, in the right hands, be used to determine *concentrations*, they rarely if ever are adequate to estimate *exposures*.

Personal sampling. The first widely used personal samplers were film badges for estimating exposures to ionizing radiation received by workers usually over a fairly long period of time. Industrial hygiene is beginning to approach health physics in this respect, with badge-like devices which sorb specific air contaminants at a rather reproducible rate and which can then be developed to provide an estimate of exposure or dose. Noise dosimeters have been available for some time but as yet little attempt has been made to use this approach with nonionizing radiation, heat or cold, ambient pressure, or aerosols.

The industrial hygiene technique of personal sampling began with the *breathing zone* air sample obtained by holding the sampler or a probe from it in the vicinity of a worker's nose and mouth while a job segment was performed. This method produced reasonable results as long as the industrial hygienist was able to obtain samples over an appreciable fraction of a shift for a few workers. Better results are now practical through use of battery-powered pumps capable of maintaining adequate airflows through samplers for a full 8-hr shift or longer. Samples of many vapors can be obtained on an adsorbent such as activated charcoal, which can later be analyzed for the material(s). Even specially calibrated detector tubes have been used in this manner to quantitate exposures where physical adsorption was not useful. The method is also used for aerosols with, perhaps, a preselector of some kind so that only respirable material is collected on a filter.

The main advantage of personal sampling is that, properly done, it gives a very good indication of the exposure during an entire shift [4]. No matter what the worker does, how he moves about, or whether he is at work or on a break, the sampler operates to determine his exposure. As equipment expense is reduced by volume manufacture, this method will have ever-increasing utility. Other than expense, its main disadvantage is that it gives few clues regarding the sources of exposure and, therefore, few hints at how to reduce an excessive exposure.

All personal monitors are subject to various kinds of interference by wearers and while this is usually not a serious problem, it can cause consternation if overlooked. Another problem that can now receive attention is that of properly defining *breathing zone*. How close to the nose or mouth must a probe be to obtain a sample truly representative of air inhaled by a worker? The answer probably varies according to the rate of airflow past the person's face, direction of airflow with relation to sources of the contaminant, the worker's body position, and perhaps even the kind of job being done. One way to obtain answers is by using biological sampling to determine dose while estimating exposure by several personal and/or general area sampling methods [8].

Another personal sampling technique available but mostly unused is the analysis of clothing especially for low-volatility materials that are absorbed through the intact skin in toxic amounts [9]. Absorbent pads under protective clothing can detect penetration by chemicals especially when the area of greatest contact is small and known in advance such as with gloves when hand exposure may be significant [10]. A possible extension is analysis of the active filtering component(s) of a respirator to determine concentrations that certainly were in the breathing zone of the wearer. These methods could be quantitated by the use of biological sampling in many instances.

Biological sampling. Urine, blood, and breath are the usual body fluids that may be examined for evidence of exposure to toxins. The concentration of lead in urine and/or blood has long been used as an index of exposure and in the 1960s the concentrations of δ-aminolevulinic acid or δ-aminolevulinic acid dehydrase were found to be even more sensitive indicators. Breath samples have been used for estimations of blood alcohol concentrations for legal purposes and more recently have found utility for the determination of CO [11, 12] and solvent [13] exposures. Biological sampling is especially important for chemical exposures where a major portion of the dose may be absorbed through the skin [14].

Saliva is another body fluid that can be used for exposure estimations; hair and nail parings are examples of tissues that have been used, especially for metals. In addition, body fat samples may be used to determine storage of fat-soluble substances, but obtaining these or bone marrow samples is more than a minor procedure.

All biological sampling methods have in common the fact that, depending on the pharmacodynamics of the material, they may represent either recent dose or the accumulated dose and hence the body burden of the substance. This is not always so; urinary mercury levels, for instance, seem to be uncorrelated with either exposure or effect [15, 16]. In many cases, utility of the method awaits a good mathematical model of uptake and excretion perhaps similar to the one for carbon monoxide described in Chapter 3.

Some of the direct effects of laser irradiation on the retina can be observed with the proper kind of examination. Similarly, audiometric examination can be used to determine the extent of temporary or permanent threshold shift, and the dynamics of the uptake of ^{86}Rb may be useful in estimations of past exposure to X or gamma rays [17]. Appearance of the lung on X-ray films is diagnostic of several pneumoconioses and in many cases correlations exist between an X-ray grade and severity of exposure [18].

Environmental sampling of various sorts can determine the exposure an individual receives to any desired degree of certainty [19]. Determination of exposure, however, is not determination of dose. In the case of absorbed chemicals, for instance, some excretion usually takes place simultaneously with exposure and *dose* must represent the manner in which the body integrates exposure. Biological sampling appears to be the best way to reconcile exposure with dose in those cases where enough is known about the uptake, excretion, and detoxification effects with time and body burden [7, 14, 20, 21]. Only when dose can be well estimated are good correlations of dose with effect likely to emerge. In the meantime, the industrial hygienist is often forced to use the next best method, to correlate exposure with effect, and many excellent examples of this have appeared in the literature [21, 22, 23, 24, 25, 26].

Probable Response Determination

Assuming that dose has been estimated (usually by estimating exposure) with as much precision as practical under the circumstances, the next step is to use that information to determine the probable response. Complicating what should be a rather straightforward application of known principles are several troublesome problems. Most exposures to stressors in the environment are not *pure* in the sense that they occur in the total absence of other stresses [27]. An exposure to a chemical may occur only (or usually) along with exposure to another as is the case with selenium and tellurium. Exposure to stresses other than chemical can also be simultaneous; the process of tunneling may involve exposures to high ambient pressure, dust, noise, biothermal stress, and carbon monoxide all at the same time, and an assessment of the effect of any of those stresses will be complicated by the presence of the others. Very little is known about how the body integrates different types of stress, even if both types are chemical. The usual assumption is that chemicals with dissimilar actions will exert their effects independently and that those with similar actions will have additive effects. In most cases the assumption is just that—an assumption—because of a paucity of data.

Another complicating factor is that even for a pure exposure there are probably little data available on the effects of that stress with man as the subject [28]. Usually the data will have been gleaned from animal experiments and can then be applied to man only with caution. In still other situations, the data relate to a different route of exposure and thus have limited utility at best. Finally, in all too many instances, no data at all are at hand and the industrial hygienist must make judgements based on analogy with similar materials or energies and hope that the safety factor used will prove large enough.

If the dose (or exposure) is at or below that known to produce little or no effect, then analysis is uncomplicated in most cases unless the flux density or concentration is high enough to produce untoward effects even at low exposures [29]. With most occupational stresses, from noise to vapor concentrations, an upper limit that should not be exceeded even for very brief times exists whether or not it is officially recognized. Irritating materials have such a limit at concentrations that cause pain even when that pain is a preferred warning property of the substance. Similarly, even brief exposures to noise in the neighborhood of 115 dbA (re 20 μPa) should be avoided. Analogous limits exist for abnormal pressure, heat, and nonionizing radiation but are not so well defined. Experimentation has not shown any particular dose-independent flux density effect at high exposure rates for ionizing radiation; this may be an exception to the general case.

If exposures or doses are not at or below some established set point value, then analysis of the situation may indicate a need for information about the effect(s) expected. These kinds of data may be most difficult to find if only because man has no way of knowing about symptoms animals experience during or after experimental exposures. Too often, then, reliance must be placed on medical reports of poisonings that appear in the literature, usually with no indication of either exposure or dose or even sure knowledge of the toxic agent.

Sources of information on safe exposures. The first *control concentration* was established for free crystalline silica in the 1920s. It actually defined two levels: one above which prolonged exposure was almost certain to result in silicosis and one which defined the maximum concentration found not to be associated with the disease even on prolonged exposure. These data were gathered from rather extensive investigations conducted in Vermont granite sheds by the U.S. Public Health Service using impingers as sample collectors. A similar survey was conducted in the early 1970s using personal mass respirable dust samplers [30], data from which were correlated with results of lung function studies and chest X rays of workers to furnish the base for a new TLV for silica [31, 32].

In the 1920s, lists of personally used maximum allowable concentration (MAC) values for gases and vapors as well as dusts began to appear and to be passed around, usually among industrial hygienists having some governmental function. Such lists were assembled and became the basis for the threshold limit values of the American Conference of Governmental Industrial Hygienists (ACGIH) after its organization in 1938. It is to the TLV list, which now contains values for physical as well as chemical agents, that most industrial hygienists have turned to find upper limits of safe exposure. Values are updated annually by a committee of experts and then published by ACGIH. In the 1960s the data on chemicals were extended to include a *C* (indicating a ceiling value rather than an average) and a *skin* (for materials where absorption through the intact skin may contribute to the exposure) notation.

The Z-37 committee of the American National Standards Institute (ANSI) has published "Acceptable Concentrations" for a number of materials. Not a list but a separate pamphlet for each material, the acceptable concentrations (AC) given are not single-valued but usually contain estimates of an acceptable ceiling concentration assuming an 8-hr day, an acceptable time-weighted average that usually corresponds closely with the TLV, and even an acceptable maximum for peaks above the ceiling. Unfortunately, only a few of these pamphlets have been published and they are not updated with any regularity.

The American Industrial Hygiene Association (AIHA) has published "Hygienic Guides" for more than 150 materials, each guide containing a summary of toxicological information, including a *maximal acceptable concentration,* which is usually identical to the TLV for the year of publication. Although there are many more hygienic guides than AC's, guides are not updated with any certain regularity, either.

Governmental agencies, particularly the U.S. National Institute for Occupational Safety and Health (NIOSH), publish summaries of information concerning chemical and physical stressors in the work environment. The NIOSH publications of most interest are called *criteria* and contain information gleaned from all possible sources; they usually include a suggested control value. Although criteria documents are more thorough than either AC's or hygienic guides, they too suffer from a lack of currency soon after publication.

The U.S. Occupational Safety and Health Administration (OSHA) adopted the 1968 TLV list as well as several publications issued by insurance organizations in its first attempt to achieve legal control of the occupational environment. Since that time it has based the new TWA (8-hr time-weighted average) values on NIOSH criteria and other judgements with some regularity. Again, however, updating is uncertain.

Many industrial hygienists are faced with the problem of staying legal as well as that of protecting the health of workers. Especially when some time has passed since the adoption of an official limit, current information may well indicate a need for more strictness to assure no injury to susceptible persons or, perhaps, it may suggest that some relaxation is in order. The industrial hygienist has a professional responsibility to provide a good, honest evaluation of the hazards of the work environment. Blind obedience is no substitute for judgement and knowledge although, of course, legal obligations must be met.

Sources of information on hazardous exposures. Many of the sources of information mentioned in the previous section also contain estimates of the consequences of overexposure. Specifically, "Hygienic Guides," ANSI "Acceptable Concentrations," and NIOSH criteria are such sources. In addition, AIHA has published a few emergency exposure limits (EEL) which contain best estimates of the highest exposures (concentrations and durations) which are not life-threatening and may be used as guides in life-saving situations.

The literature of toxicology and industrial hygiene is, in most cases, the best source of the most recent information. Access to the literature is provided most easily through the *Industrial Hygiene Digest,* an abstracting journal published monthly by the Industrial Health Foundation, Inc., 5231 Centre Avenue, Pittsburgh, Pa. 15232. Other abstracting journals including *Chemical Abstracts, Biological Abstracts,* and *Index Medicus* are, however, more likely to be found in general scientific libraries and can be very useful. Computerized abstract searching services are available, most with the disadvantage of a cutoff near 1970 with little or no information previous to that date.

Journals and newsletters are beginning to proliferate in this as well as other fields, and remaining current by casual reading is next to impossible. Nevertheless, a high proportion of articles published in the *American Industrial Hygiene Association Journal* and in the *Archives of Environmental Health* are pertinent. Each newsletter has its own advantages and disadvantages; any of them will usually match the others in coverage of major events in the field; they differ mainly in the fringes.

A good source of information on the toxic effects and hazards of materials and energies is manufacturers' literature. Many of the large chemical companies in the United States and elsewhere have a tremendous stock of information that can be tapped by means of a letter or a telephone call, or simply by reading the brochures available. In addition, manufacturers of sampling and/or analytical equipment can be very helpful. For many years the instruction book published by General Radio (*Handbook of Noise Measurement*) was the very best source of condensed information on all aspects of noise as an occupational and/or environmental hazard. Other equipment manufacturers have followed this lead; enlightened self-interest can be a powerful motivating force capable of astonishing educational effort in the interests of sales. To ignore this source of information is to handicap one's effectiveness; to rely on it exclusively is to ignore the facts of commercial life.

Hazard Control

Theory

When most people think of controlling a hazard, their thoughts are much more likely to center on reduction or elimination than control. It is in this sense that rules, regulations, and laws are promulgated demanding *zero* levels of contaminants in some portion of the environment. And, although a true concentration of zero is impossible to achieve, it does illustrate one necessity of control, that of a set point. Without something to aim at, control is unlikely and even reduction may be difficult.

To achieve control, some information about the system, source or process being controlled is necessary in addition to having a set point. Without data, the necessity of control cannot be known, and achievements cannot be verified. Information about the variable being controlled is called *feedback* if that information is used to generate an error signal whose intensity is in proportion to the deviation of the variable from its set point. That is, in order to control anything, including the time-weighted average breathing zone concentration of compound A, information must be at hand or available on not only what that concentration should be (or be below) but also on how far from that goal the concentration actually is from time to time.

The aim of much of the activity in the occupational environment is to assure that workers are never exposed to dangerous amounts of anything and, secondarily, to keep exposures to minimum practical levels. Most important then, is a set point somewhat less than the minimum exposure or dose required to produce an unwanted effect. After the desired degree of control has been attained, the set point can be reduced, if desired, to bring the dose or exposure to as near zero as is consistent with other constraints on the system.

Throughout this whole process, continuing information about the variable being controlled is necessary. Without feedback, control and even reduction may not be possible because the measures taken may have little or no effect and if that lack cannot be sensed, control is unlikely. It is for this reason that so much emphasis is put on analytical systems and devices in this field with continual improvement in accuracy, precision, sensitivity, selectivity, stability, portability, ruggedness, and cost constantly being sought.

In a history of industrial hygiene published in 1965, the following statement was made:

> Recognition of the occupational diseases led to descriptions of the symptoms and of the work environment. Later, attempts were made to prevent or minimize the diseases by controlling the environment. The control methods used were quite empirical and uncertain until the causative factors in the work environment could be identified and measured. Industrial hygiene did not emerge as a unique field of endeavor until quantitative measurements of these harmful factors could be made. Once the degree of exposure could be associated with certain medical findings, it was possible to determine safe levels of exposure and to design engineering controls to maintain these levels [33].

In other words, the field of industrial hygiene rests on a base of quantitative measurement that provides the feedback so necessary for control. To a large extent, quantitative evaluation of hazard is what distinguishes the industrial hygienist from the safety specialist (or professional) and others who also seek to reduce occupational hazards.

The set points used for control purposes in this field have been called MAC, TLV, TWA, AC, and other names. To the uninitiated, the TLV for a gas appears either to designate a line between safe and toxic concentrations or to be an expression of the relative toxicity of the material. In fact, it is neither. It is a set point for control purposes; hopefully nothing more and nothing less. The rationale behind the TLV for HCl is completely different from that behind the TLV for HCN and the reason behind the TLV for fluorocarbon-12 is completely different yet. Only one of those mentioned has to do with systemic toxicity (that for HCN) and only two of the three are related to hazard (that for fluorocarbon-12 is not). All have in common the fact that they are set points for control purposes; to use them in other ways is to ignore and perhaps pervert their derivation and purpose.

TLVs for physical agents are set on the same basis as those for chemical agents. They are not fine lines between *safe* and *dangerous* nor are they related in any way to the seriousness of effect that may be experienced on overexposure. They are *hygienic standards,* or set points, presented to the industrial hygiene community and others as primary control goals. Achievement of those goals will assure that "... nearly all workers may be repeatedly exposed day after day without adverse effect" [34].

Exposure Reduction Methods

Methods recommended by industrial hygienists and used by others to reduce exposures have been mentioned in several areas of this book and are presented in detail in Chapter 35 of *The Industrial Environment—its Evaluation & Control,* published by NIOSH in 1973, and elsewhere. Briefly, the general methods are substitution, isolation, ventilation, and education.

Substitution. Complete elimination of a hazard (or substitution of a different and hopefully lesser one) can be achieved by using a different material, piece of equipment, or process. If bicycle riding were substituted for automobile driving in the United States, for instance, overnight the annual death rate from traffic accidents would be slashed. The bumps and scrapes of bicycle disasters would be substituted for the often-fatal crashes caused by slight miscalculations at the steering wheel of a car. Desirable as this fanciful substitution scheme may seem (especially because air pollution would also be reduced considerably as would liquid fuel usage), it has consequences that reach far beyond the city street and into areas that might be labeled *dire.*

A large section of the economy of the United States is tied directly to the production of automobiles. Not only would assembly-line workers be out of work, but also employees of petroleum refineries as well as operators of garages, service stations, parking lots, and wrecking yards. Asphalt plants would go out of business, quarries would shut down as would iron mines. Steel mills would lay off large portions of their work force and rubber companies would fold. The resulting massive depression would only be alleviated slightly by the tremendously increased production of bicycles, each of which takes orders of magnitude less of everything than does the smallest automobile.

Most substitutions designed to reduce or eliminate hazards in the occupational environment can be expected to have less unfortunate consequences than can be predicted for the sudden substitution of bicycles for cars, but most will probably have some unforseen results, and not all of those will be good. Prior to suggesting any substitutions, all predictable consequences should be examined. Petroleum naptha and other gasoline- or kerosine-like hydrocarbons remove grease and oil from metal very well. These mixtures, however, burn readily so that much degreasing is done by more expensive but less flammable substitutes such as trichloroethylene or 1,1,1-trichloroethane. These are good substitutes, perhaps, but trichloroethylene may be a carcinogen in man and it does participate in the formation of photochemical smog. Both chlorinated hydrocarbons may help deplete the stratospheric layer of ozone that screens the earth's surface from UV radiation. Neither material can be used around welders or even in the vicinity of flames because of the predictable formation of hydrogen chloride and, possibly, phosgene. Perhaps a better substitute is soap or a (biodegradable) detergent and water.

Isolation. A worker can be isolated from an exposure or isolated from the consequences of an exposure by time, distance, or a physical barrier. Exposure duration may be controlled, for instance, to cause the absorbed dose to be below the set point value. This approach is called *administrative* or *work-practice* control and where it is applicable, it can work quite well. It has the major disadvantage, however, of providing a floor under exposures that cannot be further reduced easily.

Use of distance for isolation is almost automatic in many cases where the material or energy has reasonably good warning properties or else is measured thoroughly enough to enable the rational positioning of warning signs or fences.

Physical barriers may be represented by a thin film of polyethylene, or a few feet of concrete or lead, or even by the materials in a gas mask cannister. Either the process or the worker can be isolated to any degree warranted by the hazard, but in almost all cases the engineering controls that tend to isolate the process are preferred philosophically as well as practically. Anything done to tighten up equipment, reduce leaks, prevent the generation or dispersal of dust, or attenuate radiated energy can be regarded as increasing the protection offered by the barrier, whether it is a seal on a pump or the aluminum foil used as a thermal radiation shield. These measures are at once more positive and easier to maintain than personal protective devices used by workers to isolate themselves from the hazard [35]. Even though engineering physical barriers into equipment may have a much higher first cost than any personal protective equipment, the financial and other consequences of overexposing a group of people may well make that capital investment pale into insignificance [36].

Ventilation. Ventilation is obviously a method specific to reducing the hazards of chemical substances although it is also used to reduce heat stress (mainly by increasing sweat evaporation rates); it has no utility for pressure or other energy hazards. Ventilation is achieved by the directed movement of air [37] and, perhaps surprisingly, the first principle of ventilation is to use physical barriers as much as possible [38]. The cost, amortized over a reasonable time, of any barrier used in the place of a directed flow of air is considerably less than that of moving the heated, filtered, and/or conditioned air through the same area. In most cases where ventilation is used to reduce an inhalation hazard, enclosure to as great an extent as practical will be found to decrease overall cost and to increase efficiency.

Enclosure is particularly effective because local exhaust ventilation does not reach out very far to capture contaminated air and the capture effect is reduced rapidly as distance increases. An exhaust hood is a device built to cause air to move in the proper direction and also to reduce the amount of air movement necessary by enclosing the process or equipment to the greatest feasible extent. The design of exhaust hoods and of the ventilation systems connected to them is covered remarkably well in the ACGIH publication, *Industrial Ventilation: A Manual of Recommended Practice.* The latest edition of this book can be purchased from the Committee on Industrial Ventilation, P.O. Box 453, Lansing, Michigan 48902.

Another principle of exhaust ventilation is always to supply at least as much air as is exhausted. In addition to allowing building doors to be opened more easily, makeup air units filter and heat (or cool, in some instances) air much more efficiently than is done by leakage into the area, prevent drafts, and may even allow some degree of contaminant control by the direction of air movement. If 100 m^3 of air is exhausted from an enclosed volume, 100 m^3 of air must be supplied to that volume in some manner; a makeup air system not only allows control over the supply of air but also increases efficiency of exhaust ventilation.

Education. Even intelligent people can be surprisingly ignorant outside their own fields of endeavor and ignorant people are unlikely to solve problems requiring the application of knowledge. Ignorance is banished by education, formal or otherwise, to be replaced by knowledge. Hazard reduction is almost always best accomplished by people who know what they are doing, be they managers, engineers, supervisors, or workers. None of these persons is likely to solve a problem he is unaware of but any, regardless of job classification, may suggest a solution particularly if the hazard can affect him or his job.

An engineer who has never heard of the ACGIH industrial ventilation manual is unlikely to design a ventilation system well, as can be seen by inspecting the systems installed in many and perhaps most large and small manufacturing plants around the world. An engineer unaware of the problem of heat stress or of the methods used to evaluate and control it is unlikely to design workable efficient methods of hazard reduction; a chemist unaware of the hazards of a particular solvent is unlikely to suggest a substitute. Formal educational effort concerning the evaluation and control of occupational hazards is very likely to pay large dividends when chemists or engineers of any discipline are the recipients.

High-echelon supervisors and managers are usually the people who must be convinced to spend money for hazard control. Their education is an obvious necessity, all too often neglected. Although some undergraduate (and graduate) chemists and engineers become aware of a need for knowledge in the fields of industrial hygiene, very few people in training to become business managers do. They then are in a poor position to make decisions and commit resources later in life when that is their responsibility. Education of managers is at once the most difficult and potentially the most rewarding of all such efforts; in many respects it may be the most important.

REFERENCES

[1]   M. M. LUCKENS, "The Occupational Hygiene Survey: Principles, Practice, Significance," *Am. Ind. Hyg. Assn. J.* **28** (1967), 179–83.

[2]   J. E. PETERSON, R. A. COPELAND, AND H. R. HOYLE, "Health Hazards of Spraying Polyurethane Foam Out-of-Doors," *Am. Ind. Hyg. Assn. J.* **23** (1962), 345–52.

[3]   L. R. S. PATIL, ET AL., "The Health of Diaphragm Cell Workers Exposed to Chlorine," *Am. Ind. Hyg. Assn. J.* **31** (1970), 678–86.

[4]   M. K. WILLIAMS, E. KING, AND JOAN WALFORD, "An Investigation of Lead Adsorption in an Electric Accumulator Factory With the Use of Personal Samplers," *Brit. J. Ind. Med.* **26** (1969), 202–16.

[5]   L. D. PAGNOTTO AND L. M. LIEBERMAN, "Urinary Hippuric Acid Excretion as an Index of Toluene Exposure," *Am. Ind. Hyg. Assn. J.* **28** (1967), 129–34.

[6]   J. E. PETERSON, H. R. HOYLE, AND E. J. SCHNEIDER, "The Application of Computer Science to Industrial Hygiene," *Am. Ind. Hyg. Assn. J.* **27** (1966), 180–85.

[7]   C. G. KRAMER AND J. E. MUTCHLER, "The Correlation of Clinical and Environmental Measurements for Workers Exposed to Vinyl Chloride," *Am. Ind. Hyg. Assn. J.* **33** (1972), 19–30.

[8]   E. B. HAY, III, "Exposure to Aromatic Hydrocarbons in a Coke Oven By-Product Plant," *Am. Ind. Hyg. Assn. J.* **25** (1964), 386–91.

[9]   H. R. WOLFE, W. F. DURHAM, AND J. F. ARMSTRONG, "Exposure of Workers to Pesticides," *Arch. Env. Health* **14** (1967), 622–33.

[10] CHRISTINE EINERT, ET AL., "Exposure to Mixtures of Nitroglycerine and Ethylene Glycol Dinitrate," *Am. Ind. Hyg. Assn. J.* **24** (1963), 435–47.

[11] P. A. BREYSSE AND H. H. BOVEE, "Use of Expired Air-Carbon Monoxide for Carboxyhemoglobin Determinations in Evaluating Carbon Monoxide Exposures Resulting from the Operation of Gasoline Fork Lift Trucks in Holds of Ships," *Am. Ind. Hyg. Assn. J.* **30** (1969), 477–83.

[12] J. BUTT, ET AL., "Carboxyhaemoglobin Levels in Blast Furnace Workers," *Ann. Occup. Hyg.* **17** (Aug., 1974), 57–63.

[13] R. D. STEWART, C. L. HAKE, AND J. E. PETERSON, "Use of Breath Analysis to Monitor Trichloroethylene Exposures," *Arch. Env. Health* **29** (1974), 6–13.

[14] A. L. LINCH, "Biological Monitoring for Industrial Exposure to Cyanogenic Aromatic Nitro and Amino Compounds," *Am. Ind. Hyg. Assn. J.* **35** (1974), 426–32.

[15] Z. G. BELL, H. B. LOVEJOY, AND T. R. VEZENA, "Mercury Exposure Evaluations and Their Correlation with Urine Mercury Excretions. 3. Time-Weighted Average (TWA) Mercury Exposures and Urine Mercury Levels," *J. Occup. Med.* **15** (1973), 501–08.

[16] R. R. LAUWERYS AND J. P. BUCHET, "Occupational Exposure to Mercury Vapors and Biological Action," *Arch. Env. Health* **27** (1973), 65–68.

[17] K. G. SCOTT, ET AL., "Occupational X-Ray Exposure," *Arch. Env. Health* **26** (1973), 64–66.

[18] J. R. ASHFORD, J. W. J. FAY, AND C. S. SMITH, "The Correlation of Dust Exposure with Progression of Radiological Pneumoconiosis in British Coal Miners," *Am. Ind. Hyg. Assn. J.* **26** (1965), 347–61.

[19] J. R. LYNCH AND H. E. AYER, "Measurement of Dust Exposures in the Asbestos Textile Industry," *Am. Ind. Hyg. Assn. J.* **27** (1966), 431–37.

[20] M. MENZ, H. LUETKEMEIR, AND K. SACHSSE, "Long-Term Exposure of Factory Workers to Dichlorvos (DDVP) Insecticide," *Arch. Env. Health* **28** (1974), 72–76.

[21] P. L. POLANKOFF, K. A. BUSCH, AND M. T. OKAWA, "Urinary Fluoride Levels in Polytetrafluoroethylene Fabricators," *Am. Ind. Hyg. Assn. J.* **35** (1974), 99–106.

[22] H. L. WILLIAMS, "A Quarter Century of Industrial Hygiene Surveys in the Fibrous Glass Industry," *Am. Ind. Hyg. Assn. J.* **31** (1970), 362–67.

[23] H. B. ELKINS, ET AL., "Massachusetts Experience with Toluene Di-Isocyanate," *Am. Ind. Hyg. Assn. J.* **23** (1962), 265–72.

[24] J. CHOLAK, L. SCHAFER, AND D. YEAGER, "Exposures to Beryllium in a Beryllium Alloying Plant," *Am. Ind. Hyg. Assn. J.* **28** (1967), 399–407.

[25] S. J. DANZIGER AND P. A. POSSICK, "Metallic Mercury Exposure in Scientific Glassware Manufacturing Plants," *J. Occup. Med.* **15** (1973), 15–20.

[26] R. E. ROSENSTEEL, S. K. SHAMA, AND J. P. FLESCH, "Occupational Health Case Report: No. 1—Toxic Substances: Carbon Disulfide; Industry; Viscose Rayon Manufacture," *J. Occup. Med.* **16** (1974), 22–32.

[27] D. A. FRASER AND E. T. CHANLETT, "The Assessment of the Total Toxicant Exposure of Man," *Am. Ind. Hyg. Assn. J.* **24** (1963), 417–22.

[28] ANNA M. BAETJER, "Changes—Stress or Benefit?," *Am. Ind. Hyg. Assn. J.* **25** (1964), 207–12.

[29] R. L. RALEIGH AND W. A. MCGEE, "Effects of Short, High-Concentration Exposures to Acetone as Determined by Observation in the Work Area," *J. Occup. Med.* (1972), 607–10.

[30] G. P. THERIAULT, ET AL., "Dust Exposure in the Vermont Granite Sheds," *Arch. Env. Health* **28** (1974), 12–17.

[31] G. P. THERIAULT, J. M. PETERS, AND L. J. FINE, "Pulmonary Function in Granite Shed Workers of Vermont," *Arch. Env. Health* **28** (1974), 18–22.

[32] G. P. THERIAULT, J. M. PETERS, AND W. M. JOHNSON, "Pulmonary Function and Roentgenographic Changes in Granite Dust Exposure," *Arch. Env. Health* **28** (1974), 23–27.

[33] H. V. BROWN, "The History of Industrial Hygiene: A Review with Special Reference to Silicosis," *Am. Ind. Hyg. Assn. J.* **26** (1965), 212–26.

[34] *TLVs: Threshold Limit Values for Chemical Substances and Physical Agents in the Workroom Environment with Intended Changes for (current year)*, (P.O. Box 1937, Cincinnati, Ohio 45201: The American Conference of Government Industrial Hygienists).

[35] R. H. STARKEY, ET AL., "Health Aspects of the Commercial Melting of Uranium-Contaminated Ferrous Metal Scrap," *Am. Ind. Hyg. Assn. J.* **21** (1960), 178–81.

[36] H. F. SCHULTE, "Personal Protective Devices," Chapter 36 in *The Industrial Environment—Its Evaluation & Control* (Washington, D.C.: U.S. Department of Health, Education, and Welfare, Public Health Service, Center for Disease Control, National Institute for Occupational Safety and Health, 1973) 519–31.

[37] J. E. MUTCHLER, "Principles of Ventilation," Chapter 39; "Local Exhaust Systems," Chapter 41 in *The Industrial Environment—Its Evaluation & Control*, Ref. 36, pp. 573–82 and 597–608.

[38] C. L. LINDEKEN AND O. L. MEADORS, "The Control of Beryllium Hazards," *Am. Ind. Hyg. Assn. J.* **21** (1960), 245–51.

APPENDICES

A

REVIEW OF ORGANIC CHEMICAL NOMENCLATURE

Nomenclature of organic chemicals exists at several levels, from official names of the International Union of Pure and Applied Chemistry (IUPAC) to common or trivial names that may vary from building to building even within the same plant. Although this review is centered on IUPAC names of often-encountered industrial chemicals, many common names will be found here and in the Index.

Aliphatic Hydrocarbons

Aliphatic hydrocarbons are saturated (have no double or triple bonds between carbon atoms) compounds of carbon and hydrogen and are acyclic (contain no closed-ring structures). The first four members of the series are named methane, ethane, propane, and butane. All higher members have names that begin with a numerical prefix referring to the number of carbon atoms in the longest chain and end with the syllable *-ane,* for example, pentane. The name minus the suffix may be called the *stem* or *root*. Names of the first ten members of the series along with structural information are given below:

Name	Structural Formula
Methane	CH_4
Ethane	$H_3C—CH_3$
Propane	$H_3C—CH_2—CH_3$
Butane	$H_3C—CH_2—CH_2—CH_3$
Pentane	$H_3C—(CH_2)_3—CH_3$
Hexane	$H_3C—(CH_2)_4—CH_3$
Heptane	$H_3C—(CH_2)_5—CH_3$
Octane	$H_3C—(CH_2)_6—CH_3$
Nonane	$H_3C—(CH_2)_7—CH_3$
Decane	$H_3C—(CH_2)_8—CH_3$

The generic name for this group of hydrocarbons is *alkane* and the generic molecular formula is $C_nH_{(2n+2)}$. A univalent radical is denoted by substituting *-yl* for *-ane* in the name. Examples are methyl (for $H_3C—$) and hexyl [for $H_3C—(CH_3)_4—CH_2—$]; the generic name for the radicals is *alkyl.*

The compounds just named are formed in straight chains, and this configuration is called *normal* and may be designated by the prefix *n-* as *n*-butane or *n*-decane, although this prefix is implied if it is not used. Branching in the molecule is indicated by a numbering system based on the longest side chain in the structure, which then becomes the backbone. Numbers used are kept as low as possible.

EXAMPLE

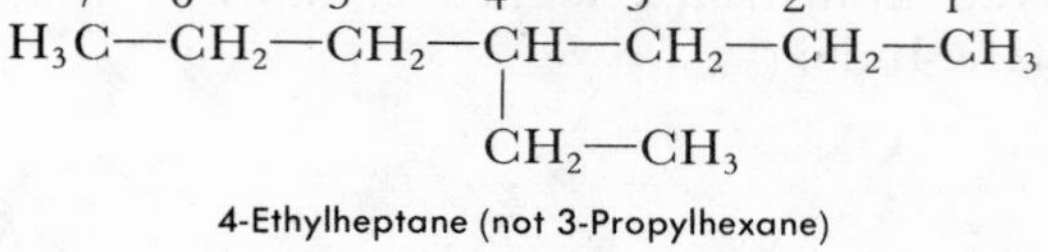

$$\overset{7}{H_3C}—\overset{6}{CH_2}—\overset{5}{CH_2}—\overset{4}{CH}—\overset{3}{CH_2}—\overset{2}{CH_2}—\overset{1}{CH_3}$$
$$|$$
$$CH_2—CH_3$$

4-Ethylheptane (not 3-Propylhexane)

This numbering system can always be used, but in a few cases another prefix will usually be found, especially for compounds having a Y structure. In these cases, the prefix *iso-* can be used. *Neo-* is now usually found only for one compound, as shown on the following page.

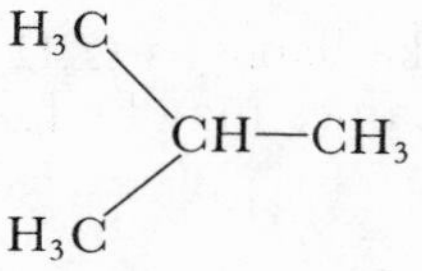

$$H_3C \diagdown CH-CH_3$$
$$H_3C \diagup$$

Isobutane

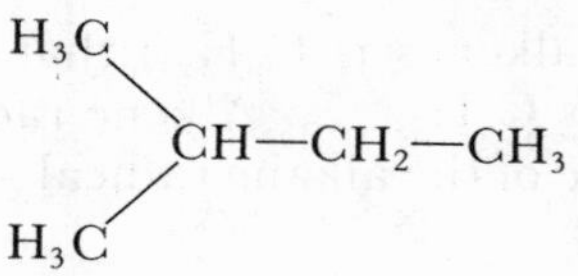

$$H_3C \diagdown CH-CH_2-CH_3$$
$$H_3C \diagup$$

Isopentane
(2-Methylbutane)

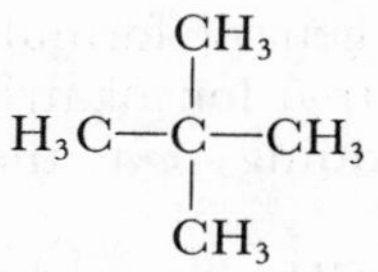

$$H_3C-\overset{\displaystyle CH_3}{\underset{\displaystyle CH_3}{\overset{|}{\underset{|}{C}}}}-CH_3$$

Neopentane
(2,2-Dimethylpropane)

Even though isobutane, isopentane, isohexane, and neopentane are the only alkanes so-named in the IUPAC, the *iso-* prefix is also used with the propyl radical to indicate that the open valence bond is on the central carbon atom:

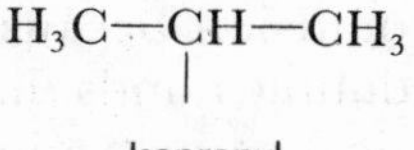

$$H_3C-\overset{|}{CH}-CH_3$$

Isopropyl

For radicals containing four or more carbon atoms, the prefixes *sec-* and *tert-* are used to indicate a free valence on an atom already having two or three, respectively, non-hydrogen substituents.

EXAMPLES

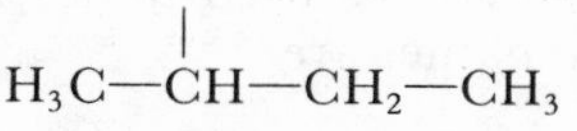

$$H_3C-\overset{|}{CH}-CH_2-CH_3$$

sec-Butyl

$$H_3C-\overset{|}{\underset{\displaystyle CH_3}{\overset{|}{C}}}-CH_3$$

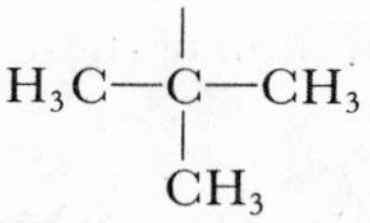

tert-Butyl

Unsaturated Aliphatic Hydrocarbons

Unsaturated acyclic aliphatic hydrocarbons with a double bond between a pair of carbon atoms are called *olefins* or *alkenes;* the suffix *-ene* is generic. Two double bonds in a molecule confer the suffix *-adiene;* three, *-atriene.* Generic names are alkadiene and alkatriene, respectively. Numbers are used to locate the double bond(s) and are kept as low as possible.

EXAMPLES

Name	Structural Formula
Ethene (ethylene)	$H_2C{=}CH_2$
Propene (propylene)	$H_2C{=}CH-CH_3$
Allene (a special name)	$H_2C{=}C{=}CH_2$
1-Butene	$H_2C{=}CH-CH_2-CH_3$
2-Butene	$H_3C-CH{=}CH-CH_3$
1,3-Butadiene	$H_2C{=}CH-CH{=}CH_2$
1,3-Pentadiene (not 2,4-Pentadiene)	$H_3C-CH{=}CH-CH{=}CH_2$

The generic formula for alkenes is C_nH_{2n}; that for alkadienes is $C_nH_{(2n-2)}$; and that for alkatrienes is $C_nH_{(2n-4)}$. Alkene radical names may be formed by adding *-ene* to the name of the alkane radical as

$$-CH_2-$$
Methylene

$$-CH_2-CH_2-$$
Ethylene

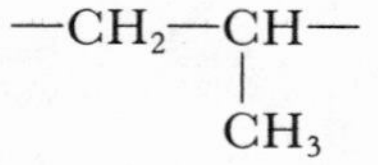

$$-CH_2-\underset{\underset{CH_3}{|}}{CH}-$$
Propylene

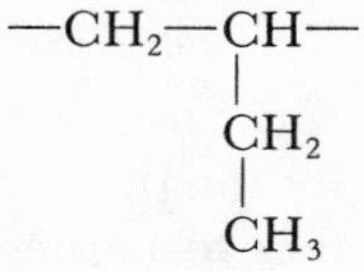

$$-CH_2-\underset{\underset{\underset{CH_3}{|}}{CH_2}}{\overset{}{CH}}-$$
Butylene

This naming system causes confusion because the radical names are also used for molecules containing double bonds that the radicals do not. A rule of the thumb is that if the chemical name begins with the name of an alkene radical (ethylene dichloride), the double bond no longer exists and the name is that of a radical. If the chemical name ends with the name of the radical (bromoethylene), then the double bond still exists and the radical is a molecule.

Because of the possible confusion associated with conventional names of alkenes, another naming system is sometimes used under the IUPAC umbrella. Most common of these radical names are

$$-CH=CH_2$$
Vinyl

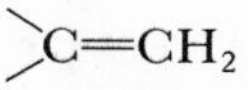

$$>C=CH_2$$
Vinylidene

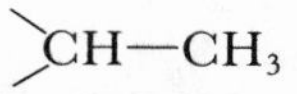

$$>CH-CH_3$$
Ethylidene

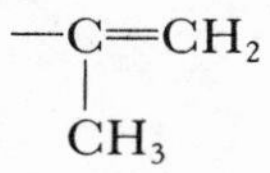

$$-\underset{\underset{CH_3}{|}}{C}=CH_2$$
Propylidene

$$-CH_2-CH=CH_2$$
Allyl

A triple bond between a pair of carbon atoms calls for the suffix *-yne* and the generic name *alkyne*. Acetylene is the name for the first member of this series of homologues, and the higher members may be named as acetylene derivatives instead of using the IUPAC numbering system.

EXAMPLES

Ethyne (acetylene)	$HC{\equiv}CH$
Propyne (methyl acetylene)	$H_3C-C{\equiv}CH$
1-Butyne (ethyl acetylene)	$H_3C-CH_2-C{\equiv}CH$
2-Butyne	$H_3C-C{\equiv}C-CH_3$

The generic formula for the alkynes is the same as that for the alkadienes, $C_nH_{(2n-2)}$.

Alicyclic Hydrocarbons

The word *alicyclic* was derived from aliphatic and cyclic. All of these compounds contain a ring structure. They are named by using the prefix *cyclo-* before the name of the analogous alkane. The generic formula is C_nH_{2n}.

EXAMPLES

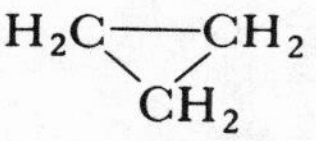

$$H_2C\underline{\qquad}CH_2$$
$$CH_2$$
Cyclopropane

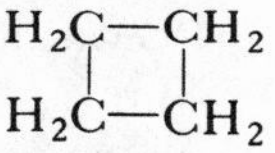

$$H_2C\underline{\quad}CH_2$$
$$H_2C\underline{\quad}CH_2$$
Cyclobutane

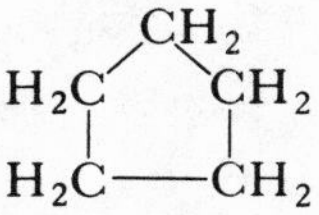

$$CH_2$$
$$H_2C \qquad CH_2$$
$$H_2C\underline{\quad}CH_2$$
Cyclopentane 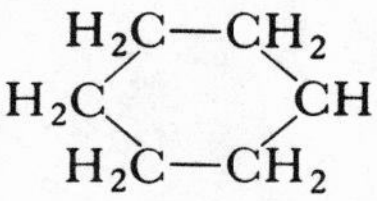 or

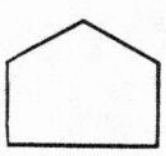

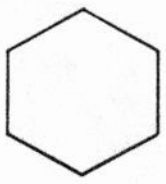

$$H_2C\underline{\quad}CH_2$$
$$H_2C \qquad CH_2$$
$$H_2C\underline{\quad}CH_2$$
Cyclohexane or

As with the alkenes, the presence of a double bond is indicated by replacing *-ane* with *-ene*, that of two double bonds with *-adiene*, etc., using numbers to indicate location if necessary.

EXAMPLES

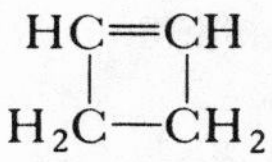

$$HC\!=\!CH$$
$$H_2C\underline{\quad}CH_2$$
Cyclobutene

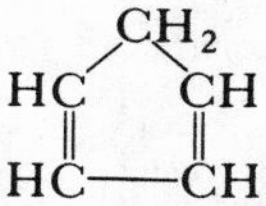

$$CH_2$$
$$HC \qquad CH$$
$$HC\underline{\quad}CH$$
1,3-Cyclopentadiene or

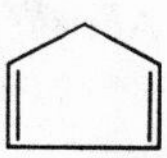

Functional Groups

Several functional groups may be attached to hydrocarbons, changing not only their names but also their physical and biological properties. The most common groups and their names are listed below:

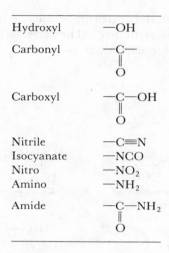

Hydroxyl	—OH
Carbonyl	$-\overset{\displaystyle \|}{\underset{\displaystyle O}{C}}-$
Carboxyl	$-\overset{\displaystyle \|}{\underset{\displaystyle O}{C}}-OH$
Nitrile	$-C\equiv N$
Isocyanate	—NCO
Nitro	$-NO_2$
Amino	$-NH_2$
Amide	$-\overset{\displaystyle \|}{\underset{\displaystyle O}{C}}-NH_2$

Halogenated Aliphatic Hydrocarbons

Special names abound in this series of compounds. Only the most common names are indicated.

Name	*Structural Formula*
Chloromethane (methyl chloride)	H_3CCl
Dichloromethane (methylene chloride, methylene dichloride)	H_2CCl_2
Trichloromethane (Chloroform)	$HCCl_3$
Tetrachloromethane (Carbon tetrachloride)	CCl_4
Chloroethane (Ethyl chloride)	H_3C-CH_2Cl
1,2-Dichloroethane (Ethylene dichloride)	ClH_2C-CH_2Cl
1,1-Dichloroethane (Ethylidene chloride)	Cl_2CH-CH_3
1,1,1-Trichloroethane (Methyl chloroform)	Cl_3C-CH_3
1,1,2-Trichloroethane	Cl_2CH-CH_2Cl
1,1,2,2-Tetrachloroethane (Acetylene tetrachloride)	$Cl_2CH-CHCl_2$

Compounds containing fluorine, bromine, and iodine are named in a similar manner. Compounds containing two or more different halogens are named, as usual, to keep the numbers as low as possible, and then with the halogens in alphabetical order.

EXAMPLES

Bromochloromethane (Methylene bromochloride)	H_2CBrCl
Dichlorodifluoromethane (Fluorocarbon-12)	CCl_2F_2
1,1,2-Trichloro-1,2,2-trifluoroethane (Fluorocarbon-113)	$CCl_2F—CClF_2$

Halogenated Alkenes

Chlorinated compounds containing double bonds most often encountered (the naming for the other halogens is similar) are

Chloroethene (Vinyl chloride)	$H_2C=CHCl$
1,1-Dichloroethene (Vinylidene chloride)	$Cl_2C=CH_2$
Trichloroethylene	$Cl_2C=CHCl$
Tetrachloroethylene (Perchloroethylene)	$Cl_2C=CCl_2$

Alcohols

Aliphatic alcohols contain a hydroxyl group attached to a carbon atom. They are named by using the stem and the suffix *-ol*.

EXAMPLES

$$H_3COH \qquad H_3C—CH_2OH \qquad H_3C—CH_2—CH_2OH$$

Methanol
(Methyl Alcohol)

Ethanol
(Ethyl Alcohol)

1-Propanol
(n-Propyl Alcohol)

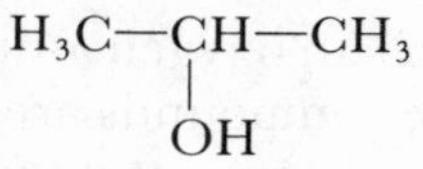

$$H_3C—\underset{\underset{\displaystyle OH}{|}}{CH}—CH_3$$

2-Propanol
(Isopropanol, Isopropyl Alcohol)

When two hydroxyl groups are present in an aliphatic compound, it is usually called a *glycol* although in the IUPAC system the suffix *-diol* is used with the full alkane name.

EXAMPLES

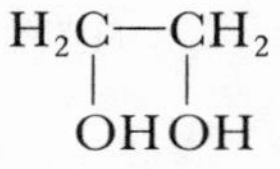

$$H_2C-CH_2$$
$$|\quad|$$
$$OH\,OH$$

1,2-Ethanediol
(Ethylene Glycol)

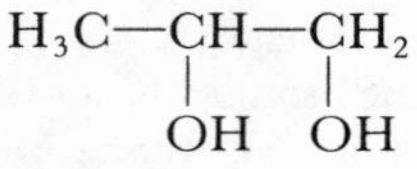

$$H_3C-CH-CH_2$$
$$|\quad\ |$$
$$OH\quad OH$$

1,2-Propanediol
(Propylene Glycol)

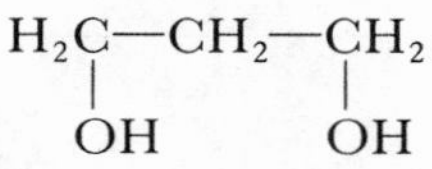

$$H_2C-CH_2-CH_2$$
$$|\qquad\quad|$$
$$OH\qquad\ OH$$

1,3-Propanediol
(Trimethylene Glycol)

The only *triol* usually encountered is 1,2,3-propanetriol, called *glycerol* or, incorrectly, glycerine.

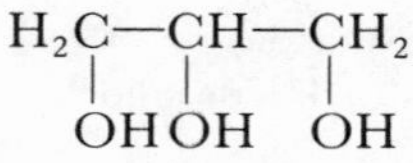

$$H_2C-CH-CH_2$$
$$|\quad\ |\quad\ |$$
$$OH\,OH\quad OH$$

1,2,3-Propanetriol
(Glycerol)

Hydroxyl groups can also appear in alkenes, alkynes, and the cyclo compounds.

EXAMPLES

$$H_2C{=}CHOH$$

Vinyl Alcohol

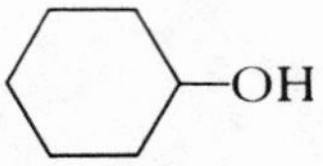

Cyclohexanol

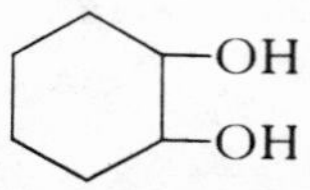

1,2-Cyclohexanediol

Ethers

Ethers are compounds in which an oxygen atom is used as a bridge between adjacent carbon atoms. These compounds are named by using the names of the two radicals attached to the oxygen atom in alphabetical order, followed by the word *ether*.

EXAMPLES

Dimethyl ether (Methyl ether)	$H_3C—O—CH_3$
Diethyl ether (Ethyl ether)	$H_3C—CH_2—O—CH_2—CH_3$
Methyl vinyl ether	$H_2C=CH—O—CH_3$

Aldehydes

Aldehydes are named by dropping the *e* from the alkane name and adding -*al* or not dropping the *e* and adding -*dial* when there are two such groups per molecule, etc.

EXAMPLES

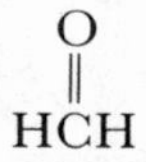

Methanal
(Formaldehyde)

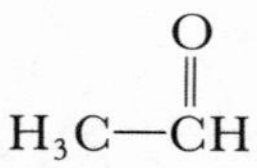

Ethanal
(Acetaldehyde)

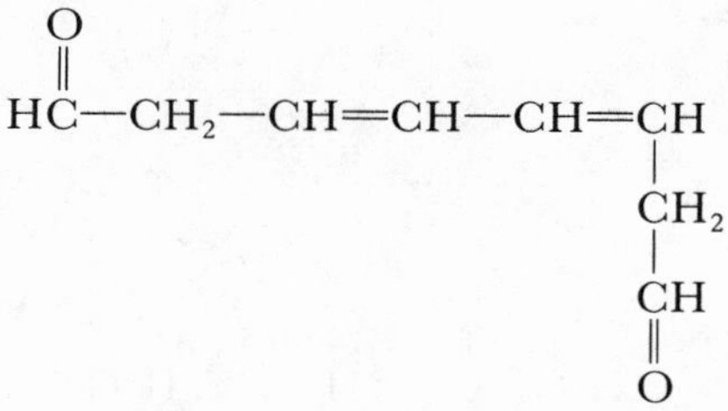

3,5-Octadienedial

Ketones

Ketones are named by dropping the *e* from the alkane name and adding the suffix -*one*. The more commonly encountered members of this family, however, have common names that are much more likely to be used. These are usually derived in a manner analogous to that for naming ethers, by using the two radical names followed by *ketone*.

EXAMPLES

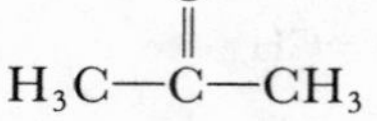

$$H_3C-\overset{\displaystyle O}{\overset{\|}{C}}-CH_3$$

Propanone
(Acetone, Dimethyl Ketone)

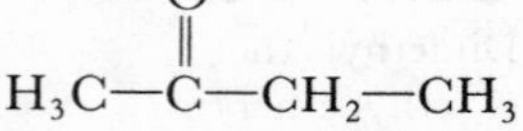

$$H_3C-\overset{\displaystyle O}{\overset{\|}{C}}-CH_2-CH_3$$

Butanone
(Methyl Ethyl Ketone)

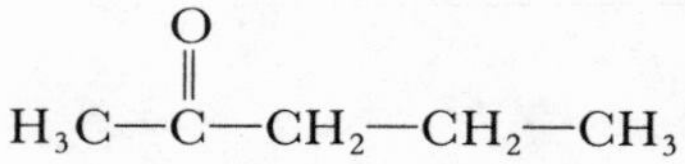

$$H_3C-\overset{\displaystyle O}{\overset{\|}{C}}-CH_2-CH_2-CH_3$$

2-Pentanone
(Methyl n-Propyl Ketone)

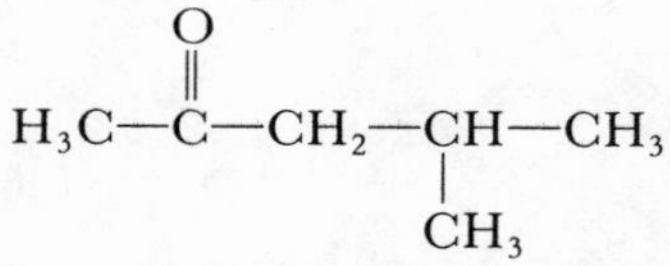

$$H_3C-\overset{\displaystyle O}{\overset{\|}{C}}-CH_2-\underset{\underset{\displaystyle CH_3}{|}}{CH}-CH_3$$

4-Methyl-2-pentanone
(Methyl Isobutyl Ketone)

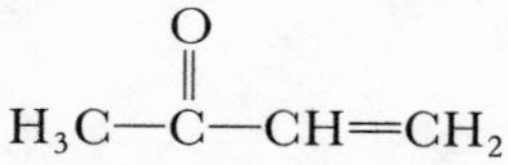

$$H_3C-\overset{\displaystyle O}{\overset{\|}{C}}-CH=CH_2$$

3-Butene-2-one
(Methyl Vinyl Ketone)

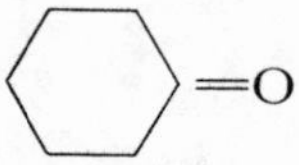

Cyclohexanone

Acids

The same root or stem as that for naming aldehydes is used for acids, followed by the word *acid*.

EXAMPLES

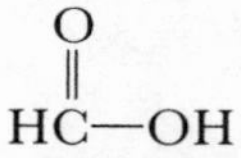

$$HC-\overset{\displaystyle O}{\overset{\|}{}}-OH$$

Methanoic Acid
(Formic Acid)

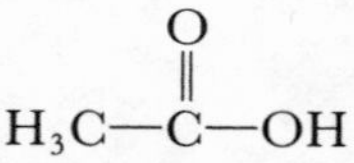

$$H_3C-\overset{\displaystyle O}{\overset{\|}{C}}-OH$$

Ethanoic Acid
(Acetic Acid)

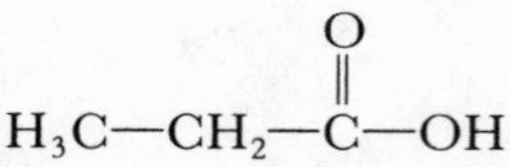

$$H_3C-CH_2-\overset{\displaystyle O}{\overset{\|}{C}}-OH$$

Propanoic Acid
(Propionic Acid)

Esters

Being the reaction products of acids and alcohols and analogous to inorganic salts, esters are named in a somewhat similar manner. The attached alcohol radical name is followed with the acid stem and a suffix *-ate*.

EXAMPLES

$$H_3C-CH_2-\overset{\overset{\displaystyle O}{\|}}{C}-O-CH_3$$

Methyl Propanate
(Methyl Propionate)

$$H_3C-\overset{\overset{\displaystyle O}{\|}}{C}-O-CH_2-CH_3$$

Ethyl Acetate

$$H\overset{\overset{\displaystyle O}{\|}}{C}-O-CH_2-CH_2-CH_3$$

n-Propyl Formate

Aromatic Hydrocarbons

Aromatic hydrocarbons are those based on the benzene ring. For the lower members of this series, common names abound and are normally used. Naming rules are similar to those for acyclic hydrocarbons.

EXAMPLES

Benzene

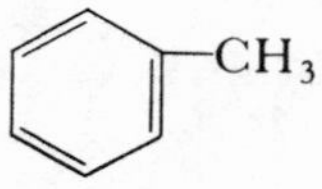

Toluene
(Methyl Benzene)

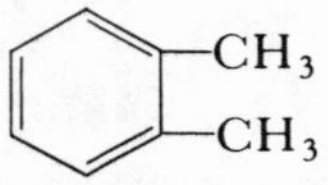

1,2-Dimethylbenzene
(*o*-Xylene)

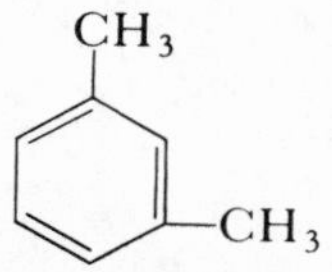

1,3-Dimethylbenzene
(*m*-Xylene)

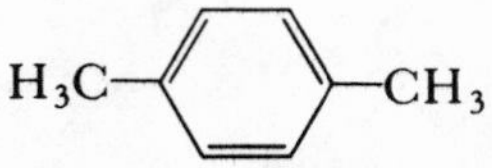

1,4-Dimethylbenzene
(*p*-Xylene)

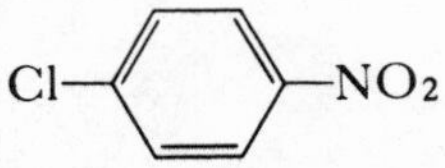

4-Nitrochlorobenzene
(*p*-Nitrochlorobenzene)

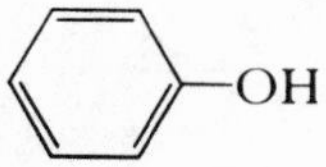

Phenol

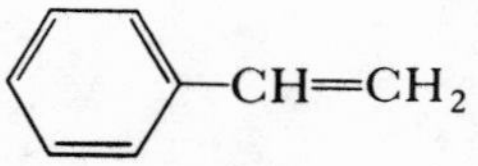

Styrene
(Vinyl Benzene, Phenylethylene)

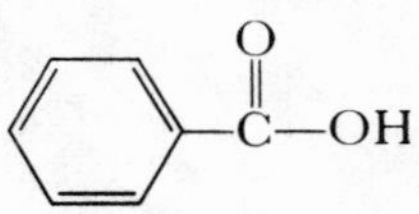

Benzoic Acid

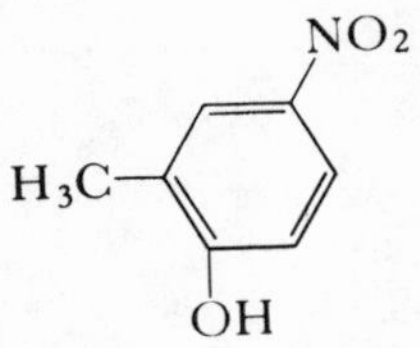

2-Hydroxy-5-nitrotoluene
(4-nitro-*o*-cresol)

B
GLOSSARY

acclimation. (*See* **acclimatization.**)

acclimatization. The process of becoming accustomed to new conditions. Also called *hardening* when used to refer to the body's reaction to chronic exposure to some irritants.

accommodation. A parameter of vision; the ability of the eye to adjust for various distances.

ACGIH. American Conference of Governmental Industrial Hygienists.

acidosis. An increase in hydrogen ion concentration in the body or an illness caused by such an increase.

acnegen. A material capable of causing the signs of acneform dermatitis.

acoustic trauma. Displacement of the bones of the middle ear by noise to the extent that sound is no longer well transmitted to the inner ear.

acroosteolysis. Disease of the bone characterized by disappearance of bone especially from the fingers.

action potential. Voltage sufficient to trigger a biological mechanism. Usually refers to impulses that travel along nerves or muscles.

317

acute.   Short-term or single dose.

aerosol.   A dispersion of solid or liquid particles in a gaseous medium.

afferent.   Going toward. Usually refers to impulses along nerves going toward the CNS.

AIHA.   American Industrial Hygiene Association.

air embolism.   Bubbles of air in blood vessels, especially when the bubbles are large enough to cause blockage of blood flow.

albuminuria.   Presence of serum albumin or serum globulin in the urine.

alkylating agent.   A chemical used to introduce an alkyl group into another molecule by reaction.

allergen.   A material capable of inducing allergy or hypersensitization.

alopecia.   Hair loss.

alveolus (plural: alveoli).   A general term meaning a small sac, but usually used to designate the sacs in the lungs through whose walls the gaseous exchange takes place.

aminoplast.   A polymeric material based on one or more monomers containing amine ($-NH_2$) groups.

amorphous.   Having no definite shape; noncrystalline.

amplitude.   Quantity. The maximum displacement from a zero value during one oscillation.

anaerobic.   Without air (oxygen).

analgesia.   An absence of sensibility to pain, designating particularly the relief of pain without loss of consciousness.

anaphylactic shock.   A condition of shock brought on by the reaction of a hypersensitized individual to the allergen that caused the hypersensitization.

anemia.   Reduction below normal in the number of red blood cells, the quantity of hemoglobin, or the volume of packed red cells in the blood (hematocrit).

anesthesia.   Specifically, loss of feeling or sensation that may or may not involve loss of consciousness.

angiosarcoma.   Cancerous (malignant) tumor of tissue like embryonic connective tissue containing many small blood vessels.

anhidrotic.   Without sweating.

anorexia.   Loss of appetite.

anoxia.   Absence or a lack of sufficient oxygen in body tissues.

antagonism.   Anticooperative action of two agents so that the total effect is less than the sum of the two effects taken independently.

anthracosis.   A form of coal-worker's pneumoconiosis caused by inhaling anthracite coal dust.

antibody.   A modified type of serum globulin produced by the body in response to the presence of an allergen.

antigen.   A material that causes or allows the formation of antibodies.

antioxidant.   A material that inhibits oxidation.

antiseptic.   A material able to inhibit the growth of microorganisms. Death of the microorganism is not implied.

anuria.   Absence of urine excretion.

aplastic.   Lack of development of new tissue.

apnea.   The (usually) transient cessation of breathing.

argyria.   Blue-gray cast of the skin caused by deposits of metallic silver beneath the surface.

asbestosis.   A fibrotic disease of the lung caused by inhaling asbestos.

aseptic bone necrosis.   Death of bone unrelated to bacteria or other biological organisms. Caused by chronic cycles of compression and decompression and probably related to caisson disease.

asphyxia.   Lack of oxygen or excess of carbon dioxide in the body, usually caused by interruption of breathing and causing unconsciousness.

asphyxiation.   Suffocation.

asthma.   A disease marked by recurring attacks of dyspnea, with wheezing cough and a sense of constriction caused by spasmotic contraction of the bronchi.

ataxia.   Failure of or irregularity of muscular coordination.

autoignition temperature.   The temperature at which a material will spontaneously burst into flame or explode.   Usually determined in air although values for some materials in oxygen can be found.

autonomic nervous system.   That part of the efferent peripheral nervous system terminating on cardiac or smooth muscle or on glands.

auxin.   Plant hormone.

Bacillus subtilis.   A common bacterium capable of forming and secreting proteolytic enzymes.

bagasse.   Sugar cane residue.   Remains of sugar cane after the sugar has been extracted.

basilar membrane.   A portion of the cochlea that serves to channel sound and perhaps to aid in pitch discrimination.

bends.   (*See* **caisson disease.**)

benign.   Not malignant.

bioconcentration.   The process of causing an increase in the concentration of a substance in a predator by ingestion of prey containing the substance in lower concentration.

biothermally neutral environment.   An environment in which a resting subject in thermal equilibrium with his surroundings has no conscious sensations of heat or cold and in which the subject's thermoregulatory system is functioning at a minimum level.

biothermal strain.   Heat or cold strain.   The displacement of a biological parameter from its value in a biothermally neutral environment produced by the action of biothermal stress.

biothermal stress.   Heat or cold stress.   The load on the thermoregulatory system of the subject due to a displacement from thermally neutral conditions caused by a change in the metabolic rate of the subject and/or by a change in the environment.

black globe thermometer.   (*See* **globe thermometer.**)

black-lung disease.   (*See* **coal-worker's pneumoconiosis.**)

blepharospasm.   Forced blink of the eyelid.

body burden.   Total amount of a substance within the body.

bolus.   A lump.

bradycardia.   Abnormal slowness of the heart beat.   Its opposite is tachycardia.

bremsstrahlung.   X rays produced by the interaction of energetic electrons and matter.

bronchiectasis.   Chronic dilitation of the bronchi or bronchioles.

bronchiole.   Small air passage within the lungs.

bronchitis.   Inflammation of the bronchi.

bronchus (plural: bronchi).   One of the larger air passages within the lungs.

byssinosis.   A pneumoconiosis produced by the chronic inhalation of dust produced during the processing, cleaning, and spinning of cotton and some other fibers.

caisson disease.   A condition characterized by pain especially in the joints and along tendons, caused by too rapid transition from a high ambient pressure to a lower one and related to the collection of bubbles of nitrogen, desorbed from blood and tissue, in many areas of the body.

cancer.   The malignant, unrestrained growth of tissue.

carboxyhemoglobin.   The reversible combination of carbon monoxide with hemoglobin.

carcinogen.   A material capable of causing cancer.

carcinoma.   A cancer made up of epithelial cells (those cells that form the covering of the body and body organs).

catalyst.   A material which accelerates (or decelerates) a chemical reaction but which remains unaltered by the reaction.

cerebellum.   The lower posterior part of the brain concerned with controlling movements.

cerebrum.   The main part of the brain.

chelate.   A chemical compound capable of binding metal ions and also the combination of such a compound plus ion(s).

chemical asphyxiant.   A material that can cause suffocation through some mechanism that involves a chemical reaction.

chemotherapy.   The use of drugs (chemicals) for treatment of disease.

chitin.   The material that forms the shell or exoskeleton of crabs and lobsters and the shards (hardened protective forewings) of beetles.

chitin inhibitor.   A material that causes chitin not to develop properly.

chokes.   A suffocating sensation caused by the collection of air (nitrogen) bubbles in the lungs, preventing blood flow through one or more vessels. Caused by too rapid decompression after a stay at high pressure.

cholinesterase inhibitor.   A material that prevents the enzyme, cholinesterase, from causing the hydrolysis of acetylcholine (or other choline compound) to choline and an acid (acetic acid in the case of acetylcholine).

chromosome.   A strand of DNA within the cell nucleus.

chronic.   Long-term or duration; repeated dose or exposure.

cilia.   Hair-like fibers attached to cells. Eyelashes are cilia but, more important, cilia are found on the inner surfaces of bronchi and bronchioles and their motion constitutes one of the ways by which particles are removed from the lungs.

ciliary action.   Cilia on the inner surfaces of the bronchi and bronchioles beat with a wave-like motion, forcing mucous toward the throat.

CNS.   Central nervous system, consisting of the several sections of the brain and of the spinal cord.

coal-worker's pneumoconiosis.   Pneumoconiosis characterized by the presence of coal dust in the lungs, a certain appearance of the lungs on X ray, the possible development of dyspnea, and possibly leading to progressive, massive fibrosis.

cochlea. A spiral structure wherein vibrations are transformed into nervous impulses. Part of the inner ear.

cochlear duct. A passage in the cochlea formed between the scala vestibuli and the scala tympani. The organ of Corti lies within the cochlear duct.

coherence. For electromagnetic radiation, the phenomenon of having all wavefronts in phase.

collagen. The protein that supports other tissues in the body.

condensation. When used in chemistry, the reaction of two materials resulting in two products, one of which has a much higher molecular weight than the other.

conductive hearing loss. Decrease in the ability to hear caused by one or more obstructions to the normal passage of sound waves from the air to the inner ear.

confound. To mingle with.

congestion. Excessive or abnormal accumulation of blood in a part.

conjunctivitis. Inflammation of the membrane that lines the eyelids.

contact dermatitis. Skin irritation resulting from direct contact of the causative agent with the skin.

contact inhibition. One of the processes that applies restraint to the growth of tissue. Physical contact with normal cells causes inhibition of the growth of epithelial cells through an obscure mechanism.

convergence. A parameter of vision. The ability of the eyes to fix on the same point.

copolymer. The result of simultaneously polymerizing two different monomers.

corrected effective temperature. The ET scale corrected for radiation effects by using the globe thermometer in place of the dry-bulb thermometer.

cortex. An external layer of an organ or other body structure. *Note:* has no relation to *core*.

cortical. Adjectival form of cortex.

cutaneous. Pertaining to the skin.

cutie pie. (*See* **ionization meter.**)

CWP. (*See* **coal-worker's pneumoconiosis.**)

cyanosis. Bluish discoloration of the skin and mucous membranes.

cytoplasm. Cell tissue or protoplasm. The liquid portion of the cell.

damping. Application of viscous mass to surfaces to alter vibrational characteristics.

debridement. Cutting away tissue. The surgical remedy to problems caused by materials that are not removed by normal body processes.

decompression sickness. (*See* **caisson disease.**)

decrement. The opposite of increment. A step down or worse.

defoliate. Remove the leaves (foliage) from.

demography. Vital and social statistics.

deoxyribonucleic acid. The material within body cells that contains the genetic code, providing for replication of the cell.

dermatitis. Inflammation of the skin.

dermis. Skin.

desensitization. The process of reducing the body's response to an allergen.

diatomaceous earth. Amorphous silica from the skeletons of microscopic diatoms.

dicotyledenous. Possessing or having seeds with two cotyledons (rudimentary leaves). Generally, broad-leafed plants.

dilation. The action of dilating or stretching; dilitation.

dilitation. The condition of being dilated or stretched beyond the normal dimensions.

diplopia. Double vision.

diuretic. A substance able to cause an increase in the secretion of urine.

DNA. (*See* **deoxyribonucleic acid.**)

dosimeter. A device usually worn on clothing and designed to respond in some manner to exposure (not dose).

dry-bulb temperature. Air temperature.

dust. An aerosol of solid particles produced by mechanical action.

dyscrasia. An imbalance of component members.

dyspnea. Difficult or labored breathing.

edema. Presence of abnormally large amounts of fluid in the intercellular tissue spaces of the body.

effective half-life. The time required for half of the original amount to disappear from the body by the processes of radioactive decay and biological reaction or elimination.

effective temperature. An empirical scale of temperature related to comfort.

efferent. Going away. Usually refers to impulses along nerves going away from the CNS.

electrolyte. Material capable of ionization in water.

electromyography. Recording of muscle action potentials.

emetic. A material that causes vomiting.

emphysema. Pulmonary emphysema is a condition of the lung characterized by an increase beyond normal in the size of air spaces beyond terminal bronchioles, either from dilitation of the alveoli or from destruction of their walls.

endemic. Present in the population at all times.

endogenous. Occurring, growing, or originating within the body.

energy. The capability of doing work.

energy density. Flux density. Amount of radiation per unit area.

environmental carcinogenesis. Production of cancer by agents in the environment external to the host.

enzyme. An organic compound, frequently a protein, that catalyzes some change in a substrate for which it is often specific.

eosinophil. A leukocyte that is readily stained with the dye eosin.

epidemiology. The field of science dealing with the relationship of the various factors that determine the frequencies and distributions of an infectious disease or a physiological state in a human community.

epidermis. Outermost layer of the skin consisting of five distinct sublayers.

epinephrine. A hormone secreted by the adrenal medulla. It is the most powerful vasopressor substance known, increasing blood pressure, stimulating the heart rate and muscle action, and increasing cardiac output.

epithelium. The covering of the internal and external surfaces of the body, including the lining of vessels and other small cavities.

erythema.   Redness of the skin.

erythrocyte.   Red blood cell.

etiology.   The cause or causes of disease; study of the causes of disease.

eucaryotic.   Having a true nucleus.

euphoria.   An exaggerated feeling of well-being.

eustachian tube.   A tube which connects the middle ear with the throat and which allows for pressure equalization between the middle ear and the ambient atmosphere.

exfoliation.   A falling off in scales or layers.

exogenous.   From outside the body.

exposure dose.   A term sometimes used instead of flux density.

external auditory canal.   The passage from the external ear to the eardrum.

extrinsic.   Coming from without. For a property, related to amount.

farmer's lung.   A pneumoconiosis caused by handling moldy hay.

feedback.   Information concerning the status of a controlled variable.

fetus.   Unborn offspring.

fibrillation.   Nonrhythmic twitching (of a muscle).

fibrogenic.   Producing or causing the formation of fibrin, a whitish, insoluble protein. Used to describe the action of some substances on tissue, especially in the lung.

fibrosis.   Formation of fibrous (scar) tissue.

flux density.   The amount of radiant energy per unit area.

fog.   An aerosol of liquid (usually water) particles produced by condensation.

fovea.   The area of the retina at the focal point of the cornea-lens system.

frequency.   Rate of recurrence.

fuller's earth.   Diatomaceous earth.

fume.   An aerosol of solid particles produced by condensation.

fungistat.   A material that prevents the growth of fungi.

ganglia.   Synapses (junctions of neurons) outside the CNS.

gastroenteritis.   Inflammation of the stomach and intestinal tract.

gauge pressure.   The difference in pressure between atmospheric (ambient) and that inside a vessel.

gene.   The unit of heredity.

genetic.   Having to do with heredity.

germicide.   A substance able to kill microorganisms.

gingival mucosa.   The membrane which forms the gums and especially which surrounds teeth and overlays unerupted teeth.

GI tract.   Gastrointestinal tract, consisting of the stomach and intestines.

globe thermometer.   A device used in the estimation of rates of radiant heat exchange. Usually a hollow metal (or other) sphere about 0.15 m in diameter painted matte black and having a thermometer bulb or other temperature-sensitive device at its center.

granuloma.   A tumor consisting of small, round, fleshy masses or granules.

granulomatous.   Composed of granules.

half-life. The time required for the disappearance of half the starting amount.

hapten. A nonproteinaceous material capable of combining with a body protein to become an antigen capable of eliciting a hypersensitization response.

heat. The form of energy that flows because of a temperature difference.

heat cramps. Painful muscle spasms caused by electrolyte imbalance brought on by over-exposure to heat stress.

heat exhaustion. Extreme fatigue, dizziness, nausea, and subnormal temperature caused by heat stress.

heat stroke. Collapse with no sweating but a high ($> 40°C$) temperature and other signs and symptoms.

heat syncope. Dizziness and possibly fainting caused by heat stress.

hematopoietic. Pertaining to or affecting the formation of blood cells.

hematuria. Discharge of blood in urine.

hemoglobin. A complex protein containing iron in the ferrous (Fe^{2+}) state and making up much of the mass of an erythrocyte.

hemolysis. Liberation of hemoglobin, consisting of separation of hemoglobin from red cells and its appearance in the fluid used for suspending the red cells.

hemolytic. Referring to destruction of red cells or hemoglobin.

histological. Dealing with the minute structure, composition, and function of the tissues. More particularly, with cell and cell structure.

homeostasis. Maintenance of internal stability.

homeotherm. An animal with constant body temperature.

homologue. One of a chemical series formed by adding a constant element to the preceding member of the series.

homopolymer. A polymer formed from a single kind of monomer.

hydride. Compound of a metal with hydrogen.

hydrolysis. Chemical addition of water to a molecule.

hygroscopic. Readily taking up and retaining moisture.

hyperemia. An excess of blood in a part.

hypergolic. Bursting into flame spontaneously upon mixing.

hyperplasia. The abnormal multiplication or increase in the number of normal cells in normal arrangement in a tissue. Usually evidenced by a thickening of the tissue.

hyperpnea. Increase in the rate and depth of breathing.

hyperpyrexia. Excessive accumulation of heat in the body leading to an elevated temperature with little sweating.

hypersensitization. The process of rendering or the condition of being abnormally sensitive, of having an ability to react with characteristic symptoms to the application or contact with certain substances (allergens) in amounts innocuous to normal individuals.

hyperventilation. A greater than normal tidal volume and/or frequency of respiration resulting in an increased amount of air per unit time at the alveolar level.

hypobaria. Pressure lower than standard atmospheric.

hyposensitization. (*See* **desensitization.**)

hypothalamus. A part of the brain beneath the thalamus and above the roof of the mouth.

hypothermia. Subnormal body temperature.

hypoventilation. Reduced frequency of respiration and/or tidal volume, resulting in a lower than normal amount of air per unit time at the alveolar level.

hypoxia. Deficiency of oxygen in inspired air.

immunocyte. A leukocyte involved in the formation of antibodies.

immunology. That science concerned with the operation of the systems that protect against a second attack of an infective agent.

impedence. Related to resistance. The ratio of the force applied to an oscillating system, to the velocity of components (particles) of that system.

incidence. An attack rate. In epidemiology, the number of new cases (of some disease) per unit of population in some (usually small) unit of time. Contrast this with prevalence.

incus. One of the bones in the middle ear. The anvil.

intrinsic. Belonging by its very nature.

in vitro. Literally, in glass. Not in the living individual but in the test tube.

in vivo. Literally, in life. Within the living body, cell, or tissue.

ionization chamber. (*See* **ionization meter.**)

ionization meter. A device for measuring the exposure rate of ionizing radiation.

iritis. Inflammation of the iris (of the eye).

irradiance. Flux density per unit of time. Power per unit of area.

isopleth. A line indicating equal concentrations.

isotope. Having the same atomic number.

lacrimation. Secretion and discharge of tears.

lactone. An internal ester formed between carboxyl and hydroxyl groups in the same molecule by splitting out water.

larynx. The upper part of the windpipe that functions as the organ of voice.

LEL. (*See* **lower explosive limit.**)

lethargy. Drowsiness.

leukocyte. A white blood cell.

leukocytosis. An increase in the number of leukocytes in the blood.

leukopenia. A reduction in the number of leukocytes in the blood.

lipid. A fat or fat-like material, insoluble in water but soluble in alcohol, ether, chloroform, and other fat solvents.

longitudinal epidemiological study. (*See* **prospective epidemiological study.**)

loudness. A characteristic of sound related to its amplitude and frequency.

lower explosive limit. The lowest concentration of the material in air that can be caused to explode under certain well-defined conditions.

lumenaire. A light fixture.

lymphocyte. A leukocyte that arises in the lymphatic system.

lymphoma. A tumor of lymphoid tissue.

MAC. Maximum (or maximal) allowable (or acceptable) concentration. A set point for control of inhalation exposure.

macula.   The area of the retina in which cone-shaped photoreceptors are found.

malignant.   Harmful.

malignant neoplasm.   Cancer.

malignant tumor.   Cancer.

malleus.   One of the bones in the middle ear. The hammer.

mandible.   Jawbone.

medulla.   The middle or innermost part, especially of an organ or structure.

melanoma.   A tumor of pigmented cells.

meninges.   The three membranes that envelope the brain and spinal cord.

mercaptan.   A hydrocarbon containing one or more sulfhydryl (—SH) groups. Analogous to alcohol where oxygen is replaced by sulfur. So-named because the material will react with mercury and not because it contains mercury, which it does not.

mesothelioma.   A cancer arising in mesothelial tissue, usually of the pleura or peritoneum.

metabolite.   A product of metabolism, the process by which a living body converts food to energy and tissue.

metalloid.   An element intermediate in properties between metals and nonmetals that has some of the properties of both. Examples are P, Si, Se, Te.

metastasis.   The spreading or transfer of disease from one organ or system to another.

methemoglobin.   A compound formed from hemoglobin by oxidation of the iron from the ferrous to the ferric state with essentially ionic bonds.

middle-ear deafness.   (*See* **acoustic trauma.**)

miticide.   A material that kills or inhibits the growth of mites.

mitosis.   The process by which cells multiply by dividing.

Monge's disease.   Chronic mountain sickness.

monochromatic.   Having a single color (or wavelength or frequency).

monocotyledenous.   Possessing or having seeds with one cotyledon (rudimentary leaf). Generally, grasses.

monomer.   An organic chemical capable of joining with like molecules to produce extremely large molecules (polymers).

morbidity.   The relative incidence of disease.

mordant.   A chemical used to react with a dye to fix the dye to the fabric. Dyes by forming an insoluble but still colored compound.

MPPCF.   Millions of particles per cubic foot of air.

mucosa.   Mucous membranes.

mutagenesis.   Causing the development of mutations (stepwise changes in structure or function of cells).

myocardium.   The middle and thickest layer of the heart wall. Composed of heart muscle.

myoglobin.   Protein similar to hemoglobin, but with 1/4 the molecular weight and with one, rather than four Fe^{2+} atoms per molecule. Found mainly in muscle.

naptha.   A blend or mixture of (usually) aliphatic hydrocarbons.

narcosis.   A reversible condition characterized by stupor and/or insensibility.

nasal septum.   Tissue that separates the nostrils.

nascent.   Freshly formed. Usually used to refer to material such as hydrogen which, by reaction, has just been set free of a chemical combination and which is especially reactive.

necrosis. Death of tissue, usually as individual cells, groups of cells, or in small localized areas.

neoplasm. Literally, *new tissue,* but used to mean the new growth of different tissue. A tumor.

nephritis. Inflammation of the kidneys.

nephron. Anatomical and functional unit of the kidney.

nephrosis. Any disease of the kidneys.

neuron. A nerve body.

neuropathy. Nonspecific disturbance in the peripheral nervous system.

neurotoxic. Causing injury to nervous tissues by the property of toxicity.

NIHL. Noise-induced hearing loss.

nitrogen narcosis. Sleepiness or drowsiness, perhaps with euphoria, caused by breathing nitrogen, neon, or argon mixed with oxygen under pressure. The effect is probably related more to the density and viscosity of the gas than to its composition.

noise. Unwanted sound.

nonionizing radiation. Electromagnetic or other radiation having an energy level of 12 eV per quantum or less.

nucleic acid. An acid found in the cell nucleus.

organ of Corti. A part of the cochlea wherein the sound waves are transformed into nervous impulses.

organometallic compound. Combination of a metal (or metalloid) and an organic compound that is neither a chelate nor the salt of an organic acid.

PAH. (*See* **polynuclear aromatic hydrocarbons.**)

palpitation. Rapid and usually strong heartbeats, felt by the person affected.

parasympathetic nervous system. That part of the autonomic nervous system having to do with maintenance of homeostasis, the normal state of the body.

partial pressure. The pressure that would be found in a container if one of the components of a mixture were present alone.

pericardium. The sac that surrounds the heart.

peripheral neuritis. Inflammation of nerves, especially of nerve endings.

peripheral neuropathy. Nonspecific disturbance in the peripheral nervous system.

peritoneum. The membrane lining the abdominal walls.

permanent threshold shift. A permanent increase in the threshold of hearing.

pesticide. A killer of one or more of man's pests. A generic term that includes insecticide, herbicide, rodenticide, etc.

phagocyte. Fixed or free-roaming cell able to ingest foreign particles and bacteria.

phagocytosis. Ingestion of particulate material of all kinds (living and dead or inert) by phagocytes.

pharynx. Back of the throat.

pheromone. A hormone-like substance secreted by insects and others that acts as a messenger to other members of the same species. Sex attractants are pheromones.

phon. A unit of loudness level.

photoallergy. Sensitization of the skin to light.

photokeratitis. Inflammation of the cornea by light.

photophobia. A condition wherein light causes pain in the eyes.

pink noise.   Noise that has been weighted, especially at the low end of the spectrum so that the energy per band (usually octave band) is about constant over the spectrum.

pinna.   The part of the external ear usually called *the ear*.

plasma.   The fluid portion of the blood in which the corpuscles are suspended. Plasma is to be distinguished from serum, which is plasma from which the fibrinogen has been separated in the process of clotting.

plasticizer.   A material added to a resin (plastic, polymer) to modify the properties of the resin. Typically, the resin is softened to become more plastic.

pleura.   The membrane that lines the chest or thoracic cavity, forming sacs around the lungs.

PMF.   (*See* **progressive massive fibrosis.**)

pneumoconiosis.   The accumulation of dust in the lungs and the tissue reaction to its presence.

pneumonitis.   Inflammation of the lungs.

polycythemia.   Excessive number of red cells in the blood.

polynuclear aromatic hydrocarbons.   Hydrocarbons with fused-ring systems, with or without aliphatic side chains.

potentiation.   The act of making another material more potent or powerful in its action. Used as a synonym of *synergism* and as an antonym of *antagonism*.

power density.   Irradiance. Power per unit area.

presbycusis.   Loss of hearing with age.

prevalence.   In epidemiology, at some instant in time the total number of cases at all stages of some disease per unit (usually 100,000) of population. Contrast this with incidence.

primary (skin) irritation.   Skin irritation produced on first contact with the material.

progressive massive fibrosis.   Rapid development of large amounts of fibrotic tissue, usually in the lungs.

prophylaxis.   Preventive treatment.

prospective epidemiological study.   A study done by examining a group of subjects from the present on into the future, at each stage relating their state of health to one or more aspects of the environment.

prostration.   Extreme exhaustion.

proteinuria.   The presence of protein in the urine.

proteolytic enzyme.   One of the biological catalysts with the ability to decompose proteinaceous materials.

protista.   One-celled plants or animals.

pruritis.   Itching.

psychoacoustical.   Having to do with the interaction of sound and mind.

PTS.   (*See* **permanent threshold shift.**)

pulmonary hypertension.   Increased pressure in the lung blood vessels.

punctate.   Points or dots.

purpura.   The presence of red-to-purple splotches or bruises beneath the skin surface.

pyrolysis.   Thermal decomposition.

pyrophoric (pyroforic).   Bursting into flame spontaneously on exposure to air.

rad.   *Radiation absorbed dose:* 1 rad equals 100 erg g^{-1} or 1.0×10^{-2} J kg^{-1}.

radiant exposure. A term sometimes used to mean flux density.

radiation intensity. Irradiance. Power per unit area.

radiomimetic. Mimics the action of radioactivity; specifically, the spectrum of actions by which radioactivity causes injury to the organism or its individual cells.

range. For a charged particle, the linear distance between origin and final capture.

rate meter. (*See* **ionization meter**.)

Raynaud's phenomenon. A condition of the hands (usually) characterized by pallor or cyanosis.

rem. Roentgen equivalent man (or mammal).

renal. Pertaining to the kidneys.

retrospective epidemiological study. A study done by examining subjects in the present and relating those findings to their histories.

rhinitis. Inflammation of the mucous membranes in the nose. Runny nose.

ribonucleic acid. A material used by the cell to replicate DNA and other proteins.

RNA. (*See* **ribonucleic acid**.)

roentgen. A unit of radiation exposure: 2.58×10^{-4} coulomb kg^{-1}.

sarcoma. A tumor of tissue similar to normal connective tissue. Usually malignant.

scala tympani. A passage in the cochlea. (*See* **scala vestibuli**.)

scala vestibuli. A passage in the cochlea formed between the vestibular membrane and bone. Sound waves enter through the oval window, travel through this passage, and then go out through the scala tympani to the round window.

senescence. Aging.

senile. Old.

sensitization. (*See* **hypersensitization**.)

serum. The clear portion of any animal liquid separated from its more solid elements, especially the clear liquid that separates in the clotting of blood from the clot and corpuscles.

Shaver's disease. A pneumoconiosis named for its discoverer, Dr. Cecil G. Shaver. Its most characteristic signs are a marked emphysema with a diffuse rather than nodular fibrosis visible on X ray. Several deaths have occurred, usually from spontaneous pneumothorax. The cause is obscure but is associated with bauxite (Al_2O_3), an ore of aluminum.

sign. Any objective evidence of disease.

silicosis. Fibrotic disease of the lungs caused by inhaling crystalline silica.

simple asphyxiant. A gas that causes suffocation by dilution of oxygen.

slimicide. A material that kills or inhibits the growth of slime molds.

somatic. Referring to the body or to skeletal muscle.

somnolence. Sleepiness.

sone. A unit of loudness.

spasm. A sudden, violent, involuntary contraction of a muscle or a group of muscles, or a sudden but transitory constriction of a passage, canal, or orifice.

specific activity. The number of disintegrations per unit time in a unit weight of material. Usually, curies per gram.

specific ionization. Energy loss per unit distance traveled.

stapes. One of the bones in the middle ear. The stirrup.

stereoacuity. A parameter of vision; the ability to detect depth; depth perception.

subcutaneous. Beneath the skin.

subliminal. Below the range of sensation.

subsonic. Below the hearing range in frequency.

sympathetic nervous system. That part of the autonomic nervous system having to do with the fight or flight response.

symptom. Any functional or subjective evidence of disease.

synapse. The region of contact between adjacent neurons. This is the place where the impulse is transferred from one (or more) neuron(s) to another (or several).

syncope. Dizziness and possibly fainting.

syndrome. A set of symptoms and/or signs that occur together.

synergism. Cooperative action of agents so that the total effect is greater than the sum of the two effects taken independently.

systole. Contraction of the left ventricle of the (human) heart.

tachycardia. Rapid heart rate; opposite of bradycardia.

tautomer. An organic compound that exists in two isomeric forms, usually in equilibrium.

temporary threshold shift. An impermanent increase in the threshold of hearing.

teratogenic. Pertaining to the production of physical defects in the unborn offspring.

thermoluminescence. The emission of light upon heating; the amount of light being proportional to the amount of ionizing radiation absorbed by the material.

thermoplastic. Becoming soft and easily deformable on the application of sufficient heat.

thermosetting. Becoming hard, rigid, or infusible on the application of sufficient heat.

thesaurosis. The (perhaps hypothetical) pneumoconiosis resulting from inhalation of hair spray.

thoracic cavity. In man, inside the chest.

threshold limit value. A concentration in air or an exposure rate or dose rate of an energy felt by ACGIH to be safe for repeated exposure of workers. A set point for control of a hazard.

thymus gland. A ductless gland of vertebrate animals, found near the base of the neck in man but disappearing in the adult. Its function relates to the development of immunity.

tidal volume. The amount of air inhaled or exhaled during normal (or any other) activity.

timbre. A characteristic of sound related to the strengths of frequency components. The quality of sound that allows its source to be inferred.

TL. (*See* **thermoluminescence.**)

TLV. (*See* **threshold limit value.**)

toxic health hazard. Liklihood of toxic injury.

toxicity. The ability of a material to injure a living organism by other than mechanical means.

toxicology. Study of the effects of chemicals on biologic systems, with emphasis on the mechanisms of harmful effects, and the conditions under which those effects occur.

trachea. Windpipe.

transducer. A device for receiving and retransmitting energy, usually changing the form of the energy in the process.

TTS. Temporary (hearing) threshold shift.

tumor.   A swelling of tissue.

tympanic membrane.   The ear drum.

ultrasonic.   Above hearing range in frequency.

upper explosive limit.   The highest concentration of the material in air that can be caused to explode under certain well-defined conditions.

urticaria.   A vascular reaction of the skin marked by a transient appearance of smooth, slightly elevated patches that are redder or paler than the surrounding skin and often attended by severe itching.

vascularity.   Presence of blood vessels.

vasoconstriction.   The narrowing of blood vessels, especially arterioles, leading to decreased blood flow.

vasodilitation.   A state of increased caliper of the blood vessels.

vasomotor.   Regulating the diameter of certain blood vessels.

ventricular arrhythmia.   A nonrhythmic heartbeat.

ventricular fibrillation.   A condition characterized by twitching of the heart muscle.

vesiculation.   Presence of inflammation and blisters.

PREFIXES

a-	negation
an-	negation
cardi-	heart
hem(at)-	blood
hepat-	liver
hist-	web, tissue
hyper-	above, beyond
hypo-	under, below
nephr-	kidney
ot-	ear
per-	to the greatest possible extent
peri-	around
pne-	breathing
pneum(at)-	breath, air
pneumo(n)-	lung
poly-	many
vas-	vessel

SUFFIXES

-ase	an enzyme
-cyte	cell
-emia	appearing in blood
-itis	inflammation
-lysis	destruction, decomposition
-oma	tumor, neoplasm
-osis	a process, often a disease, and sometimes an abnormal increase
-pnea	breathing
-uria	appearing in urine

INDEX

INDEX

B